Applications of Nanomaterials in Energy Systems

Applications of Nanomaterials in Energy Systems

Editors

Eleftheria C. Pyrgioti
Ioannis F. Gonos
Diaa-Eldin A. Mansour

Basel • Beijing • Wuhan • Barcelona • Belgrade • Novi Sad • Cluj • Manchester

Editors
Eleftheria C. Pyrgioti
University of Patras
Patras, Greece

Ioannis F. Gonos
National Technical University
of Athens
Athens, Greece

Diaa-Eldin A. Mansour
Tanta University
Tanta, Egypt

Editorial Office
MDPI
St. Alban-Anlage 66
4052 Basel, Switzerland

This is a reprint of articles from the Topic published online in the open access journals *Energies* (ISSN 1996-1073), *Materials* (ISSN 1996-1944), and *Applied Sciences* (ISSN 2076-3417) (available at: https://www.mdpi.com/topics/nanomaterials).

For citation purposes, cite each article independently as indicated on the article page online and as indicated below:

Lastname, A.A.; Lastname, B.B. Article Title. *Journal Name* **Year**, *Volume Number*, Page Range.

ISBN 978-3-0365-9678-5 (Hbk)
ISBN 978-3-0365-9679-2 (PDF)
doi.org/10.3390/books978-3-0365-9679-2

© 2023 by the authors. Articles in this book are Open Access and distributed under the Creative Commons Attribution (CC BY) license. The book as a whole is distributed by MDPI under the terms and conditions of the Creative Commons Attribution-NonCommercial-NoDerivs (CC BY-NC-ND) license.

Contents

About the Editors vii

Preface ix

Imran Abbas, Shahid Hasnain, Nawal A. Alatawi, Muhammad Saqib and Daoud S. Mashat
Thermal Radiation Energy Performance on Stagnation-Point Flow in the Presence of Base Fluids Ethylene Glycol and Water over Stretching Sheet with Slip Boundary Condition
Reprinted from: *Energies* **2022**, *15*, 7965, doi:10.3390/en15217965 1

Marco Milanese, Francesco Micali, Gianpiero Colangelo and Arturo de Risi
Experimental Evaluation of a Full-Scale HVAC System Working with Nanofluid
Reprinted from: *Energies* **2022**, *15*, 2902, doi:10.3390/en15082902 25

Karolina Wenelska, Martyna Trukawka, Wojciech Kukulka, Xuecheng Chen and Ewa Mijowska
Co-Existence of Iron Oxide Nanoparticles and Manganese Oxide Nanorods as Decoration of Hollow Carbon Spheres for Boosting Electrochemical Performance of Li-Ion Battery
Reprinted from: *Materials* **2021**, *14*, 6902, doi:10.3390/ma14226902 39

Yujia Chen, Jiaqi Wang, Xiaohu Wang, Xuelei Li, Jun Liu, Jingshun Liu, et al.
Constructing High-Performance Carbon Nanofiber Anodes by the Hierarchical Porous Structure Regulation and Silicon/Nitrogen Co-Doping
Reprinted from: *Energies* **2022**, *15*, 4839, doi:10.3390/en15134839 49

Gang Liu, Xiaoyi Zhu, Xiaohua Li, Dongchen Jia, Dong Li, Zhaoli Ma and Jianjiang Li
Flexible Porous Silicon/Carbon Fiber Anode for High−Performance Lithium−Ion Batteries
Reprinted from: *Materials* **2022**, *15*, 3190, doi:10.3390/ma15093190 61

Alejandra Martínez-Lázaro, Luis A. Ramírez-Montoya, Janet Ledesma-García, Miguel A. Montes-Morán, Mayra P. Gurrola, José. Angel Menéndez, et al.
Facile Synthesis of Unsupported Pd Aerogel for High Performance Formic Acid Microfluidic Fuel Cell
Reprinted from: *Materials* **2022**, *15*, 1422, doi:10.3390/ma15041422 71

Chenchen Ji, Haonan Cui, Hongyu Mi and Shengchun Yang
Applications of 2D MXenes for Electrochemical Energy Conversion and Storage
Reprinted from: *Energies* **2021**, *14*, 8183, doi:10.3390/en14238183 87

Zhaoliang Xing, Chong Zhang, Mengyao Han, Ziwei Gao, Qingzhou Wu and Daomin Min
A Comparison of Electrical Breakdown Models for Polyethylene Nanocomposites
Reprinted from: *Appl. Sci.* **2022**, *12*, 6157, doi:10.3390/app12126157 111

Vladimir Bordo and Thomas Ebel
Theory of Electrical Breakdown in a Nanocomposite Capacitor
Reprinted from: *Appl. Sci.* **2022**, *12*, 5669, doi:10.3390/app12115669 127

Konstantinos N. Koutras, Ioannis A. Naxakis, Eleftheria C. Pyrgioti, Vasilios P. Charalampakos, Ioannis F. Gonos, Aspasia E. Antonelou and Spyros N. Yannopoulos
The Influence of Nanoparticles' Conductivity and Charging on Dielectric Properties of Ester Oil Based Nanofluid
Reprinted from: *Energies* **2020**, *13*, 6540, doi:10.3390/en13246540 137

Roberto D'Amato, Anna Donnadio, Chiara Battocchio, Paola Sassi, Monica Pica, Alessandra Carbone, et al.
Polydopamine Coated CeO_2 as Radical Scavenger Filler for Aquivion Membranes with High Proton Conductivity
Reprinted from: *Materials* **2021**, *14*, 5280, doi:10.3390/ma14185280 . . . . **153**

Zhiyu Tong, Linfeng Li, Yuanyuan Li, Qingmeng Wang and Xiaomin Cheng
The Effect of In Situ Synthesis of MgO Nanoparticles on the Thermal Properties of Ternary Nitrate
Reprinted from: *Materials* **2021**, *14*, 5737, doi:10.3390/ma14195737 . . . . **169**

Iman E. Shaaban, Ahmed S. Samra, Shabbir Muhammad and Swelm Wageh
Design of Distributed Bragg Reflectors for Green Light-Emitting Devices Based on Quantum Dots as Emission Layer
Reprinted from: *Energies* **2022**, *15*, 1237, doi:10.3390/en15031237 . . . . **183**

Jie Liu, Fangchao Li, Cheng Zhong and Wenbin Hu
Clean Electrochemical Synthesis of Pd–Pt Bimetallic Dendrites with High Electrocatalytic Performance for the Oxidation of Formic Acid
Reprinted from: *Materials* **2022**, *15*, 1554, doi:10.3390/ma15041554 . . . . **201**

Jianzhong Wang, Suo Tian, Xiaoze Liu, Xiangtao Wang, Yue Huang, Yingchao Fu and Qingfa Xu
Molecular Dynamics Simulation of the Oil–Water Interface Behavior of Modified Graphene Oxide and Its Effect on Interfacial Phenomena
Reprinted from: *Energies* **2022**, *15*, 4443, doi:10.3390/en15124443 . . . . **215**

About the Editors

Eleftheria C. Pyrgioti

Eleftheria C. Pyrgioti was born in Kanalia, Karditsa, Greece, in 1958. She received a diploma in Electrical Engineering and a Ph.D. from the Electrical and Computer Engineering Dept., University of Patras, Greece, in 1981 and 1991, respectively. She is a professor at the Electrical and Computer Engineering Dept., University of Patras. Her research interests concern high-voltage systems, lightning protection, high-voltage insulation, dielectric liquids' distributed generation, and renewable energy.

Ioannis F. Gonos

Ioannis F. Gonos was born in Artemisio, Arcadia, Greece, in 1970. He received a diploma in Electrical Engineering and a Ph.D. from the National Technical University of Athens (NTUA) in 1993 and 2002, respectively. He is a professor at NTUA. His research interests concern grounding systems, dielectric liquids, high voltages, measurements, and insulators. He is the author of more than 200 papers in scientific journals and conference proceedings.

Diaa-Eldin A. Mansour

Diaa-Eldin A. Mansour was born in Tanta, Egypt, in 1978. He received his B.Sc. and M.Sc. degrees in Electrical Engineering from Tanta University, Tanta, Egypt, in 2000 and 2004, respectively, and a Ph.D. in Electrical Engineering from Nagoya University, Nagoya, Japan, in 2010. Since 2000, he has been with the Department of Electrical Power and Machines Engineering, Faculty of Engineering, Tanta University, Egypt, where he has been a professor since 2020. He is currently working as a professor at the Electrical Power Engineering Department, Egypt–Japan University of Science and Technology (E-JUST), Alexandria, Egypt, whilst on leave from Tanta University. His research interests include high-voltage engineering, the condition monitoring of electrical power equipment, IoT applications in electrical power systems, and applied superconductivity. He received the Best Presentation Award twice from the IEE of Japan, in 2008 and 2009, Prof. Khalifa's Prize from the Egyptian Academy of Scientific Research and Technology in 2013, the Tanta University Encouragement Award in 2016, and the Egypt-State Encouragement Award in the field of engineering sciences in 2018. Recently, he has been listed among the world's top 2% scientists by Stanford University, USA, from 2020 to 2023.

Preface

Recently, there have been remarkable advancements in the field of energy materials, especially for energy conversion and storage. The development of nanomaterials has been the main driver of these advancements. Nanomaterials have shown great potential in enhancing the performance and efficiency of various energy-related systems due to their unique properties and tailored structures at the nanoscale. This book presents a collection of research studies that explore the exciting developments and applications of nanomaterials in energy-related systems. The book covers various aspects from the material level to the device level. It presents a detailed description of the structural and compositional characteristics of the used nanomaterials. In addition, it includes diverse methodologies, either based on experimental measurements or based on molecular dynamic simulations. For experimental measurements, different properties are evaluated, such as electrical, thermal, dielectric, and electrochemical properties. Overall, the book serves as a valuable resource for researchers, scientists, and engineers involved in the fields of energy-related systems and nanotechnology.

Eleftheria C. Pyrgioti, Ioannis F. Gonos, and Diaa-Eldin A. Mansour
Editors

Article

Thermal Radiation Energy Performance on Stagnation-Point Flow in the Presence of Base Fluids Ethylene Glycol and Water over Stretching Sheet with Slip Boundary Condition

Imran Abbas [1,†], Shahid Hasnain [2,†], Nawal A. Alatawi [2,†], Muhammad Saqib [3,†] and Daoud S. Mashat [2,*]

1. Department of Mathematics, Air University, Islamabad 44000, Pakistan
2. Department of Mathematics and Science, King Abdulaziz University, Jeddah 21589, Saudi Arabia
3. Department of Mathematics, Khwaja Fareed University of Engineering & Information Technology, Rahim Yar Khan 48800, Pakistan
* Correspondence: dmashat@kau.edu.sa
† These authors contributed equally to this work.

Abstract: Nanoparticles are useful in improving the efficiency of convective heat transfer. The current study addresses this gap by making use of an analogy between Al_2O_3 and $\gamma\text{-}Al_2O_3$ nanoparticles in various base fluids across a stretched sheet conjunction with f. Base fluids include ethylene glycol and water. We address, for the first time, the stagnation-point flow of a boundary layer of $\gamma\text{-}Al_2O_3$ nanofluid over a stretched sheet with slip boundary condition. Al_2O_3 nanofluids employ Brinkman viscosity and Maxwell's thermal conductivity models with thermal radiations, whereas $\gamma\text{-}Al_2O_3$ nanofluids use viscosity and thermal conductivity models generated from experimental data. For the boundary layer, the motion equation was solved numerically using the fourth-order Runge–Kutta method and the shooting approach. Plots of the velocity profile, temperature profile, skin friction coefficient and reduced Nusselt number are shown. Simultaneous exposure of the identical nanoparticles to water and ethylene glycol, it is projected, would result in markedly different behaviors with respect to the temperature profile. Therefore, this kind of research instills confidence in us to conduct an analysis of the various nanoparticle decompositions and profile structures with regard to various base fluids.

Keywords: Al_2O_3; $\gamma\text{-}Al_2O_3$; stagnation-point flow; nanofluids; stretching sheet; slip boundary condition; shooting method

1. Introduction

Water, ethylene glycol and mineral oils are traditional heat transfer fluids with low thermal conductivities which may restrict their effectiveness in many industrial domains such as chemical processing, generation of power, air conditioning systems, microelectronics and transportation. Solid particle suspensions and fluids have a strong ability to enhance heat transfer. One may classify particles in a number of ways including metallic, non-metallic and polymeric. However, there are issues that arise when industries use macro-sized suspensions such as heat transfer erosion and flow channel blockage owing to poor suspension stability and a gradual but steady decrease in pressure. Therefore, researchers and engineers have been hammering away to conquer this fundamental barrier by dispersing particles as small as millimeters or micrometers in liquids since Maxwell (1873). Large particles quickly settle in fluids which is an issue. Extended surface technology has reached its limits in thermal management system designs; thus, innovations that might increase interest in which nanofluids are nanotechnological heat transfer fluids. Nanofluids are suited for engineering applications and have various benefits over convectional suspensions which include improved stability, high thermal conductivity and negligible pressure loss. Thus, nanofluid technology will emerge as a promising and interesting field of study in the twenty-first century [1–9].

Flow and heat transfer caused by the stretching sheet is a prominent process in many industrial applications, such as metallic sheet cooling, crystal growth in cooling baths, plastics and rubber sheets manufacturing, paper and glass fiber production, and the eviction of polymer and metals. Although enhancement of thermal and electrical conductivity can be achieved by using metallic particles was first proposed by Maxwell (1873), Choi learned from his work with micrometer-sized particle and fiber suspensions in the 1980s in which traditional particles in micro-channel flow passages are unusable [9–11]. However, advancements in nanotechnology can process and manufacture materials with average crystal-lite sizes below 50 nm. Thermal conductivity is key to developing energy-efficient heat transfer fluids. Because of the growing level of competitiveness on a worldwide scale, a number of different sectors are in desperate need of innovative heat transfer fluids that have much greater thermal conductivities than those that are already on the market [12]. In recent years, several studies have investigated the nanofluids boundary layer flow across a stretched surface using a wide range of metal and oxide nanoparticles [13–17]. Using a reworked version of Buongiorno's model with verified thermo-physical correlations which examine the impact of Darcy–Forchheimer and Lorentz forces on radiative alumina-water nanofluid flows across a slippery curved geometry subject to numerous convective restrictions [18]. Later on, Saima et al. [19] used a finite volume method, a numerical approach to study micropolar nanofluids flow through a lid-driven cavity. The development of innovative hybrid 2D-3D graphene oxide diamond micro composite polyimide films to alleviate electrical and thermal conduction. It is believed to be a useful option for the thermal dissipation of the electronic components of electric machines as a result of its high and outstanding thermal conductivity [20]. A numerical investigation for two-dimensional Sutterby fluid flow which is bounded at a stagnation point with an inclined magnetic field and thermal radiation was conducted by Sabir et al. [21]. Meanwhile, another computational approach was performed on stagnation point pseudo-plastic nano-liquid flow towards a flexible Riga sheet by Azad et al. [22]. The stagnation-point flow of an incompressible non-Newtonian fluid over a non-isothermal stretching sheet is investigated by Rashidi et al. [23]. Baag et al. [24] investigated numerical methods into the flow of MHD micro-polar fluids toward a stagnation point on a vertical surface with a heat source and a chemical reaction.

In light of this, a comparison study was carried out on the flow with velocity and thermal boundary layers (BL) of Al_2O_3 and γ-Al_2O_3 nanofluids along various base fluids across a stretching sheet. We have been successful in developing a conjuncture between stagnation-point flow and nanofluids flow over the stretching sheet with slip boundary conditions, by the numerical technique. Models of viscosity and thermal conductivity are developed from data through experiments. For Al_2O_3, Maxwell's thermal conductivity with radiation and Brinkman viscosity models are used in nanofluids flow. Therefore, this is a comparative new study and its contribution to the existing body of research will be significant.

1.1. Theoretical to Experimental Perspective on Nanofluids

The thermal properties of nanoparticles are explained through nanofluid theory, which supports physics and chemistry-based predictive models. The thermal conductivity of the nanofluids has not been explained for numerous reasons. First, nanofluids behave differently from solid-to-fluid suspensions or typical solid-to-solid composites. Reducing nanoparticle size increases nanofluid thermal conductivity. Second, nanofluids and traditional solid-to-liquid suspensions differ in thermal conductivity, concentration of the particles, and size. Thirdly, nanofluids are a new, highly multidisciplinary field that spans engineering, material science, physics, chemistry and colloidal science. Nanofluids hence need expertise in each field, which shows the difficulty to build a nanofluid theory [17,25–27]. Therefore, predictions are poorer when the nanoparticles are suspended in a liquid because these interactions include electromagnetic or particle-to-lattice heat transfer in addition to the lattice vibrational heat transfer predicted by the liquid models. Thus, a theory for nanofluid thermal conductivity can be developed by taking into account two crucial components, static and dynamic mechanisms. Near field radiation in nanofluids seems like a promising theory [26,28].

The most recent advances in fabrication technology have opened up exciting new possibilities for actively processing materials at nano-scale sizes. Materials that are either nanostructured or nanophase are composed of nanometer-sized components. Because of this, particles of a size less than 100 nm have characteristics that are distinct from those of traditional solids. The remarkable qualities that are associated with nanophase materials are a direct result of the relatively high surface area/volume ratio that these materials possess. This ratio is made possible by the presence of a significant number of constituent atoms that are located at the grain boundaries. Nanophase materials have superior thermal, mechanical, optical, magnetic, and electrical capabilities compared to traditional materials that have coarse grain patterns. These qualities include magnetism and electrical conductivity. As a consequence of this, the exploration of nanophase materials in research and development has attracted a significant amount of attention from both material scientists and engineers [29]. Several kinds of nanoparticles that are employed in nanofluids can be constructed out of a wide variety of materials, including oxide ceramics (Al_2O_3, CuO), carbide ceramic, metals (Cu, Ag, Au), semiconductors, composite materials and alloyed nanoparticles are some examples of advanced materials. Whereas, in the development of nanofluids a wide variety of liquids, including water, ethylene glycol and oil, have been employed as base fluids [29,30].

1.1.1. Volume Fraction and the Particle Size

Many experiments on nanofluid thermal conductivity have been described in recent years. Table 1 summarizes published experimental studies on nanofluids at room temperature. So, nanofluids with thermal conductivity higher than their base fluids, even at low nanoparticle concentrations, grow well with nanoparticle volume fraction. Below are several nanofluid thermal conductivity investigations.

Eastman et al. [31] first reported nanofluids' increased effective thermal conductivity. Al_2O_3 and CuO nanoparticles dispersed in water increased thermal conductivity by 29% to 60% for 5% nanoparticle volume fraction which later findings found a moderate increase in thermal conductivity for Al_2O_3 and CuO nanoparticles in water and ethylene glycol. Li et al. [32] recently studied the thermal conductivity of CuO and Al_2O_3 nanoparticles boosted waters thermal conductivity by 52% and 22% at 6% fractional part of volume at a temperature of 34 °C. Choi et al. [33] investigated multi-walled carbon nanotube-containing oil suspensions' thermal conductivity. Thermal conductivity doubled with 1% volumetric loading. Even at low volume fractions, nanotube addition increases conductivity non-linearly. Strong thermal field interactions between fibers might be to cause. TiO_2 nanoparticles are used to evaluate the thermal conductivity in deionized water by Murshed et al. [34]. Their findings demonstrated, for the very first time, that there was no anomalous improvement in the thermal conductivity of nanofluids containing a very low volume proportion of particles. This conclusion is in direct opposition to Patel et al.'s [35] unusual finding which shows the incorrect Patel's hypothesis. According to a comparison of the studies that have been conducted, the increases in thermal conductivities of various types of nanofluids are distinct from one another. The size and composition of the nanoparticles, in addition to the base fluids, both have an effect on the thermal conductivity of nanofluids. Particle size is essential to optimize results and build a relation to volume fraction which causes nano-scale mechanism in the suspensions. According to theoretical evidence, with a reduction in particle size, the effective thermal conductivity of nanofluids improved [36,37].

Table 1. An overview of various nanofluids and thermal conductivity.

Researchers	Nanoparticle Size/Base-Fluid	Measurement	Thermal Conductivity (k) ↑
Eastman et al. [31]	Al_2O_3 (33)/water	transient hot-wire	29% for 5 vol %
	CuO (36)/water		60% for 5 vol %
Murshed et al. [34]	TiO_2/deionised-water	transient hot-wire	30% for 5 vol %
Li and Peterson [32]	Al_2O_3 (36)/water	SS method	52% for 6 vol %
Xuan and Li [37]	Cu (10)/water	transient hot-wire	70% for 3 vol %
Hwang et al. [38]	CuO (35.4)/water/EG	transient hot-wire	9% for 1 vol %
Liu et al. [39]	CuO (29)/EG	transient hot-wire	23% for 5 vol%
Kri. et al. [40]	Al_2O_3(20)/water	Unspecified	16% for 1 vol %
Wen and Ding [41]	TiO_2 (34)/water	transient hot-wire	6% for 0.66 vol %

1.1.2. The Influence of Temperature on Nanofluid

The temperature may affect nanofluid thermal conductivity. Despite the fact that nanofluids can be used at different temperatures, few studies have examined the temperature effect on their thermal conductivity. Table 2 summarizes published studies on nanofluids in which thermal conductivity dependency on temperature. The nanofluids which comprise Al_2O_3 and CuO nanoparticles in water are used to measure the thermal conductivity which is investigated by Das et al. [42]. In their investigation, they employed a temperature oscillation technique to measure thermal diffusivity. They found that the thermal conductivity enhancement of these nanofluids increased by a factor of two to four over the temperature range of 21° to 51° Celsius. The different sizes of nanoparticles of Al_2O_3 dispersing in water to investigate the thermal conductivity of nanofluids are carried out by Chon and Kihm [43]. The nanoparticles ranged from 47 to 150 nanometers in size. They found that there was a slight increase in thermal conductivity in relation to temperature. When the temperature of the fluid was raised from 31 °C to 51 °C, there was a 6–11% increase in the thermal conductivity of nanofluids. It is interesting to note a little reduction in the thermal conductivity as the temperature increases, which is the opposite tendency seen in nanofluids containing spherical nanoparticles.

Table 2. The Influence of Temperature on Nanofluids.

Researchers	Nanoparticle Size/Base-Fluid	Measurement	Thermal Conductivity (k) ↑
Das et al. [42]	Al_2O_3 (38.4)/water	Temperature Oscillations	For 4 vol %: 16% at 36 °C and 25% at 51 °C
Li and Peterson [32]	Al_2O_3 (36)/water	SS method	For 2 vol %: 7% at 27.5 °C and 23% at 36 °C
Chon and Kihm [43]	Al_2O_3 (47)/water	transient hot-wire	6% at 31 °C and 11% at 51 °C
	Al_2O_3 (150)/water		3% at 51 °C and 8.5% at 51 °C
Murshed et al. [44]	Al_2O_3 (150)/DIW	transient hot-wire	For 1 vol %: 11.4% at 60 °C

2. Mathematical Formulation

Assume a two-dimensional laminar boundary layer incompressible flow with steady characteristics in which Al_2O_3 and γ-Al_2O_3 nanofluids move across a stretched sheet of various base fluids, such as water and $C_2H_6O_2$, see from Figure 1. The flow that is related to nanofluids is produced as a result of the sheet being stretched along the x-axis by two forces that are identical in magnitude but act in opposite directions. The stretching velocity $u_w(x)$

is used to suppose that the flow is confined to $y > 0$ and external velocity to boundary layer flow is U_∞, such that $U_\infty = \tilde{a}x$, where $\tilde{a}$ is constant. It is observed that the ratio of the stretching surface velocity to the inviscid flow at the stagnation point determines the shape of a boundary layer produced in a stagnation-point flow of an incompressible viscous fluid towards a stretching surface. Therefore, the stagnation-point flow is represented by the term, $\tilde{U}_\infty \dfrac{d\tilde{U}_\infty}{dx}$ in the momentum equation of the fluid flow where $\tilde{U}_\infty$ is the velocity distribution for the free stream which is far away from the surface. The temperature profile at the surface which is being stretched is $T_w = T_\infty + \tilde{b}x^2$, with $\tilde{b}$ is constant and T_∞ is the ambient temperature. In addition, it is assumed that the nanoparticles and the base fluids are in a state of thermal equilibrium and there is a slip between them [21,23,45]. Table 3 summarizes the considered thermo-physical features of nanofluids.

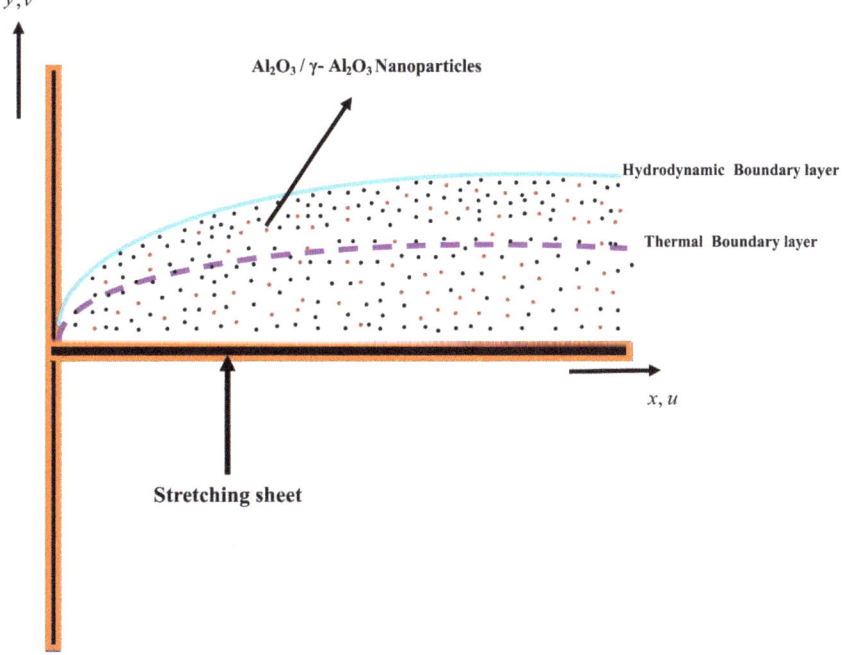

Figure 1. Physical description to mathematical model.

Table 3. Thermophysical characteristics of Nanofluids.

	Density (kg/m³)	C_p (J/Kg K)	k (W/mK)	P_r
H$_2$O (Pure Water)	998.3	4182	0.60	6.96
C$_2$H$_6$O$_2$ (Ethylene glycol)	1116.6	2382	0.249	204
Al$_2$O$_3$ (Alumina)	3970	765	40	-

Under these conditions, we can write down the steady boundary equation that controls the convective flow and heat transfer of nanofluids as:

$$\tilde{u}_x + \tilde{v}_y = 0, \qquad (1)$$

$$\tilde{u}\tilde{u}_x + \tilde{v}\tilde{u}_y = \dfrac{\mu_{nf}}{\rho_{nf}} \dfrac{\partial^2 \tilde{u}}{\partial y^2} + \tilde{U}_\infty \dfrac{d\tilde{U}_\infty}{dx} \qquad (2)$$

$$\tilde{u}\tilde{T}_x + \tilde{v}\tilde{T}_y = \frac{k_{nf}}{(\rho C_p)_{nf}} \frac{\partial^2 \tilde{T}}{\partial y^2} - \frac{1}{\tilde{\rho}\tilde{C}_p} \frac{\partial \tilde{q}_r}{\partial y}, \tag{3}$$

where $\tilde{u}$ and $\tilde{v}$ are velocity components along x and y directions, respectively. Moreover, $\tilde{\rho}$ is the density of the fluid and $\tilde{C}_p$ represents specific heat at constant pressure. For the reason that the intensity of radiant emission increases with increasing absolute temperature a very crucial factor in the heat transfer process. Therefore, the term $\frac{1}{\tilde{\rho}\tilde{C}_p} \frac{\partial \tilde{q}_r}{\partial y}$ represents the thermal radiation in which $\tilde{q}_r$ is the radiative heat flux. Such heat flux is defined by Rosseland approximation,

$$\tilde{q}_r = -\left(\frac{4\tilde{\sigma}}{3\tilde{k}}\right) \frac{\partial T^4}{\partial y}, \tag{4}$$

where $\tilde{k}$ is the Rosseland coefficient of mean absorption and $\tilde{\sigma}$ is the Stefan Boltzmann constant. For the construction of the linear function of temperature T^4, apply the Taylor series about the free stream temperature T_∞ and neglect the higher power, we have the following equation,

$$T^4 \approx (4T - 3T_\infty)T_\infty^3. \tag{5}$$

Therefore, the radiative heat flux can be expressed as;

$$\tilde{q}_r = -\left(\frac{16\tilde{\sigma}}{3\tilde{k}}\right) T_\infty^3 \frac{\partial T}{\partial y}. \tag{6}$$

Boundary conditions are:

$$\tilde{u} = u_w + \check{a}\frac{\partial u}{\partial y}, \; \tilde{v} = 0, \; \tilde{T} = T_\omega(T_\omega = T_\infty + \tilde{b}x^2) \; at \; y = 0, \tag{7}$$

$$\tilde{u} \to 0, \; \tilde{T} \to T_\infty \; as \; y \to \infty \tag{8}$$

where, u is the tangential velocity of the free fluid which is exterior normal to the stretching sheet and $\check{a} = \frac{\sqrt{\check{K}}}{\check{\alpha}}$ in which $\check{K}$ is the permeability, $u_w(x)$ stretching velocity and $\check{\alpha}$ is a dimensionless parameter which depends only on the properties of the fluid and permeable material.

3. Thermophysical Characteristics of Al$_2$O$_3$ and γ-Al$_2$O$_3$ Nanofluids

The heat capacitance $(\rho C_p)_{nf}$ and the effective dynamic density ρ_{nf} of the nanofluids have been provided by the following expressions

$$\left.\begin{aligned}\rho_{nf} &= (1-\varphi)\rho_f + \varphi\rho_s, \\ (\rho C_p)_{nf} &= (1-\varphi)(\rho C_p)_f + \varphi(\rho C_p)_s,\end{aligned}\right\} \tag{9}$$

where the solid volume fraction of nanofluids is denoted by φ whereas the nanofluid's dynamic viscosity is characterized by

$$\left.\begin{aligned}\frac{\mu_{nf}}{\mu_f} &= (1-\varphi)^{-2.5}(for \; Al_2O_3 - water) \\ \frac{\mu_{nf}}{\mu_f} &= 123\varphi^2 + 7.3\varphi + 1 (for \; \gamma - Al_2O_3 - water) \\ \frac{\mu_{nf}}{\mu_f} &= 306\varphi^2 - 0.19\varphi + 1 (for \; \gamma - Al_2O_3 - C_2H_6O_2).\end{aligned}\right\} \tag{10}$$

An expression for the nanofluid's effective thermal conductivity is

$$\left.\begin{array}{l}\dfrac{k_{nf}}{k_f} = \dfrac{k_s + 2k_f - 2\varphi(k_f - k_s)}{k_s + 2k_f + \varphi(k_f - k_s)} \quad (for \ Al_2O_3 - water) \\[2mm] \dfrac{k_{nf}}{k_f} = 4.97\varphi^2 + 2.72\varphi + 1 \quad (for \ \gamma - Al_2O_3 - water) \\[2mm] \dfrac{k_{nf}}{k_f} = 28.905\varphi^2 + 2.87\varphi + 1 \quad (for \ \gamma - Al_2O_3 - C_2H_6O_2)\end{array}\right\} \quad (11)$$

4. Non-Dimensionalization through Similarity Transformation

By making use of similarities, transformation is achieved by

$$\eta = \sqrt{\dfrac{a}{v_f}}\, y, \ \tilde{u} = axf'(\eta), \ \tilde{v} = -(av_f)^{\frac{1}{2}} f(\eta), \ \theta = \dfrac{\tilde{T} - T_\infty}{T_w - T_\infty} \quad (12)$$

4.1. Momentum Equations

The Equation (2) regulate the boundary layer into non-dimensional ordinary differential equations which can be written in the following way,

$$\left.\begin{array}{l} f''' - (1-\varphi)^{-2.5}\left(1 - \varphi\left(\dfrac{\rho_s}{\rho_f}\right)\right)(f' - ff'') + \varepsilon^2 = 0 \ (for \ Al_2O_3 - H_2O) \\[3mm] f''' - \left(\dfrac{1 - \varphi\left(\dfrac{\rho_s}{\rho_f}\right)}{306\varphi^2 - 0.19\varphi + 1}\right)(f' - ff'') + \varepsilon^2 = 0 \ (for \ \gamma - Al_2O_3 - water) \\[3mm] f''' - \left(\dfrac{1 - \varphi\left(\dfrac{\rho_s}{\rho_f}\right)}{123\varphi^2 + 7.3\varphi + 1}\right)(f' - ff'') + \varepsilon^2 = 0 \ (for \ \gamma - Al_2O_3 - C_2H_6O_2). \end{array}\right\} \quad (13)$$

4.2. Energy Equations

The temperature Equation (3) regulate the boundary layer into non-dimensional ordinary differential equations which can be explained as,

$$\left.\begin{array}{l} \theta''\left(\dfrac{\tilde{k}_{nf}}{\tilde{k}_f} + \dfrac{4}{3}R\right) - p_r\left((1-\varphi)^{2.5} + \varphi\left(\dfrac{(\rho C_p)_s}{(\rho C_p)_f}\right)\right)(f\theta' - 2\theta f') = 0 \ for \ Al_2O_3 - H_2O, \\[3mm] \theta''\left(\dfrac{\tilde{k}_{nf}}{\tilde{k}_f} + \dfrac{4}{3}R\right) - p_r\left((1-\varphi)^{2.5} + \varphi\left(\dfrac{(\rho C_p)_s}{(\rho C_p)_f}\right)\right)(f\theta' - 2\theta f') = 0 \ for \ \gamma - Al_2O_3 - water, \\[3mm] \theta''\left(\dfrac{\tilde{k}_{nf}}{\tilde{k}_f} + \dfrac{4}{3}R\right) - p_r\left((1-\varphi)^{2.5} + \varphi\left(\dfrac{(\rho C_p)_s}{(\rho C_p)_f}\right)\right)(f\theta' - 2\theta f') = 0 \ for \ \gamma - Al_2O_3 - C_2H_6O_2. \end{array}\right\} \quad (14)$$

4.3. Boundary Conditions

The related boundary conditions are as follows:

For Momentum equation

$$f(0) = 0, \ f'(0) = 1 \ \& \ f'(\infty) = 0. \quad (15)$$

For Energy equation

$$\theta(0) = 1 \ \& \ \theta(\infty) = 0. \quad (16)$$

4.4. Solution Methodology

In this part, an overview of the solution to the non-linear ODE with regard to the boundary conditions is offered. A well-known method known as the shooting approach is used in order to show the solution for these non-linear ODEs. MATLAB's built-in solver called bvp4c is used to do an analysis of the numerical results, with the domain set at zero to η_{max}.

Pseudo-Algorithm

To accomplish the transformation from BVP to IVP, the representations f by Y_1 and θ by Y_4 have been put into action:

$$\left.\begin{array}{ll} Y_1' = Y_2, & Y_1(0) = 0 \\ Y_2' = Y_3, & Y_2(0) = 1 \\ Y_3' = (A_1 \times A_1)(Y_2 - Y_1)Y_3 - \epsilon^2, & Y_3(0) = I_1 \\ Y_4' = Y_5, & Y_4(0) = 1 \\ Y_5' = \left(\dfrac{A_5}{(A_6 + (4/3)\lambda)}\right) * P_r * (2 * Y_4 Y_2 - Y_1)Y_5, & Y_5(0) = I_2 \end{array}\right\} \quad (17)$$

The preceding IVP is supplied by making use of a well-known shooting strategy in conjunction with the Runge–Kutta scheme. The starting conditions that are not present are represented by the notation $Y_3(0) = I_1$ and $Y_5(0) = I_2$, respectively. Newton's technique, in its usual form, may be used to fill in missing beginning circumstances and still obtain accurate results. The numerical strategy of the shooting is applied to the missing values of I_1 and I_2 until it is unable to satisfy the tolerance ζ, which is $max\{|Y_3(\eta_{max})|, |Y_5(\eta_{max})|\} < \zeta$ [20,21].

5. Important Physical Characteristics

The skin friction coefficient (shear stress rate) and the Nusselt number (rate of heat transfer) are two physical characteristics of importance in engineering problems. The skin friction coefficient C_f is used to calculate the shear stress at the stretched sheet which can be defined as:

$$c_f = -\dfrac{-2\mu_{nf}}{\rho_f u_w^2}(u_y)_{y=0}. \quad (18)$$

By utilizing Equation (18) into Equation (13), we have the following expressions:

$$\left.\begin{array}{l} \dfrac{1}{2}Re_x^{\frac{1}{2}} c_f = -(1-\varphi)^{-2.5} f''(0), \ (for\ Al_2O_3 - H_2O) \\ \dfrac{1}{2}Re_x^{\frac{1}{2}} c_f = -(123\varphi^2 + 7.3\varphi + 1)f''(0), \ (for\ \gamma - Al_2O_3 - H_2O) \\ \dfrac{1}{2}Re_x^{\frac{1}{2}} c_f = -(306\varphi^2 - 0.19\varphi + 1)f''(0), \ (for\ \gamma - Al_2O_3 - C_2H_6O_2) \end{array}\right\} \quad (19)$$

where the local Reynolds number is denoted by $Re_x = \dfrac{x u_w(x)}{v_f}$, which completely base on the stretching velocity $U_\infty(x)$ and local skin friction coefficient $Re_x^{\frac{1}{2}} c_f$. The Nusselt number Nu_x can be defined as

$$Nu_x = \dfrac{x \tilde{q}_w}{\tilde{k}_f (T_\omega - T_\infty)}, \quad (20)$$

where, the local surface heat flux is $\tilde{q} = k_{nf}(T_y)_{y=0}$. Based on Equation (12), we obtain the following Nusselt number,

$$Re_x^{-\frac{1}{2}}Nu_x = \left(\frac{k_s + 2k_f - 2\varphi(k_f - k_s)}{k_s + 2k_f + \varphi(k_f - k_s)}\right)(-\theta'(0)) \ (for \ Al_2O_3 - H_2O)$$
$$Re_x^{-\frac{1}{2}}Nu_x = (4.97\varphi^2 + 2.72\varphi + 1)(-\theta'(0)) \ (for \ \gamma - Al_2O_3 - H_2O) \quad (21)$$
$$\frac{k_{nf}}{k_f} = (28.905\varphi^2 + 2.87\varphi + 1)(-\theta'(0)) \ (for \ \gamma - Al_2O_3 - C_2H_6O_2).$$

6. Discussion

The graphical representation of the numerical findings is shown for the nanoparticles of Al_2O_3 and γ-Al_2O_3 combined with water and ethylene glycol as the base fluids. An investigation into the effects of a number of disparate parameters, including the stagnation point parameter ϵ, the dimensionless slip parameter $\grave{\alpha}$, the solid volume fraction φ, Prandtl number Pr, reduced skin friction coefficient $Re_x^{\frac{1}{2}}c_f$, reduced Nusselt number $Re_x^{-\frac{1}{2}}Nu_x$, shear stress $f''(0)$, velocity profile $f'(\eta)$ and temperature profile $\theta(\eta)$ have been analyzed with regard to the slip boundary conditions on a stretching sheet conjunction with stagnation-point flow phenomena.

Table 4 presents the results of a comparison between the values of $-f''(0)$ that were reported for Al_2O_3-water by Hamad et al. [16] and Vishnu et al. [45]. We are confident in continuing to utilize the existing code since the results indicate great agreement. Through the use of the figures, the authors examine the impact that the nanoparticle volume friction has on the velocity profile, temperature profile, skin friction coefficient and the reduced Nusselt number.

Table 4. Results comparison with earlier findings for Al_2O_3 nanofluids.

φ	Hamid et al. [16]	Vishnu et al. [45]	Present Work
0.05	1.00538	1.00537	1.00530
0.10	0.99877	0.99877	0.98866
0.15	0.98185	0.98184	0.97132
0.2	0.95992	0.95591	0.94581

The influence of the nanoparticle volume fraction on the velocity profile is described via the Figures 2–4 for Al_2O_3-water, Al_2O_3-ethylene glycol and γ-Al_2O_3-water nanofluids. It has been determined, based on the findings presented here, that higher values of nanoparticle volume fraction result in an increase in the velocity of oxide nanofluids. It may be said that the velocity of the γ-Al_2O_3 nanofluids is greater than that of the Al_2O_3 nanofluids [16,45]. This is because the thickness of the momentum boundary layer in γ-Al_2O_3 nanofluids is much greater than in Al_2O_3 nanofluids. When comparing nanofluids according to different base fluids, those based on ethylene glycol have a faster velocity than those based on water. Nanofluids based on water have a smaller momentum boundary barrier thickness compared to nanofluids based on ethylene glycol. The results which are depicted from Figures 2–4 show that the Al_2O_3-water mixture has a lower velocity but the γ-Al_2O_3-Water mixture has a greater velocity.

Figures 5–7 show the impact of the nanoparticle volume fraction on the temperature profile of the nanofluids. It has been observed that when the values of the nanoparticle volume fraction rise, the temperatures of both γ-Al_2O_3 and Al_2O_3 nanofluids rise as well. Therefore, nanofluids based on water have a steeper temperature profile compared to those on ethylene glycol (shallower profile) [16,45–47]. This is owing to the fact that the thermal diffusivity of water is significantly greater than that of ethylene glycol whereas, the Prandtl number P_r of water is substantially lower than that of ethylene glycol [34,40,45]. Comparing nanoparticles, Al_2O_3-water has a higher temperature profile than γ-Al_2O_3-water, while γ-Al_2O_3-ethylene glycol is higher. It is possible to draw the conclusion that the Al_2O_3 nanoparticles and the γ-Al_2O_3 nanoparticles have opposing impacts on the temperature profile when used in conjunction with various base fluids such as water and ethylene glycol.

To achieve cooling effects, it is possible to make use of nanofluids that include ethylene glycol functioning as the basis fluid [45].

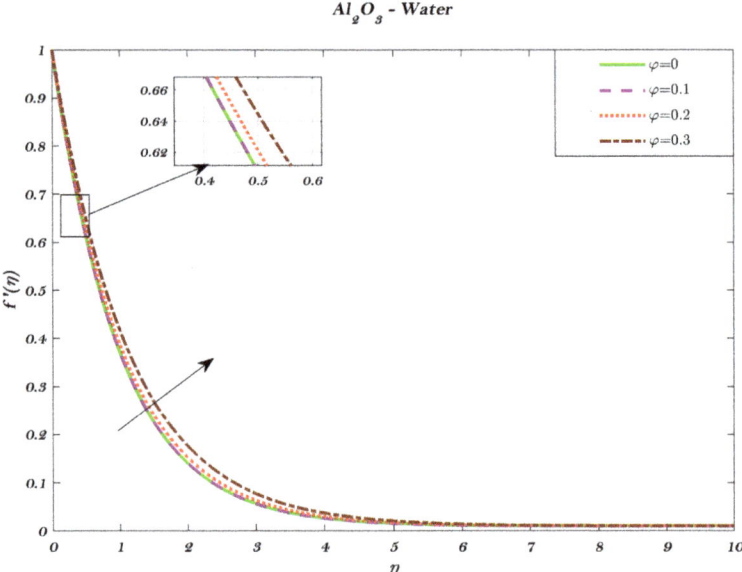

Figure 2. Velocity profile as a function of nanoparticle volume fraction φ with fix parameters $\epsilon = 0.02$ $\check{a} = 0.01$ and $P_r = 6.96$.

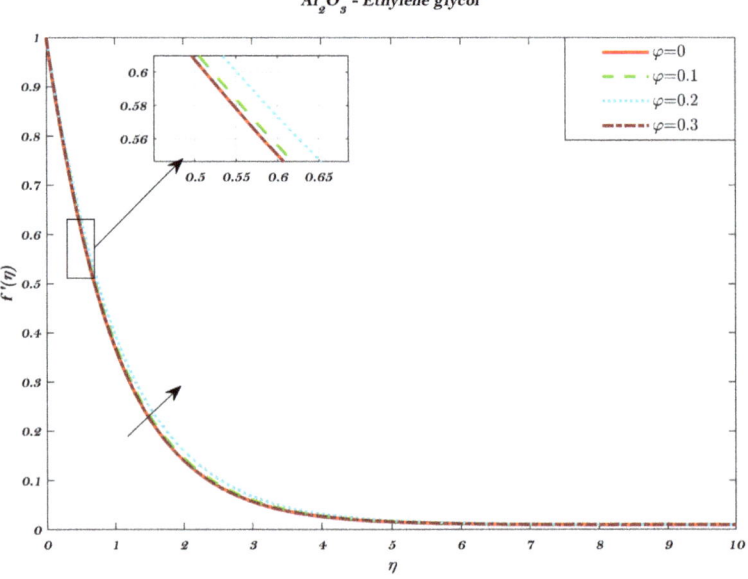

Figure 3. Velocity profile as a function of nanoparticle volume fraction φ with fix parameters $\epsilon = 0.02$ $\check{a} = 0.01$ and $P_r = 204$.

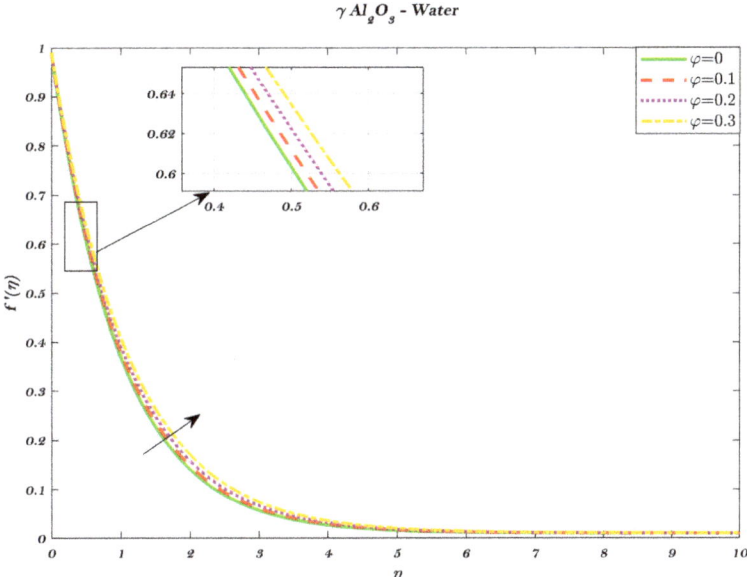

Figure 4. Velocity profile as a function of nanoparticle volume fraction φ with fix parameters $\epsilon = 0.02$ $\breve{a} = 0.01$ and $P_r = 6.96$.

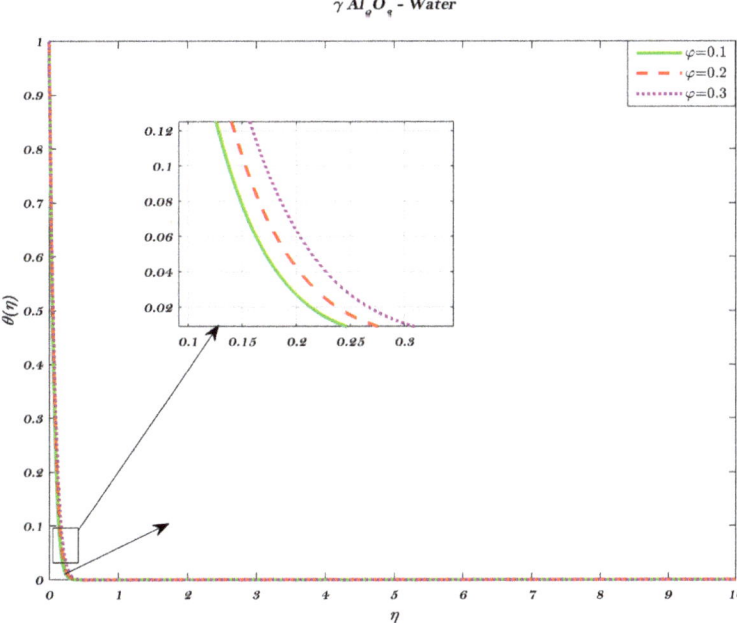

Figure 5. Temperature profile as a function of nanoparticle volume fraction φ with fix parameters $\epsilon = 0.02$ $\breve{a} = 0.01$, $P_r = 6.96$ and $R = 1$.

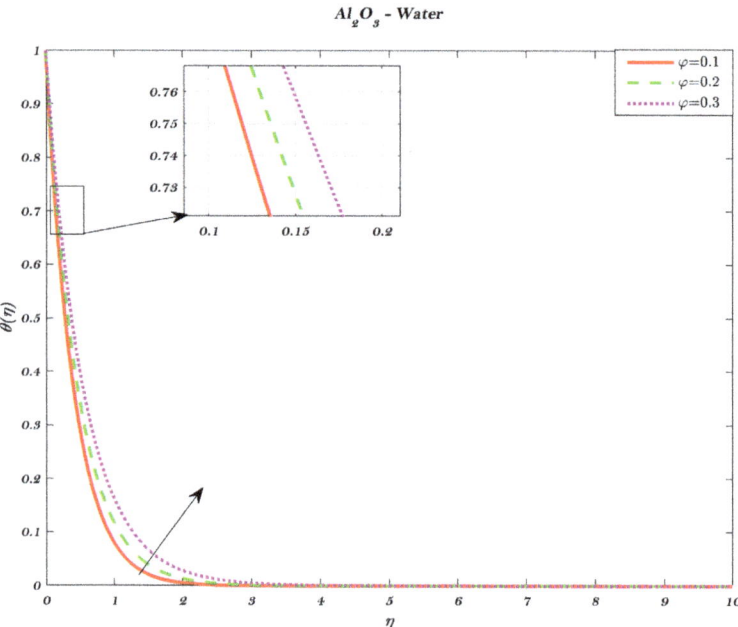

Figure 6. Temperature profile as a function of nanoparticle volume fraction φ with fix parameters $\epsilon = 0.02$ $ă = 0.01$, $P_r = 204$ and $R = 1$.

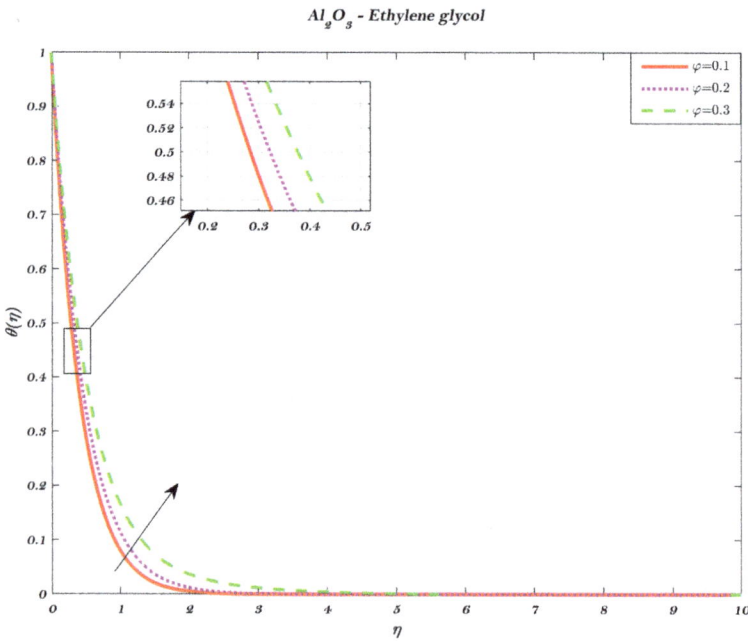

Figure 7. Temperature profile as a function of nanoparticle volume fraction φ with fix parameters $\epsilon = 0.02$ $ă = 0.01$, $P_r = 6.96$ and $R = 1$.

The impacts of different physical parameters in the flow model on the velocity of the nanofluids within the boundary layer are depicted in Figures 8–10, while Figures 11–13 show the results for temperature profile. The influence of the slip parameter $ă$ against the similarity variable η on the velocity and temperature profiles show that an increase in the value of a parameter known as the slip velocity results in a reduction in the velocity of the nanofluids also an increase in the value of the same parameter results in a rise in the temperature [45,47].

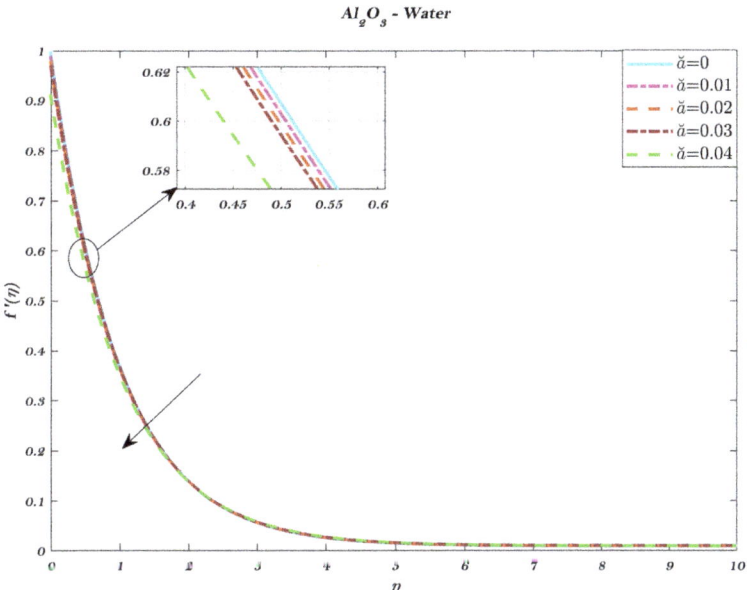

Figure 8. The influence of the slip parameter $ă$ against η on the velocity profile with φ = 3%, ϵ = 0.01 and R = 1.

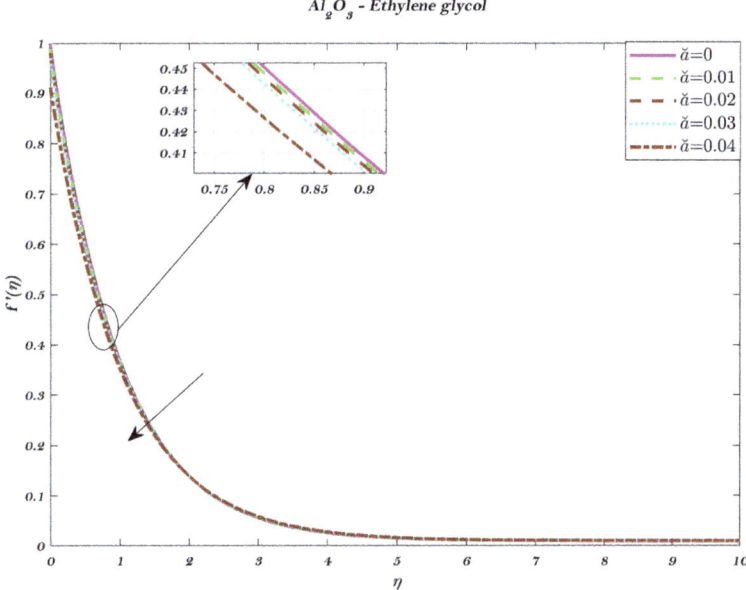

Figure 9. The influence of the slip parameter $ă$ against η on the velocity profile with φ = 3%, ϵ = 0.01 and R = 1.

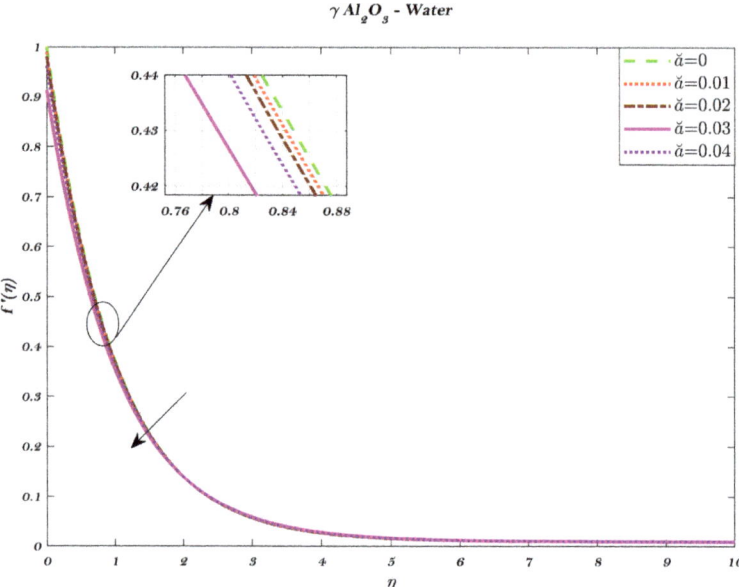

Figure 10. The influence of the slip parameter $ă$ against η on the velocity profile with $\varphi = 3\%$, $\epsilon = 0.01$ and $R = 1$.

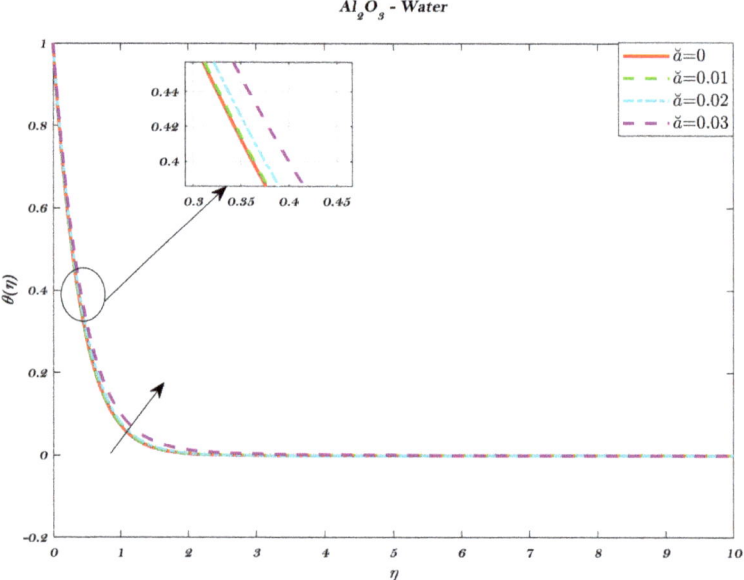

Figure 11. The slip parameters $ă$ against η on the temperature profile with $\varphi = 2\%$, $\epsilon = 0.01$ and $R = 1$.

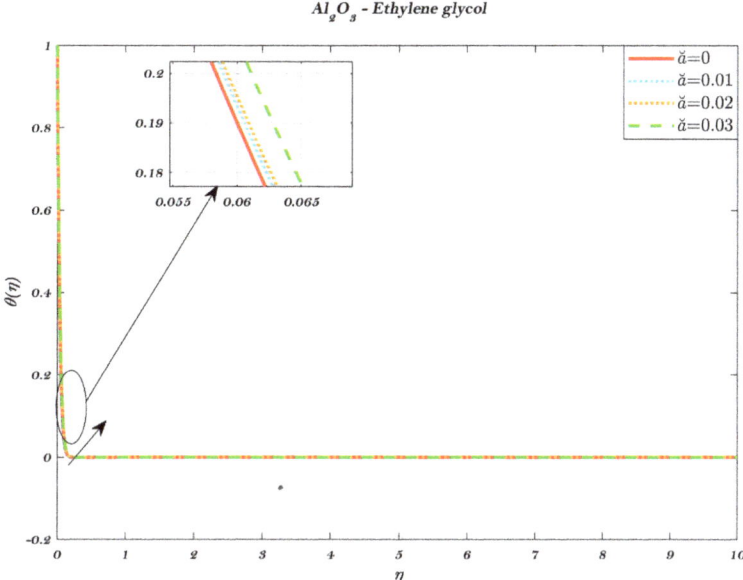

Figure 12. The slip parameters $ă$ against η on the temperature profile with $\varphi = 2\%$, $\epsilon = 0.01$ and $R = 1$.

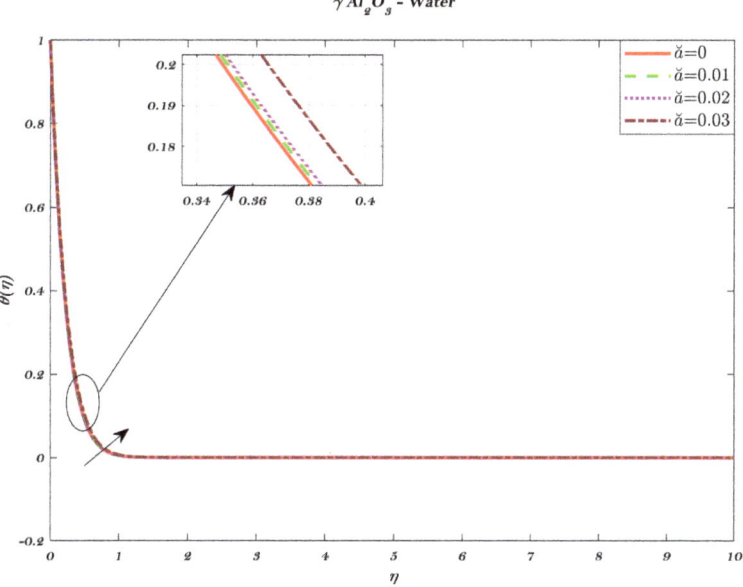

Figure 13. The slip parameters $ă$ against η on the temperature profile with $\varphi = 2\%$, $\epsilon = 0.01$ and $R = 1$.

Figures 14–16 illustrate the effect that the stagnation parameter ϵ has on the velocity profile in which the value of ϵ rises, the velocity of the nanofluids also rises but the temperature and the nanoparticle volume fraction fall [3,13,46]. If we assume that the velocity of the stream remains the same then an increase in the value of ϵ will result in a decrease in the stretch velocity.

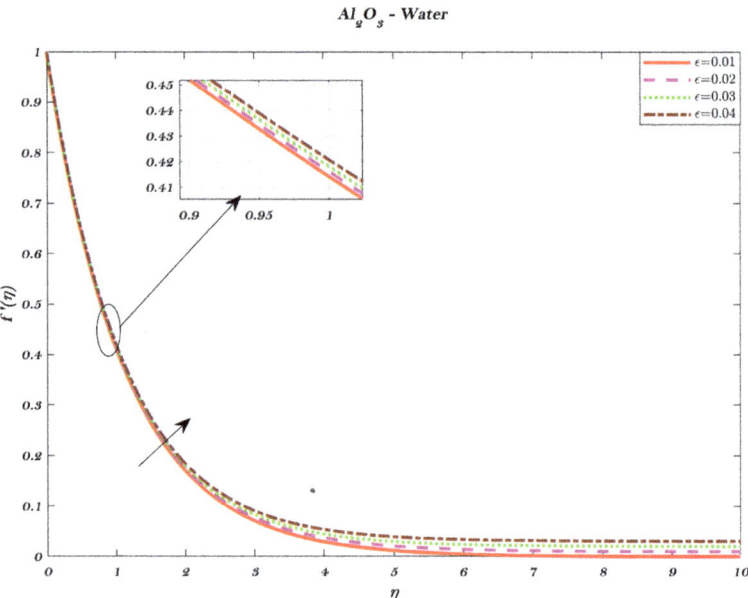

Figure 14. The effect of the stagnation parameter ϵ on the velocity profile with fix parameters $\varphi = 3\%$, $P_r = 6.96$ and $ă = 0.01$.

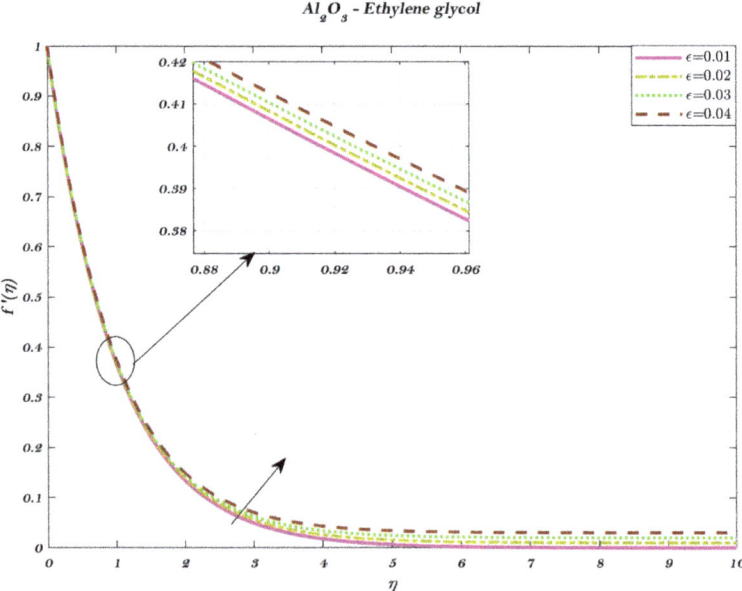

Figure 15. The effect of the stagnation parameter ϵ on the velocity profile with fix parameter $\varphi = 3\%$, $P_r = 204$ and $ă = 0.01$.

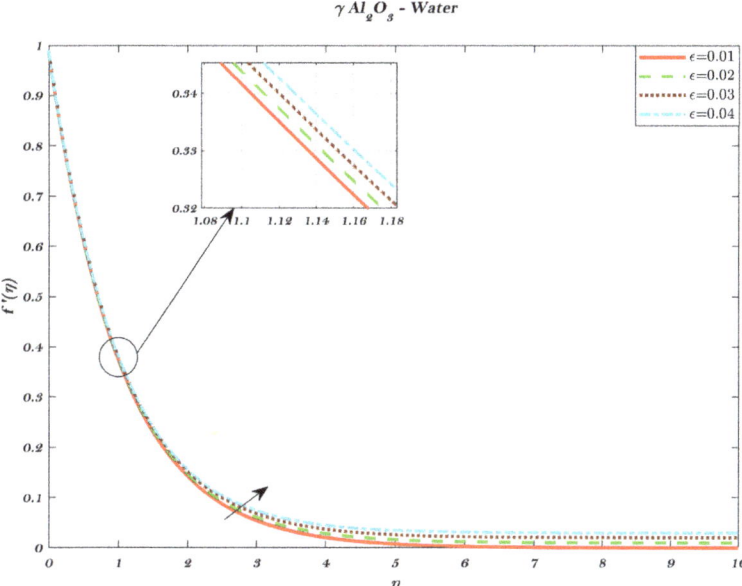

Figure 16. The effect of the stagnation parameter ϵ on the velocity profile with fix parameters $\varphi = 3\%$, $P_r = 6.96$ and $\check{a} = 0.01$.

Figures 17–19 show the result for different values of radiative heat flux parameter R. When radiative heat flux increases, the temperature profile shows enhancement whereas the velocity profile shows a downfall. Even though the velocity profile shows a trend that is slightly decreasing as well as a trend that is slightly increasing as the values of thermal radiation increased [12,16,34]. These results explain that an increase in the value of the parameter R, thermal radiation, has an insignificant impact on the fluid velocity. This is the case despite the fact that the velocity profile shows a slight decreasing trend [17,29,47]. In point of fact, the fluid viscosity has a propensity to grow with increased resistance to distortion, which results in a reduction within the velocity profile. On the other hand, it has a propensity to drop as internal heat production and thermal radiation both increase.

Figures 20 and 21 illustrate the value fluctuation $\frac{1}{2} Re_x^{\frac{1}{2}} c_f$ which is the local (reduced) skin friction coefficient, as well as $Re_x^{-\frac{1}{2}} Nu_x$, located along the y-axis and the nanoparticle volume fraction φ along the x-axis. It has been shown that an increase in φ values, the skin friction coefficient and Nusselt number are found to increase as well [46,48,49]. Therefore, the skin friction coefficient is greater for nanofluid γ-Al_2O_3 than it is for Al_2O_3. For different base fluids, the amount of skin friction coefficient by Al_2O_3 with base fluid as water is greater than that produced by Al_2O_3 ethylene glycol as base fluid. However, γ-Al_2O_3 nanoparticles have been shown to exhibit the reverse tendency. Nanofluids that are based on ethylene glycol have a greater Nusselt number than nanofluids that are based on water. When compared to other nanoparticles, the Nusselt number for γ-Al_2O_3 nanoparticles is much greater.

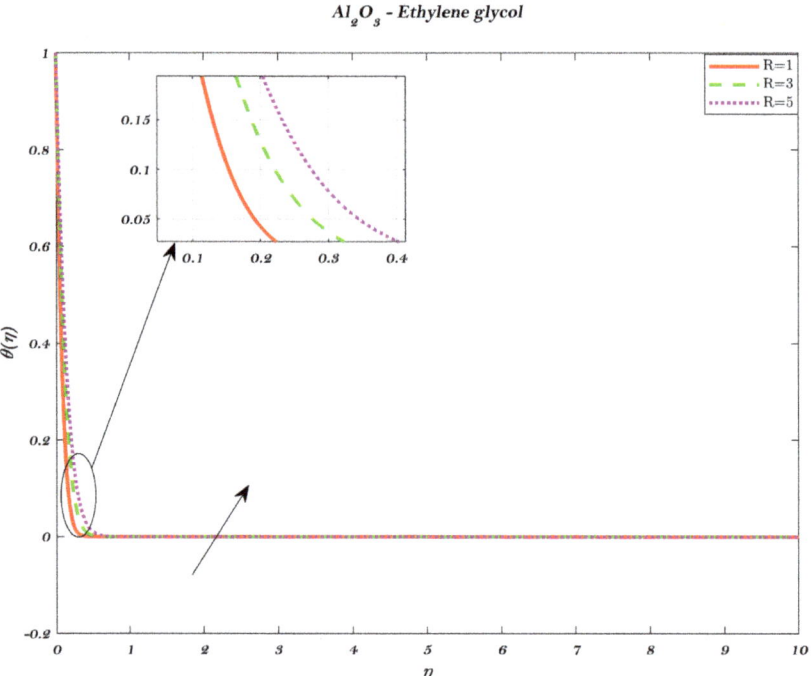

Figure 17. The radiative heat flux R against η with fix parameters $\varphi = 2\%$, $ǎ = 0.02$ and $P_r = 204$.

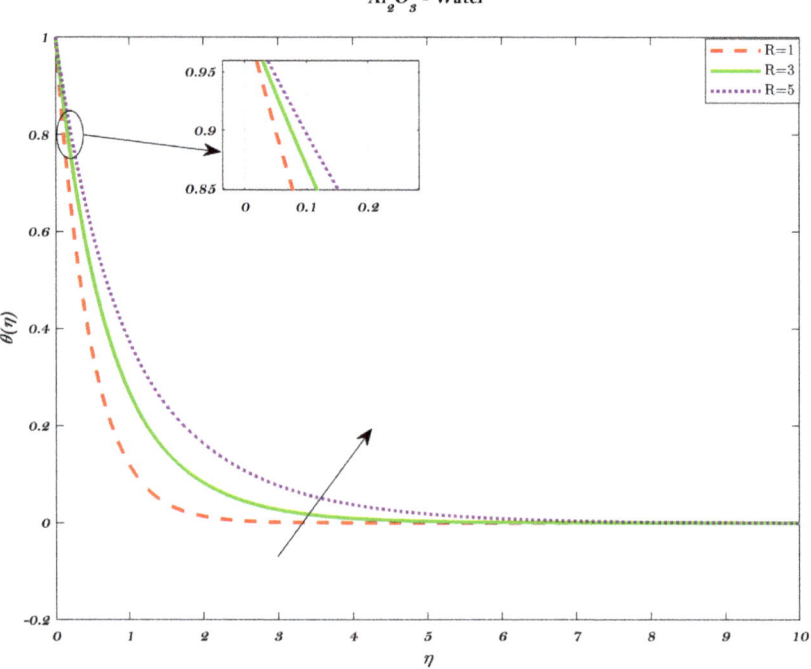

Figure 18. The radiative heat flux R against η with fix parameters $\varphi = 2\%$, $ǎ = 0.02$ and $P_r = 6.96$.

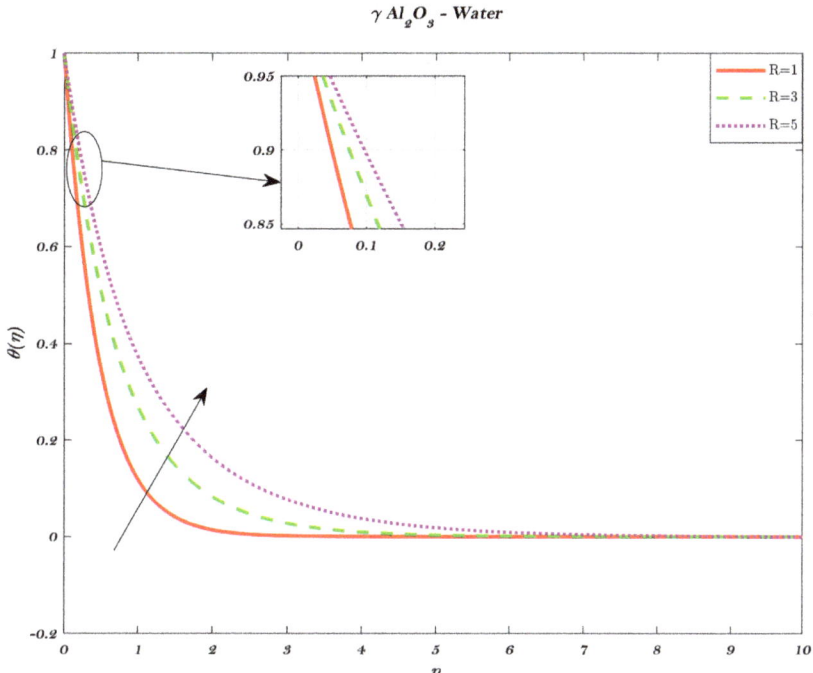

Figure 19. The radiative heat flux R against η with fix parameters $\varphi = 2\%$, $\tilde{a} = 0.02$ and $P_r = 6.96$.

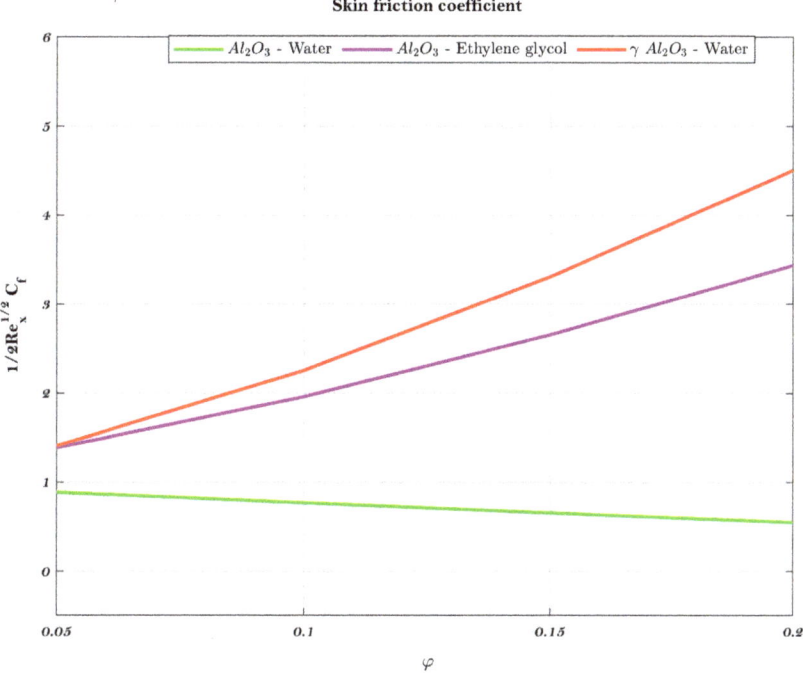

Figure 20. Frictional efficiency of skin as a function of φ.

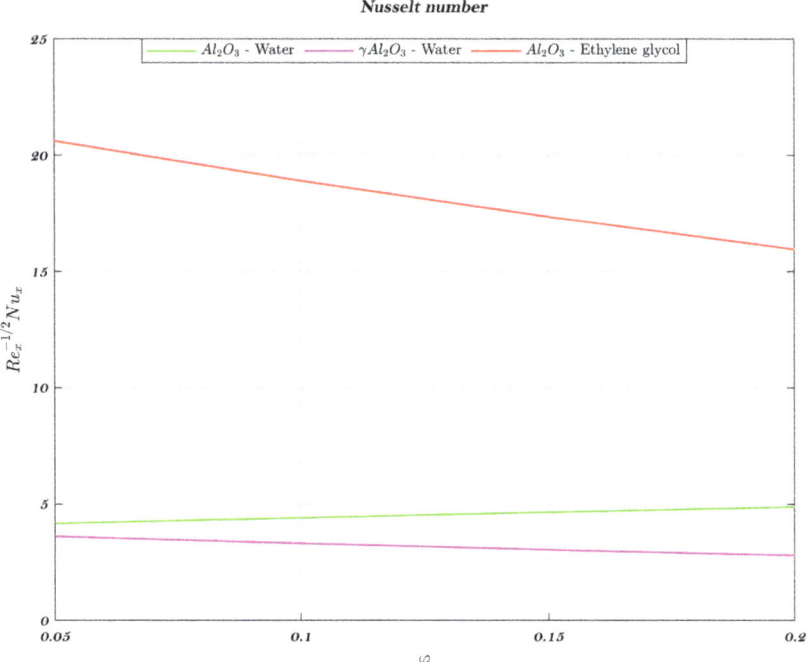

Figure 21. Impact of φ on Nusselt number.

7. Conclusions

The boundary layer flow of nanofluids of Al_2O_3 and γ-Al_2O_3 with various base fluids is carried out on a stretching sheet. We have been successful in developing a numerical solution for the steady BL flow and heat transfer at the stagnation point of nanofluids with a slip boundary condition. The velocity of the sheet's stretching in its own plane u_w, is different from the velocity of the external flow U_∞. The similarity ordinary differential equations that were produced are solved using the shooting method in conjunction with the $RK-4$ approach. For γ-Al_2O_3 nanofluids, we use models of viscosity and thermal conductivity that are built from actual experiments. For Al_2O_3, the viscosity model from Brinkman and the thermal conductivity model from Maxwell, are used along thermal radiation. The following are some particular findings that has be drawn from this investigation.

- Both the momentum and the thermal boundary layer, thickness increase with increasing nanoparticle volume fraction of Al_2O_3 and γ-Al_2O_3 nanoparticles. The velocity of nanofluids γ-Al_2O_3, is greater than that of Al_2O_3, in a comparison of nanoparticles [16,45–47]. Nanofluids made from ethylene glycol move more quickly than those made from water.
- However, in terms of the temperature distribution, γ-Al_2O_3-$C_2H_6O_2$ (ethylene glycol) is greater than that of Al_2O_3-ethylene glycol, whereas that of Al_2O_3-water is more than that of γ-Al_2O_3-water. The temperature profile increased with increasing values of parameter R, indicating that the thermal boundary layer thickness increases with increasing thermal radiation. As a consequence, a higher radiation parameter causes more heat to be produced in the flow, resulting in an increase in the temperature profile of the fluid. Nanofluids that are based on ethylene glycol have a lower temperature profile than those based on water. There is potential for cooling applications of nanofluids based on ethylene glycol [16,45,47].
- When compared to Al_2O_3 nanofluids, γ-Al_2O_3 nanofluids have greater skin friction. The skin friction is greater for Al_2O_3-water than it is for Al_2O_3-ethylene glycol. How-

ever, γ-Al$_2$O$_3$ nanoparticles exhibit the opposite tendency. The Nusselt number of ethylene glycol-based nanofluids is greater than that of water-based nanofluids. The Nusselt number for γ-Al$_2$O$_3$ nanoparticles is greater than that of other nanoparticles.

Author Contributions: Writing–review & editing, I.A.; Methodology and Resources, S.H. and N.A.A.; Data curation and visualization, M.S.; Supervision, D.S.M. All authors have read and agreed to the published version of the manuscript.

Funding: This research received no external funding.

Institutional Review Board Statement: Not applicable.

Informed Consent Statement: Not applicable.

Data Availability Statement: Not applicable

Conflicts of Interest: The authors declare no conflict of interest.

Abbreviations

The following abbreviations are used in this manuscript:

x, y	Coordinate $\perp$ to sheet (m)
$\tilde{u}, \tilde{v}$	Velocity components along x & y directions (ms^{-1})
T	Fluid's local temperature (K)
T_∞	Temperature far away from the sheet
φ	Solid Volume fraction
ρ_{nf}	Effective Nanofluid density (kg m^{-3})
ρ_f	Pure fluid density (kg m^{-3})
ρ_s	Nanoparticles density (kg m^{-3})
μ_{nf}	Nanofluid's effective dynamic viscosity (kg m^{-1} s^{-1})
μ_f	Dynamic viscosity of pure fluid (kg m^{-1} s^{-1})
k_{nf}	Nanofluid's thermal conductivity (W m^{-1} K^{-1})
k_f	Base fluid's thermal conductivity (W m^{-1} K^{-1})
k_s	Nanoparticles thermal conductivity (W m^{-1} K^{-1})
P_r	Prandtl number
$\check{a}$	Slip parameter
$\check{K}$	Permeability
$\check{\alpha}$	Dimensionless parameter
$\tilde{U}_\infty \frac{d\tilde{U}_\infty}{dx}$	stagnation-point flow term
$\tilde{q}_r$	Radiative heat flux
C_f	Skin Friction coefficient
$Re_x^{1/2} C_f$	Reduced Skin Friction coefficient
$Re_x^{-1/2} Nu_x$	Reduced Nusselt number

References

1. Das, S.K.; Choi, S.U.; Yu, W.; Pradeep, T. *Nanofluids: Science and Technology*; John Wiley & Sons: Hoboken, NJ, USA, 2007.
2. Wang, X.Q.; Mujumdar, A.S. A review on nanofluids-part I: Theoretical and numerical investigations. *Braz. J. Chem. Eng.* **2008**, *25*, 613–630. [CrossRef]
3. Choi, S.; Eastman, J. *Enhancing Thermal Conductivity of Fluids with Nanoparticles* (No. ANL/MSD/CP-84938; CONF-951135-29); Argonne National Lab.: Argonne, IL, USA, 1995.
4. Turkyilmazoglu, M. Dual and triple solutions for MHD slip flow of non-Newtonian fluid over a shrinking surface. *Comput. Fluids* **2012**, *70*, 53–58. [CrossRef]
5. Kandelousi, M.S. KKL correlation for simulation of nanofluid flow and heat transfer in a permeable channel. *Phys. Lett. A* **2014**, *378*, 3331–3339. [CrossRef]
6. Sheikholeslami, M.; Abelman, S.; Ganji, D.D. Numerical simulation of MHD nanofluid flow and heat transfer considering viscous dissipation. *Int. J. Heat Mass Transf.* **2014**, *79*, 212–222. [CrossRef]
7. Sheikholeslami, M.; Abelman, S. Two-phase simulation of nanofluid flow and heat transfer in an annulus in the presence of an axial magnetic field. *IEEE Trans. Nanotechnol.* **2015**, *14*, 561–569. [CrossRef]

8. Turkyilmazoglu, M. An analytical treatment for the exact solutions of MHD flow and heat over two–three dimensional deforming bodies. *Int. J. Heat Mass Transf.* **2015**, *90*, 781–789. [CrossRef]
9. Turkyilmazoglu, M. A note on the correspondence between certain nanofluid flows and standard fluid flows. *J. Heat Transf.* **2015**, *137*, 024501. [CrossRef]
10. Sheikholeslami, M.; Ellahi, R. Three dimensional mesoscopic simulation of magnetic field effect on natural convection of nanofluid. *Int. J. Heat Mass Transf.* **2015**, *89*, 799–808. [CrossRef]
11. Jamaludin, A.; Naganthran, K.; Nazar, R.; Pop, I. Thermal radiation and MHD effects in the mixed convection flow of Fe_3O_4–water ferrofluid towards a nonlinearly moving surface. *Processes* **2020**, *8*, 95. [CrossRef]
12. Jamaludin, A.; Naganthran, K.; Nazar, R.; Pop, I. MHD mixed convection stagnation-point flow of $Cu-Al_2O_3$/water hybrid nanofluid over a permeable stretching/shrinking surface with heat source/sink. *Eur. J. Mech.-B/Fluids* **2020**, *84*, 71–80. [CrossRef]
13. Farajollahi, B.; Etemad, S.G.; Hojjat, M. Heat transfer of nanofluids in a shell and tube heat exchanger. *Int. J. Heat Mass Transf.* **2010**, *53*, 12–17. [CrossRef]
14. Rehman, R.; Wahab, H.A.; Khan, U. Heat transfer analysis and entropy generation in the nanofluids composed by Aluminum and γ-Aluminum oxides nanoparticles. *Case Stud. Therm. Eng.* **2022**, *31*, 101812. [CrossRef]
15. Soltanipour, H.; Choupani, P.; Mirzaee, I. Numerical analysis of heat transfer enhancement with the use of γ-Al_2O_3/water nanofluid and longitudinal ribs in a curved duct. *Therm. Sci.* **2012**, *16*, 469–480. [CrossRef]
16. Hamad, M. Analytical solution of natural convection flow of a nanofluid over a linearly stretching sheet in the presence of magnetic field. *Int. Commun. Heat Mass Transf.* **2011**, *38*, 487–492. [CrossRef]
17. Bridgman, P.W. The thermal conductivity of liquids. *Proc. Natl. Acad. Sci. USA* **1923**, *9*, 341–345. [CrossRef]
18. Algehyne, E.A.; Wakif, A.; Rasool, G.; Saeed, A.; Ghouli, Z. Significance of Darcy-Forchheimer and Lorentz forces on radiative alumina-water nanofluid flows over a slippery curved geometry under multiple convective constraints: A renovated Buongiorno's model with validated thermophysical correlations. *Waves Random Complex Media* **2022**, 1–30. [CrossRef]
19. Batool, S.; Rasool, G.; Alshammari, N.; Khan, I.; Kaneez, H.; Hamadneh, N. Numerical analysis of heat and mass transfer in micropolar nanofluids flow through lid driven cavity: Finite volume approach. *Case Stud. Therm. Eng.* **2022**, *37*, 102233. [CrossRef]
20. Bhutta, M.S.; Xuebang, T.; Akram, S.; Yidong, C.; Ren, X.; Fasehullah, M.; Rasool, G.; Nazir, M.T. Development of novel hybrid 2D-3D graphene oxide diamond micro composite polyimide films to ameliorate electrical & thermal conduction. *J. Ind. Eng. Chem.* **2022**, *114*, 108–114.
21. Sabir, Z.; Imran, A.; Umar, M.; Zeb, M.; Shoaib, M.; Raja, M.A.Z. A numerical approach for 2-D Sutterby fluid-flow bounded at a stagnation point with an inclined magnetic field and thermal radiation impacts. *Therm. Sci.* **2021**, *25*, 1975–1987. [CrossRef]
22. Rehman, A.; Hussain, A.; Nadeem, S. Assisting and opposing stagnation point pseudoplastic nano liquid flow towards a flexible Riga sheet: A computational approach. *Math. Probl. Eng.* **2021**, *2021*, 6610332. [CrossRef]
23. Ramesh, G.; Prasannakumara, B.; Gireesha, B.; Rashidi, M. Casson fluid flow near the stagnation point over a stretching sheet with variable thickness and radiation. *J. Appl. Fluid Mech.* **2016**, *9*, 1115–1022. [CrossRef]
24. Baag, S.; Mishra, S.; Dash, G.; Acharya, M. Numerical investigation on MHD micropolar fluid flow toward a stagnation point on a vertical surface with heat source and chemical reaction. *J. King Saud Univ.-Eng. Sci.* **2017**, *29*, 75–83. [CrossRef]
25. Gleiter, H. Nanocrystalline Materials and Nanometer-Sized Glasses. *Europhys. News* **1989**, *20*, 130–133. [CrossRef]
26. Keblinski, P.; Phillpot, S.; Choi, S.; Eastman, J. Mechanisms of heat flow in suspensions of nano-sized particles (nanofluids). *Int. J. Heat Mass Transf.* **2002**, *45*, 855–863. [CrossRef]
27. Hamilton, R.L.; Crosser, O. Thermal conductivity of heterogeneous two-component systems. *Ind. Eng. Chem. Fundam.* **1962**, *1*, 187–191. [CrossRef]
28. Horrocks, J.; McLaughlin, E. Thermal conductivity of simple molecules in the condensed state. *Trans. Faraday Soc.* **1960**, *56*, 206–212. [CrossRef]
29. Duncan, A.; Peterson, G. Review of microscale heat transfer. *Appl. Mech. Rev.* **1994**, *47*, 397–428. [CrossRef]
30. Luo, M.; Zhao, J.; Liu, L. Normal heat diffusion in many-body system via thermal photons. *arXiv* **2020**, arXiv:2011.07588.
31. Eastman, J.A.; Choi, S.; Li, S.; Yu, W.; Thompson, L. Anomalously increased effective thermal conductivities of ethylene glycol-based nanofluids containing copper nanoparticles. *Appl. Phys. Lett.* **2001**, *78*, 718–720. [CrossRef]
32. Li, C.H.; Peterson, G. Experimental investigation of temperature and volume fraction variations on the effective thermal conductivity of nanoparticle suspensions (nanofluids). *J. Appl. Phys.* **2006**, *99*, 084314. [CrossRef]
33. Lee, S.; Choi, S.S.; Li, S.; Eastman, J. Measuring thermal conductivity of fluids containing oxide nanoparticles. *J. Heat Transf.* **1999**, *121*, 280–289. [CrossRef]
34. Murshed, S.; Leong, K.; Yang, C. Enhanced thermal conductivity of TiO2—Water based nanofluids. *Int. J. Therm. Sci.* **2005**, *44*, 367–373. [CrossRef]
35. Patel, H.E.; Das, S.K.; Sundararajan, T.; Sreekumaran Nair, A.; George, B.; Pradeep, T. Thermal conductivities of naked and monolayer protected metal nanoparticle based nanofluids: Manifestation of anomalous enhancement and chemical effects. *Appl. Phys. Lett.* **2003**, *83*, 2931–2933. [CrossRef]
36. Yu, W.; Xie, H.; Li, Y.; Chen, L. Experimental investigation on thermal conductivity and viscosity of aluminum nitride nanofluid. *Particuology* **2011**, *9*, 187–191. [CrossRef]

37. Xuan, Y.; Li, Q. Investigation on convective heat transfer and flow features of nanofluids. *J. Heat Transf.* **2003**, *125*, 151–155. [CrossRef]
38. Hwang, Y.; Ahn, Y.; Shin, H.; Lee, C.; Kim, G.; Park, H.; Lee, J. Investigation on characteristics of thermal conductivity enhancement of nanofluids. *Curr. Appl. Phys.* **2006**, *6*, 1068–1071. [CrossRef]
39. Liu, M.S.; Lin, M.C.; Huang, I.T.; Wang, C.C. Enhancement of thermal conductivity with CuO for nanofluids. *Chem. Eng. Technol. Ind. Chem.-Plant Equip.-Process Eng.-Biotechnol.* **2006**, *29*, 72–77. [CrossRef]
40. Krishnamurthy, S.; Bhattacharya, P.; Phelan, P.; Prasher, R. Enhanced mass transport in nanofluids. *Nano Lett.* **2006**, *6*, 419–423. [CrossRef]
41. Wen, D.; Ding, Y. Natural convective heat transfer of suspensions of titanium dioxide nanoparticles (nanofluids). *IEEE Trans. Nanotechnol.* **2006**, *5*, 220–227.
42. Das, S.K.; Putra, N.; Thiesen, P.; Roetzel, W. Temperature dependence of thermal conductivity enhancement for nanofluids. *J. Heat Transf.* **2003**, *125*, 567–574. [CrossRef]
43. Chon, C.; Kihm, K. Thermal conductivity enhancement of nanofluids by Brownian motion. *Trans.-Am. Soc. Mech. Eng. J. Heat Transf.* **2005**, *127*, 810. [CrossRef]
44. Murshed, S.; Leong, K.; Yang, C. Investigations of thermal conductivity and viscosity of nanofluids. *Int. J. Therm. Sci.* **2008**, *47*, 560–568. [CrossRef]
45. Ganesh, N.V.; Hakeem, A.A.; Ganga, B. A comparative theoretical study on Al_2O_3 and γ-Al_2O_3 nanoparticles with different base fluids over a stretching sheet. *Adv. Powder Technol.* **2016**, *27*, 436–441. [CrossRef]
46. Devi, S.; Kandasamy, R. Analysis of nonlinear two dimensional laminar natural flow and mixed convection over variable surface with free stream conditions. *J. Comput. Appl. Math.* **2002**, *3*, 107–116.
47. Reza-E-Rabbi, S.; Ahmmed, S.F.; Arifuzzaman, S.; Sarkar, T.; Khan, M.S. Computational modelling of multiphase fluid flow behaviour over a stretching sheet in the presence of nanoparticles. *Eng. Sci. Technol. Int. J.* **2020**, *23*, 605–617. [CrossRef]
48. Dogonchi, A.; Alizadeh, M.; Ganji, D. Investigation of MHD Go-water nanofluid flow and heat transfer in a porous channel in the presence of thermal radiation effect. *Adv. Powder Technol.* **2017**, *28*, 1815–1825. [CrossRef]
49. Maiga, S.E.B.; Palm, S.J.; Nguyen, C.T.; Roy, G.; Galanis, N. Heat transfer enhancement by using nanofluids in forced convection flows. *Int. J. Heat Fluid Flow* **2005**, *26*, 530–546. [CrossRef]

Article

Experimental Evaluation of a Full-Scale HVAC System Working with Nanofluid

Marco Milanese *, Francesco Micali, Gianpiero Colangelo and Arturo de Risi

Department of Engineering for Innovation, University of Salento, 73100 Lecce, Italy; francesco.micali@unisalento.it (F.M.); gianpiero.colangelo@unisalento.it (G.C.); arturo.derisi@unisalento.it (A.d.R.)
* Correspondence: marco.milanese@unisalento.it; Tel.: +39-0832-299-438

Abstract: Nowadays, energy saving is considered a key issue worldwide, as it brings a variety of benefits: reducing greenhouse gas emissions and the demand for energy imports and lowering costs on a household and economy-wide level. Researchers and building designers are looking to optimize building efficiency by means of new energy technologies. Changes can also be made in existing buildings to reduce the energy consumption of air conditioning systems, even during operational conditions without dramatically modifying the system layout and have as low an impact as possible on the cost of the modification. These may include the usage of new heat transfer fluids based on nanofluids. In this work, an extended experimental campaign (from February 2020 to March 2021) has been carried out on the HVAC system of an educational building in the Campus of University of Salento, Lecce, Italy. The scope of the investigation was comparing the COP for the two HVAC systems (one with nanofluid and the other one without) operating concurrently during winter and summer: simultaneous measurements on the two HVAC systems show that the coefficient of performance (COP) with nanofluid increased on average by 9.8% in winter and 8.9% in summer, with average daily peaks of about 15%. Furthermore, the comparison between the performance of the same HVAC system, working in different comparable periods with and without nanofluids, shows a mean increase in COP equal to about 13%.

Keywords: heat transfer fluid; nanofluid; heating; ventilation and air conditioning system; experimental test; coefficient of performance

1. Introduction

Decreasing the energy consumption of heating, ventilation and air conditioning (HVAC) systems is a very important issue, due to their high environmental and energy costs, together with a significant actual increase in their demand from the market. Several studies have described various technologies and techniques that can be used to reduce HVAC energy consumption, one of which is nanofluids [1,2].

Nanofluids are engineered heat transfer fluids which can be used to improve heat transfer, thus increasing energy efficiency in a variety of applications based on chillers, heat pumps and other hydronic HVAC systems. They are suspensions of nanoparticles dispersed in a liquid that are formulated to achieve higher heat transfer performance than their basefluids. Numerous studies have demonstrated that nanofluid thermal conductivity can be improved based on some variables, such as nanoparticle volume concentration, size, morphology, etc.

In early experiments on nanofluids, Lee et al. [3] measured thermal conductivity with the transient hot-wire method, demonstrating that a small amount of nanoparticles was enough to increase the thermal conductivity of the base fluid.

Beck et al. [4] presented data for the thermal conductivity enhancement in seven nanofluids containing 8–282 nm diameter alumina nanoparticles in water or ethylene glycol.

They found that the thermal conductivity enhancement in these nanofluids decreases as the particle size decreases below about 50 nm. This finding could be attributed to phonon scattering at the solid–liquid interface.

To confirm this result, Colangelo et al. [5,6] designed, built and tested a new experimental setup to investigate the physical phenomena involved in the thermal conductivity enhancement of nanofluids.

In recent years, experimental investigations on the effects of nanofluids on convective heat transfer coefficients in laminar and turbulent conditions were developed, demonstrating a significant improvement with respect to conventional heat transfer fluids [7].

Balla et al. [8] studied several suspensions of Cu and Zn nanoparticles with a size of 50 nm in water base fluid. They found that the heat transfer coefficient of nanofluids was higher than its base fluid. Similar results were found by Kai et al. [9] studying nanofluid heat transfer in a mini-tube using SiO_2 nanoparticles.

Recently, several numerical and experimental studies on nanofluids and their applications have been developed, such as solar thermal systems. Lee et al. [10] and Alsalame et al. [11] studied photovoltaic thermal systems based on nanofluids. Colangelo et al. [12,13] experimentally investigated the use of nanofluids in flat solar thermal collectors: tests on traditional solar flat panels revealed some technical issues, due to the nanoparticles' sedimentation. Therefore, the modification of the panel shape allowed this problem to be fixed. Flat plate solar collectors based on nanofluids were also studied by Chaji et al. [14]. Furthermore, the application of nanofluids on different solar thermal energy conversion systems was investigated in [15–18].

Different studies have been carried out to increase the performance of internal combustion engines. Zhang et al. [19] improved the heat-transfer performance of a diesel-engine cylinder head by means of a nanofluid coolant.

Micali et al. [20] developed an experimental campaign related to the use of CuO nanofluid as the coolant in a biodiesel four-strokes engine. They measured reductions in temperature up to 13.6% on the exhaust valve seat and up to 4.1% on the exhaust valve spindle.

Further studies related to the use of nanofluid within electronic devices [21], geothermal heat exchangers [22,23], and a cooling system for wind turbines [24], demonstrated a significant increase in heat transfer performance versus traditional fluids.

Considering these thermal performance improvements, the use of fluids containing suspended solid particles in HVAC systems is expected to show significant enhancements of their efficiency.

Devdatta et al. [25] observed that the use of nanofluids inside the heating system of the building is a suitable solution to reduce the size of the heat transfer system, and, in particular, the size of the heat exchangers, heat pumps and other components as well. This will reduce energy consumption and will, thus, indirectly reduce environmental pollution.

Ahmed and Ahmed Khan [26] used nanofluids in the external cooling jacket around the condenser of an air conditioner. In particular, they studied the benefits of two types of nanofluids, made of copper and aluminum oxide, respectively, on the performance of an air/water conditioner. Their experimental results showed a significant enhancement in the coefficient of performance (COP), up to 22.1% with Al_2O_3 nanofluid and 29.4% with CuO nanofluid.

Hatami et al. [27] experimentally tested three types of nanoparticles (SiO_2, TiO_2 and Carbon Nanotubes), dispersed in water inside HVAC systems. They found the best result, in terms of energy consumption reduction, with SiO_2-based nanofluid.

In order to use nanofluids as a heat transfer fluid within full-scale HVAC systems, different problems have to be solved, such as nanoparticle stability in suspension [28] and increment of viscosity [29]. Regarding the first issue and according to Awais et al. [29], sedimentation and agglomeration of nanoparticles within nanofluids can produce fouling on heat transfer surfaces and, therefore, higher pressure drops and damages in ducts, pumps, etc. On the other hand, the use of nanoparticles, having an optimal shape, size and volume fraction in the base-fluid coupled with the addition of surfactants can improve the suspension stability, avoiding the above problems [30].

The electrical potential at the shear slippage plane is called the zeta potential (ZP), and its value aids in evaluating nanoparticles' (NPs) stability in suspension [31,32], according to Bogdan [33] and Lee [34]: indeed, particles in colloidal suspension tend to develop a surface charge by the adsorption of ions from the base fluid. This superficial charge is double-layer-structured, with a sliding surface located beyond the first layer. In the nanofluid formulation used in this investigation, an anionic surfactant was used in order to improve the stability. The stability of the aluminum oxide suspension depends on the dispersant to modify the ZP and the surface repulsion between particles. In cases with anionic dispersant, a ZP value higher than 25 mV (absolute value) is necessary to achieve enough repulsion forces.

In the case of a sample at rest, the settling occurs over a long time, with 50% of particles settled in 6 months.

Regarding the second issue (viscosity), it is important to remark that the variation in nanofluid viscosity is directly proportional to the particles' concentration in suspension. In the nanofluid formulation, a volume of only 2% nanoparticles has been added to achieve a very limited viscosity increment. Furthermore, the test campaign was carried out in a large HVAC system, where the relevant diameter of the pipes and relevant size of the heat exchangers limited the impact of viscosity increment on the pressure drop in the system. Pantzali et al. [35] studied nanofluid use in industrial applications, mainly focusing on the pressure drop increment related to the viscosity of the nanofluid. They concluded that in the case of industrial heat exchangers and large pipes with turbulent flow, usually developed inside, the substitution of conventional fluids by nanofluids had no relevant incidence on pressure drops in the system.

In order to overcome the above discussed problems, this work was based on a nanofluid composed of water–glycol and aluminum oxide (Al_2O_3) nanoparticles, having a controlled size distribution (D_{v90} = 617 nm) and good stability, that deliver efficient, reliable, and consistent performance over a wide temperature range, with little effect on viscosity, and, therefore, on system fluid pumping energy. Particularly, in [36] Colangelo et al. developed dynamic simulations in order to compare the efficiency of two full-scale HVAC systems (installed at the educational building "Corpo O" in the Campus of University of Salento, Lecce, Italy), working with a traditional water–glycol mixture and with Al_2O_3-nanofluid, they found a numerical increment in efficiency of about 10%. As a follow-up to that study, the objective of this work was to carry out an extended experimental campaign on the same building in order to quantify, over a long period of time and under real operating conditions, the increase in performance of the HVAC system due to the use of nanofluids. These results will also allow the validation of the numerical results previously found, verifying their congruence with the experimental measurements.

2. Test Conditions and Experimental Apparatus

The experimental campaign was carried out on an educational building, named "Corpo O" (Figure 1), at the Campus of University of Salento, which is in Lecce, Italy at latitude 40°21′ and longitude 18°10′.

The building consists of two symmetrical wings (left and right wings), each of which has its own HVAC system. Each wing is composed of three floors: ground floor, first floor and second floor, with a total area of 2400 m² and a total volume of 13,163 m³. The HVAC systems are used for air conditioning of offices and labs inside the building.

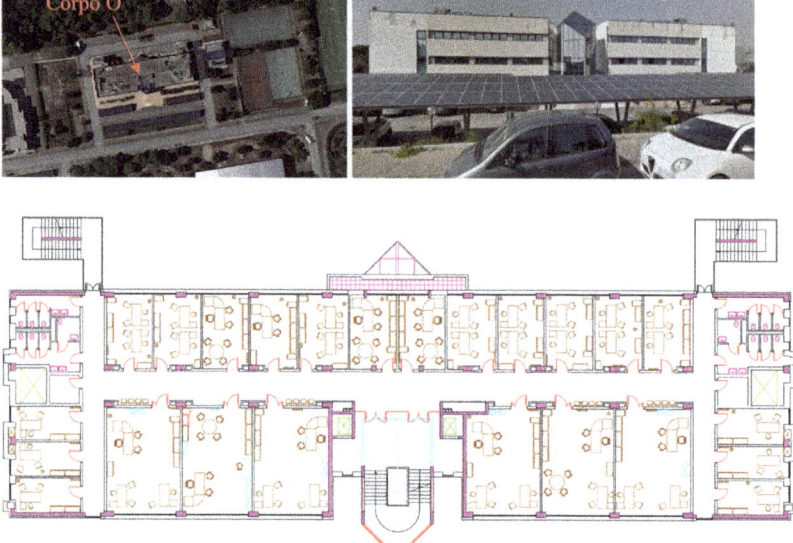

Figure 1. Building "Corpo O" at Campus Ecotekne of the University of Salento in Lecce, Italy.

The experimental test campaign was focused on data acquisition during winter (heating mode) and summer (cooling mode).

2.1. Description of the Thermal System

Two symmetrical HVAC systems (named HVAC-1 and HVAC-2 in the following) are installed on the roof of the building, and each system is used for thermal conditioning of half of the building. Both systems are equipped with heat pumps (model CLIVET WSAN-XEE 302 [37]), having the following technical specifications (Table 1).

Table 1. Characteristics of the heat pump CLIVET WSAN-XEE 302 [37].

Characteristics	Value
Compressor	
Type	2 Scroll
Refrigerant charge	8.28 L
Internal heat exchanger	
Water flow	3.4 L/s
Maximum water flow	5.4 L/s
Pressure drop	41.9 kPa
Useful pump discharge	131 kPa
External heat exchanger	
Fans	6
Standard air flow	6971 L/s
Installed power unit	0.18 kW
Expansion vessel	
Capacity	5 l
Maximum pressure on the water	550 kPa
Storage tank	
Inertial tank	130 L

Finally, each HVAC unit supplies three pipelines, through which the heat transfer fluid is pumped:

- Fan coils and radiator lines;
- AHU line.

2.2. Nanofluid Characteristics

In order to evaluate the increase in performance due to the use of a nanofluid as a heat transfer fluid, in this study an aluminum oxide-based nanofluid has been loaded inside the HVAC-1 system. This nanofluid has been chosen taking into account its high stability and its low viscosity, which are comparable to the base fluid ones. Table 2 summarizes the main specifications of the nanofluid.

Table 2. Specifications of the nanofluid.

Characteristics	Value
Composition (% by weight):	
Propylene Glycol	37
Performance Additives	2
Water	61
Color	White
Odor	Odorless
pH	10
Specific Weight (kg/m^3) at 25 °C	1.079
Operating Range (°C)	−22 to 65
Freeze Point (°C)	−22
Burst Point (°C)	−51
Boiling Point (°C)	105
Thermal Conductivity (W/m K) at 20 °C	0.471
Specific Heat (kJ/kg K) at 20 °C	3.51
Viscosity (mPa s) at 20 °C	4.74

The nanofluid was made of aluminum oxide nanoparticles at 2% in volume concentration and size distribution with Dv_{90} = 617 nm, with a density of 1079 g/L. Density is strictly related to the particle concentration, therefore sample density has been measured over the test campaign to monitor sedimentation phenomena inside the system. According to regulation and restrictions, the use of propylene glycol in the composition of the nanofluid for the test campaign comes from the necessity to avoid a toxic grade of glycol, as ethylene glycol is.

The remaining 2% in weight are dispersants, anti-corrosion inhibitors and aluminum oxide nanoparticles.

2.3. Test Instrumentation and Data Acquisition System

The coefficient of performance (COP) of each HVAC unit was calculated as the ratio between thermal (E_{th}) and electrical (E_{el}) energy:

$$COP = E_{th}/E_{el} \qquad (1)$$

Therefore, the energy monitoring system required the installation of several instrumentations, such as electricity meters, temperature sensors and mass flow rate meters:

- electricity meters were used to measure current, tension, and electrical power absorbed by the HVAC systems. Data were collected for all pumps and heat pumps by means of the energy meter IME—NEMO D4—three phase (Figure 2). It measures active energy and energy/power, with an accuracy of ±1% for active energy, conforming to IEC62053-23, and ±2% for reactive energy, conforming to IEC61557-12;
- thermal energy was evaluated by measuring the heat transfer fluid mass flow rate and its temperature at the heat pump inlet and outlet. In particular, the energy meter Caleffi—Conteca Easy is an ultrasonic direct heat meter with two temperature probes, with an accuracy of ≤0.05 °C and one flowmeter with an accuracy of ±2% according to EN 1434 (Figure 3);

- environmental measurements (indoor and outdoor air temperature and RH) were carried out in order to analyze the heat pump performance under different meteorological conditions;
- all data were recorded using a dedicated PLC to acquire and to store the data, sampled every 60 s by sensors, through the communication bus. Using the integrated web interface, it was possible to monitor consumption and other data and review data history (Figure 4).

Figure 2. Acquisition system for the electrical energy measurements IME—NEMO D4.

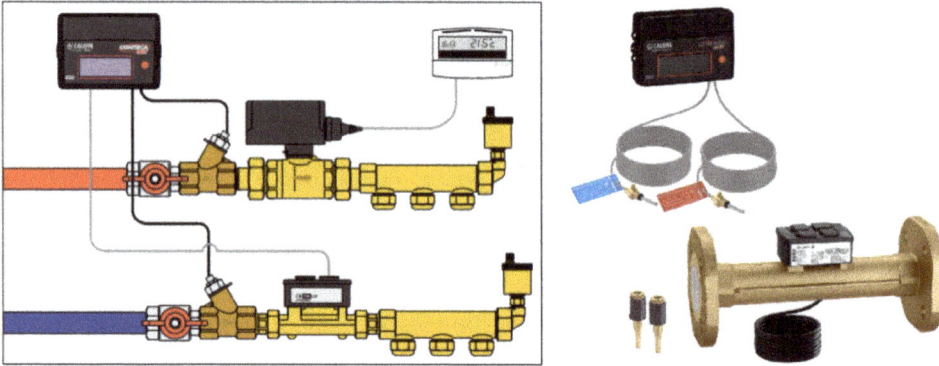

Figure 3. Drawing of the acquisition system for thermal energy Caleffi—Conteca Easy—Ultra.

2.4. Nanofluid Loading Procedure

The volume of heat transfer fluid within each HVAC thermal line was 1460 L. As the final step in preparing the building for experimental testing, the concentration of propylene glycol within the HVAC system (left and right wings of the building) was measured: it was the same in both systems and equal to 30% vol.

Therefore, the nanofluid was loaded within the HVAC-1 system (left wing) only, up to reach a nanoparticle concentration of 2% vol, leaving the right wing of the building (HVAC-2) loaded with the conventional water–propylene glycol heat transfer fluid: this approach allowed comparison of the performance of two identical systems, working with different heat transfer fluids at the same time.

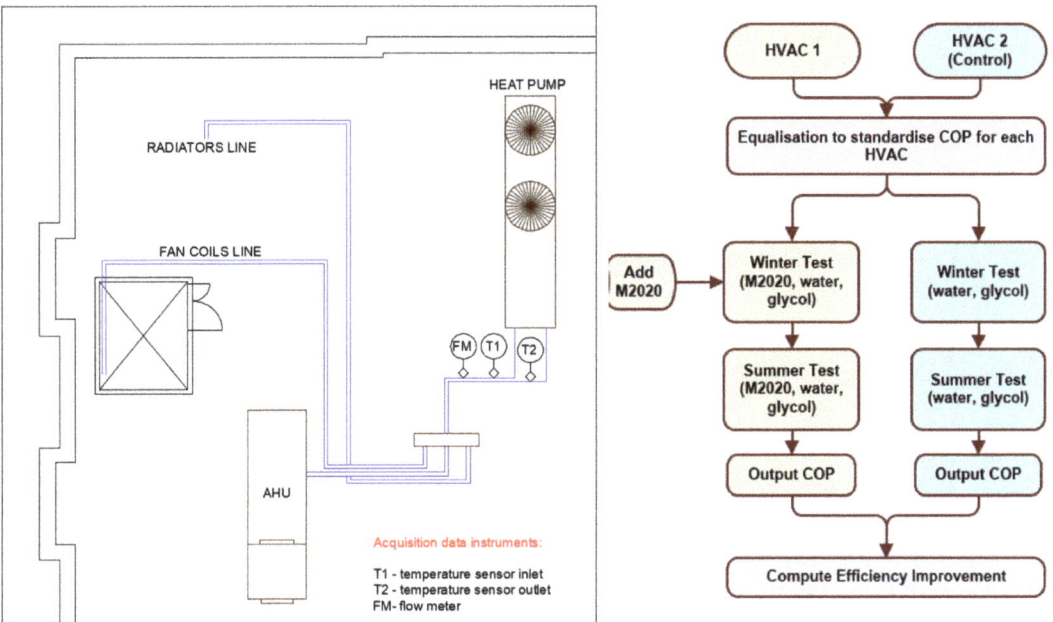

Figure 4. HVAC system with data sensors positions and flowchart of test methodology.

3. Experimental Results

This paper summarizes and compares the heat pump performance, recorded in three experimental tests, carried out from February 2020 to March 2021, according to the following scheduling:

- first test (heating mode): 10 February 2020–9 March 2020;
- second test (cooling mode): 7 September 2020–25 September 2020;
- third test (heating mode): 27 November 2020–9 March 2021.

As this study referred to a real building, the parameters which can affect the performance of the HVAC systems cannot be fully controlled. For this reason, in order to minimize their effects, in this study the performance has been evaluated over a long period of time, balancing as much as possible, oscillations related to stochastic variables.

3.1. February–March 2020 Results

In the first week of monitoring (from 10 February 2020 to 14 February 2020), the HVAC-1 and HVAC-2 systems were loaded with the same heat transfer fluid (water–glycol 30% vol). During this period, a short database was acquired, to be used as reference data for the experiment. Then, on 17 February, nanofluid was loaded in the HVAC-1 and the next 7 working days were used to balance both the heat pumps by performing preliminary tests. After such balance, energy consumption comparison between both the machines was restarted on 26 February and continued until 9 March.

Figure 5 shows the hourly COP comparison between HVAC-1 and HVAC-2, while Figure 6 shows the average daily COP comparison and the daily COP ratio between HVAC-1 and HVAC-2.

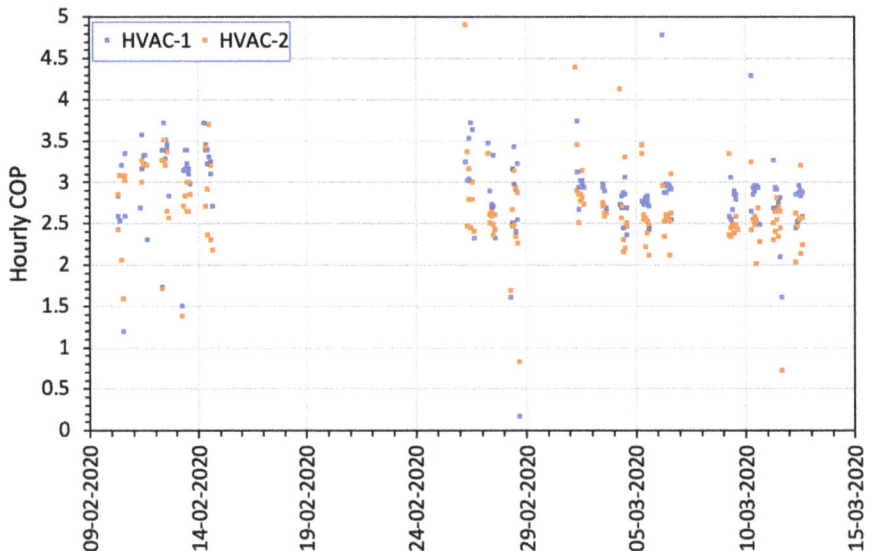

Figure 5. Hourly COP comparison between HVAC-1 and HVAC-2 (10 February 2020–9 March 2020).

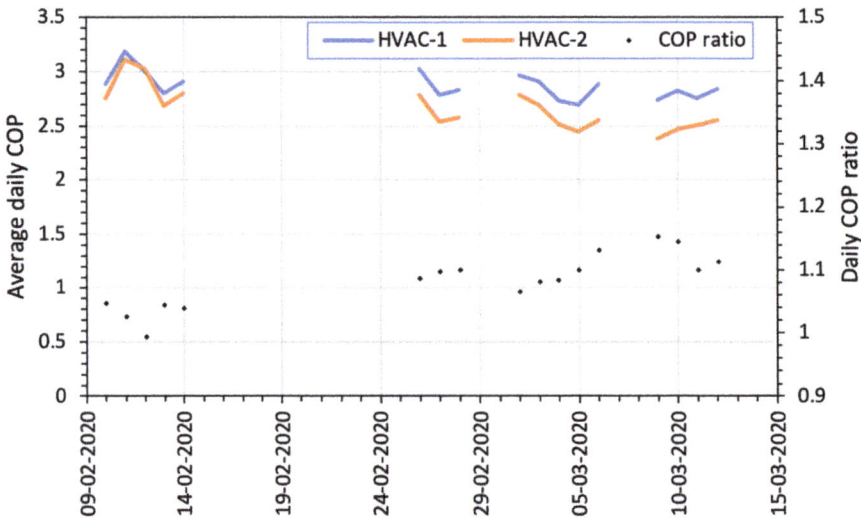

Figure 6. Average daily COP comparison and daily COP ratio between HVAC-1 and HVAC-2 (10 February 2020–9 March 2020).

In the first 3 days of monitoring (from 26 February to 28 February), after nanofluid loading, the mean increase in performance was 9.36%. However, the best performances were achieved in the last two weeks of experimental data acquisition, when the average increase in performance was 10.8%. On 12 March 2020, the HVAC systems were shut off.

During the acquisition period, the density of the nanofluid was measured weekly by sampling from the system in operation. Since it was 1079 g/L over the test period, a constant concentration of 2% of nanoparticles was ensured inside the system fluid.

3.2. September 2020 Results

Figures 7 and 8 show the experimental results in terms of mean hourly COP and average daily COP obtained in September 2020. In this period, the HVAC machines worked as chillers in cooling mode.

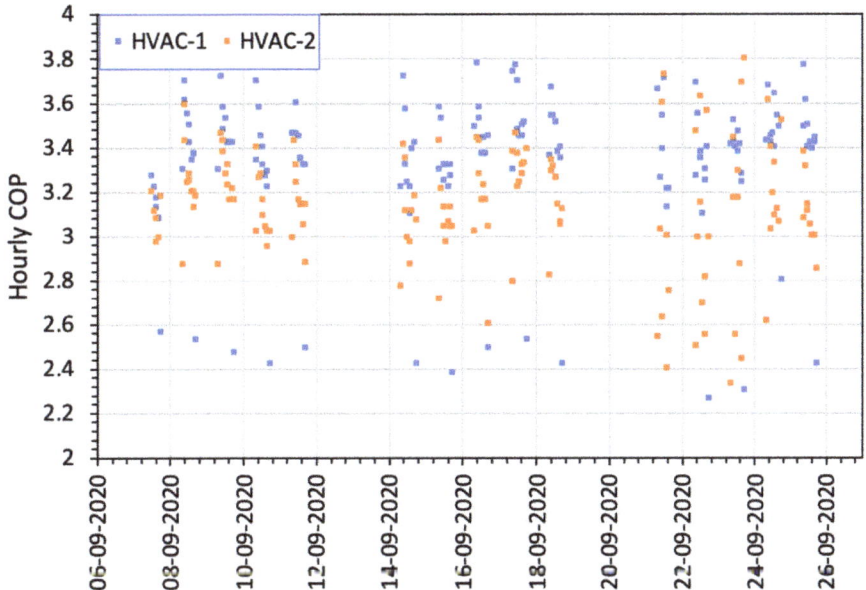

Figure 7. Hourly COP comparison between HVAC-1 and HVAC-2 (7 September 2020–25 September 2020).

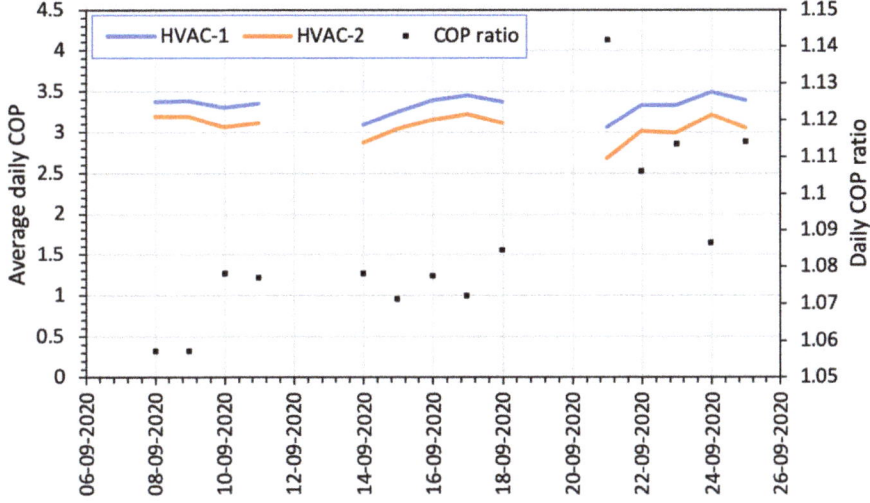

Figure 8. Average daily COP comparison and daily COP ratio between HVAC-1 and HVAC-2 (7 September 2020–25 September 2020).

It can be seen that increments in terms of COP were recorded during the entire period of experimentation, with lower values in the first week (average weekly increment equal to 6.7%), middle values in the second week (average weekly increment equal to 7.7%), and maximum values in the final week (average weekly increment equal to 11.2%), with a mean value of 8.9% over the entire period.

Nanoparticle concentration was measured before the data acquisition, in September. In fact, before the test period, the system was stopped from April to September. In that period, nanoparticle sedimentation occurred in the system. Density measured before the test campaign was 1065 g/L, therefore concentration was 1.7% in volume. Total re-dispersion of the nanoparticles was achieved after 29 h of the pumps being in operation, and the density was again 1079 g/L, therefore the concentration was again 2% (Figure 9).

(a)

(b)

Figure 9. Nanofluid dispersion during weight measurements in lab. (**a**) Density = 1065 g/L before pumps switched on. Concentration (1.7%); (**b**) Density = 1079 g/L after 29 h of running pumps. Concentration (2.0%).

3.3. February–March 2021 Results

In order to confirm the results acquired during winter 2020, the previous heat transfer fluid (water–glycol 30% vol) was reloaded within the HVAC-1 system and a long data set was acquired, in order to compare the performance of the two systems over a long time period (from 27 November 2020 to 5 February 2021). Therefore, on 8 February, the nanofluid was reloaded again within the HVAC-1 system and the performance was monitored until 9 March 2021. Figure 10 shows the hourly COP comparison between HVAC-1 and HVAC-2, from 1 January 2021 to 9 March 2021.

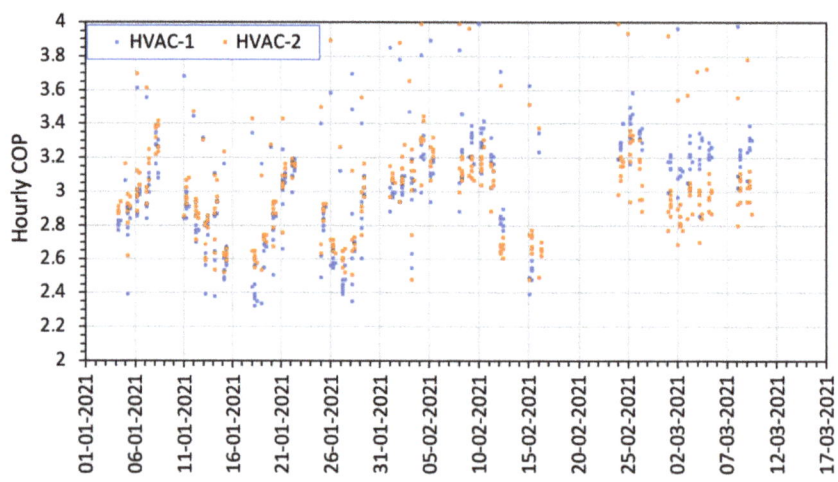

Figure 10. Hourly COP comparison between HVAC-1 and HVAC-2 (1 January 2021–9 March 2021).

It can be seen that the hourly COP values related to the HVAC-2 (orange dots), until 5 February are mainly higher than the HVAC-1 values. After loading the nanofluid (8 February) the trend reverses, with blue dots over orange ones.

This effect is particularly visible in Figure 11, where a comparison of the instantaneous COP is shown, following the loading of the nanofluid around noon. A significant increase in HVAC-1 performance is evident, due to the heat transfer fluid change.

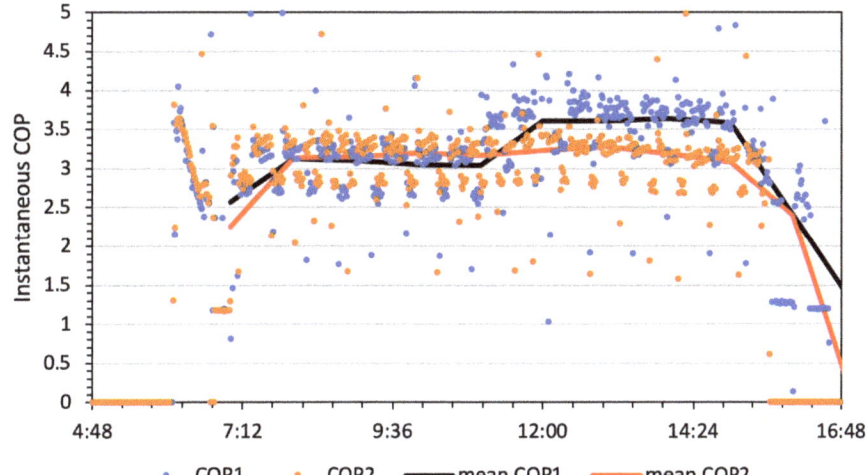

Figure 11. Instantaneous COP comparison between HVAC-1 and HVAC-2. Data acquired on 8 February 2021.

Finally, Figure 12 shows the average daily COP comparison and the daily COP ratio between HVAC-1 and HVAC-2 from 27 November 2020 to 9 March 2021. Clearly, it can be observed that during the entire experimental period, in which the two plants operated with the same fluid (from 27 November to 5 February), the performance of HVAC-1 was worse than HVAC-2, with an average COP1/COP2 ratio of 0.970. After the nanofluid loading, for the period from 9 February to 9 March 2021, the COP1/COP2 ratio was 1.078, with a peak of 1.121 and an average increase of 10.5%.

All the results shown in the above graphs demonstrate that the increased performance of the HVAC-2 system is not due to favorable environmental conditions, but only to the positive action of the nanofluid. The above discussed experimental results essentially agree with the numerical results found by Colangelo et al. [36].

In order to better understand the results described above, the performances of the HVAC-1 working with the base fluid and nanofluid have been investigated in depth and compared. In particular, the data related to the period January–March 2021 have been collected as a function of outside air temperature and fluid temperature at the inlet of HVAC-1. These parameters have been chosen since they directly affect the heat pump COP. Figure 13 shows the results.

As it can be seen, the COP was strongly influenced by the operating conditions of the heat pump. Nevertheless, the performance growth between the base fluid (data until 5 February) and nanofluid (data related to the following days) seems quite constant during the whole period of experimentation and equal to 13% on average.

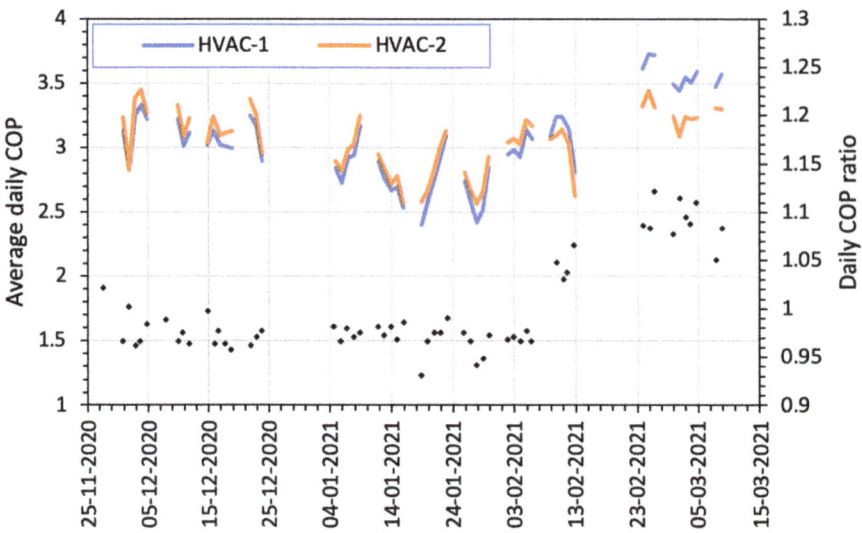

Figure 12. Average daily COP comparison and daily COP ratio between HVAC-1 and HVAC-2 (27 November 2020–9 March 2021).

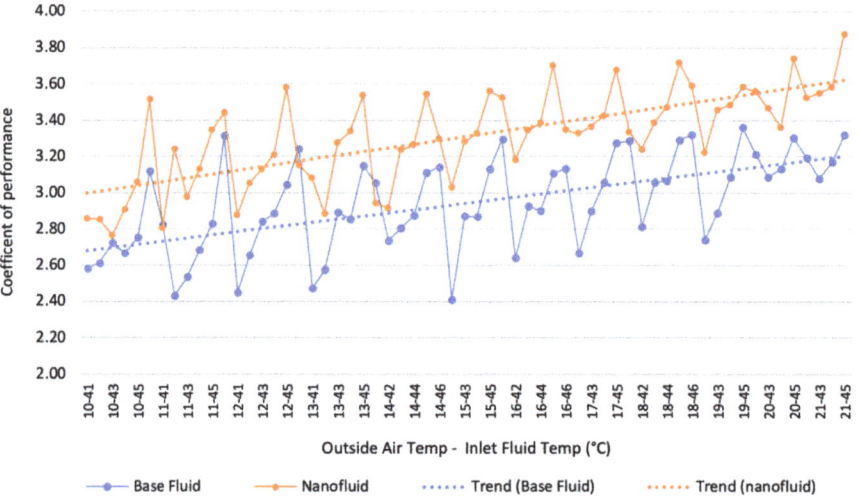

Figure 13. Average COP of HVAC-1 working with base fluid and nanofluid (January 2021–March 2021).

3.4. Practical Significance/Usefulness

The results of this work suggest that the use of nanofluids within hydronic HVAC systems can have a big impact from an environmental, energetic and economic perspective. In fact, taking into account the annual energy consumption of the building "Corpo O" [36], it was possible to calculate annual energy savings equal to 50.2 MWh. According to the Italian CO_2 emission factor [38], it was possible to preliminarily evaluate an annual avoided CO_2 emission equal to 21.8 tons related to the use of the nanofluid.

Finally, it is important to remark that the replacement of the traditional heat transfer fluid with a high performance nanofluid does not require important modification to the HVAC plant, resulting in an easy and effective retrofitting of old systems.

4. Conclusions

In this work, we investigated the performances of two full-scale HVAC systems, installed at the educational building "Corpo O" in the Campus of University of Salento, Lecce, Italy, working with a conventional water–glycol mixture and with Al_2O_3-nanofluid. In particular, the nanofluid was composed of water–glycol and 2% vol aluminum oxide (Al_2O_3) nanoparticles, having a controlled size distribution between 100 nm and 600 nm and a controlled stable suspension during the operation in the system. Long term stability of the nanofluid caused reliable and consistent thermal performance over a wide temperature range with limited effects on viscosity.

The results obtained in three experimental campaigns allowed both to quantify the performance increase due to the use of nanofluid instead of the conventional water and glycol mixture:

(1) under real operating conditions, the increase in energy efficiency due to the nanofluid of HVAC-1 with respect to HVAC-2, working simultaneously, has been on average equal to 9.8% in winter and 8.9% in summer, with average daily peaks of about 15%.

(2) the comparison between the performance of the same HVAC system, working in different comparable periods with and without nanofluid, shows a mean increase in COP equal to about 13%

(3) nanofluid density was monitored over the period of the test. Constant density was measured, and therefore a stable suspension of the nanofluid was found inside the distribution system of HVAC-1.

Although the relationship between the HVAC system performance and the use of nanofluids needs to be better investigated, the results of this work suggest that nanofluids can significantly improve the performance of air conditioning systems.

Author Contributions: Conceptualization, M.M. and A.d.R.; methodology, M.M., F.M. and G.C.; validation, M.M. and G.C.; writing—original draft preparation, M.M. and F.M.; writing—review and editing, M.M. and G.C.; supervision, A.d.R.; project administration, A.d.R.; funding acquisition, A.d.R. All authors have read and agreed to the published version of the manuscript.

Funding: This research received no external funding.

Conflicts of Interest: The authors declare no conflict of interest.

References

1. Saidur, R.; Leong, K.Y.; Mohammed, H.A. A review on applications and challenges of nanofluids. *Renew. Sustain. Energy Rev.* **2011**, *15*, 1646–1668. [CrossRef]
2. Lomascolo, M.; Colangelo, G.; Milanese, M.; de Risi, A. Review of heat transfer in nanofluids: Conductive, convective and radiative experimental results. *Renew. Sustain. Energy Rev.* **2015**, *43*, 1182–1198. [CrossRef]
3. Lee, S.; Choi, S.U.S.; Li, S.; Eastman, J.A. Measuring thermal conductivity of fluids containing oxide nanoparticles. *J. Heat Transf.* **1999**, *121*, 280–289. [CrossRef]
4. Beck, M.P.; Yuan, Y.; Warrier, P.; Teja, A.S. The effect of particle size on the thermal conductivity of alumina nanofluids. *J. Nanopart. Res.* **2009**, *11*, 1129–1136. [CrossRef]
5. Colangelo, G.; Milanese, M.; Iacobazzi, F.; De Risi, A. Experimental setup for low temperature thermal conductivity analysis of micro and nano suspensions. *AIP Conf. Proc.* **2019**, *2191*, 020050.
6. Colangelo, G.; Favale, E.; Miglietta, P.; Milanese, M.; de Risi, A. Thermal conductivity, viscosity and stability of Al_2O_3-diathermic oil nanofluids for solar energy systems. *Energy* **2016**, *95*, 124–136. [CrossRef]
7. Colangelo, G.; Diamante, N.F.; Milanese, M.; Starace, G.; de Risi, A. A critical review of experimental investigations about convective heat transfer characteristics of nanofluids under turbulent and laminar regimes with a focus on the experimental setup. *Energies* **2021**, *14*, 6004. [CrossRef]
8. Balla, H.; Abdullah, S.; Faizal, W.M.W.; Zulkifli, R.; Sopian, K. Enhancement of heat transfer coefficient multi-metallic nanofluid with ANSIS modeling for thermophysical properties. *Therm. Sci.* **2015**, *9*, 1613–1620. [CrossRef]
9. Kai, L.C.; Abdullah, M.Z.; Ismail, M.Z.; Mamat, H. Enhancement of nanofluid heat transfer in a mini-tube using SiO_2 nanoparticles. *Adv. Mater. Process. Technol.* **2019**, *5*, 607–616. [CrossRef]
10. Lee, J.H.; Hwang, S.G.; Lee, G.H. Efficiency improvement of a photovoltaic thermal (PVT) system using nanofluids. *Energies* **2019**, *12*, 3063. [CrossRef]

11. Colangelo, G.; Favale, E.; Miglietta, P.; de Risi, A. Innovation in flat solar thermal collectors: A review of the last ten years experimental results. *Renew. Sustain. Energy Rev.* **2016**, *57*, 1141–1159. [CrossRef]
12. Colangelo, G.; Favale, E.; de Risi, A.; Laforgia, D. A new solution for reduced sedimentation flat panel solar thermal collector using nanofluids. *Appl. Energy* **2013**, *111*, 80–93. [CrossRef]
13. Alsalame, H.A.M.; Hee Lee, J.; Hyun Lee, G. Performance Evaluation of a Photovoltaic thermal (PVT) system using nanofluids. *Energies* **2021**, *14*, 301. [CrossRef]
14. Milanese, M.; De Risi, A.; Colangelo, G. Energy simulation of a nanofluid solar cooling system in Italy. *Proc. Inst. Civ. Eng. Eng. Sustain.* **2018**, *172*, 32–39.
15. Marefati, M.; Huang, W. Energy, exergy, environmental and economic comparison of various solar thermal systems using water and Thermic Oil B base fluids, and CuO and Al_2O_3 nanofluids. *Energy Rep.* **2020**, *6*, 2919–2947.
16. Sattar, A.; Faroop, M.; Amjad, M.; Saeed, M.; Nawaz, S.; Mujataba, M.A.; Anwar, S.; Ali, Q.; Imran, M.; Pettinau, A. Performance evaluation of a direct absorption collector for solar thermal energy conversion. *Energies* **2020**, *13*, 4956. [CrossRef]
17. Colangelo, G.; D'Andrea, G.; Franciosa, M.; Milanese, M.; De Risi, A. Dynamic simulation of a solar cooling HVAC system with nanofluid solar collector. In Proceedings of the International Conference ZEMCH, Lecce, Italy, 22–24 September 2015.
18. Chaji, H.; Ajabshirchi, Y.; Esmaeilzadeh, E.; Heris, S.Z.; Hedayatizadeh, M.; Kahani, M. Experimental study on thermal efficiency of flat plate solar collector using TiO_2/Water nanofluid. *Appl. Sci.* **2013**, *7*, 60–70. [CrossRef]
19. Zhang, Z.D.; Zheng, W.; Su, Z.G. Study on diesel cylinder-head cooling using nanofluid coolant with jet impingement. *Therm. Sci.* **2015**, *19*, 2025–2037.
20. Micali, F.; Milanese, M.; Colangelo, G.; de Risi, A. Experimental investigation on 4-strokes biodiesel engine cooling system based on nanofluid. *Renew. Energy* **2018**, *125*, 319–326. [CrossRef]
21. Colangelo, G.; Favale, E.; Milanese, M.; de Risi, A.; Laforgia, D. Cooling of electronic devices: Nanofluids contribution. *Appl. Therm. Eng.* **2017**, *127*, 421–435. [CrossRef]
22. Du, R.; Jiang, D.; Wang, Y. Numerical investigation of the effect of nanoparticle diameter and sphericity on the thermal performance of geothermal heat exchanger using nanofluid as heat transfer fluid. *Energies* **2020**, *13*, 1653. [CrossRef]
23. Hui Sun, X.; Yan, H.; Massoudi, M.; Chen, Z.H.; Wu, W.T. Numerical simulation of nanofluid suspension in a geothermal heat exchanger. *Energies* **2018**, *11*, 1919. [CrossRef]
24. De Risi, A.; Milanese, M.; Colangelo, G.; Laforgia, D. High Efficiency Nanofluid Cooling System for Wind Turbines. *Therm. Sci.* **2014**, *18*, 543–554.
25. Kulkarni, D.P.; Das, D.K.; Vajjha, R.S. Application of nanofluids in heating buildings and reducing pollution. *Appl. Energy* **2009**, *86*, 2566–2573. [CrossRef]
26. Ahmed, F.; Khan, W.A. Efficiency enhancement of an air-conditioner utilizing nanofluid: An experimental study. *Energy Rep.* **2021**, *7*, 575–583. [CrossRef]
27. Hatami, M.; Domairry, G.; Mirzababaei, S.N. Experimental investigation of preparing and using the H_2O based nanofluids in the heating process of HVAC system model. *Int. J. Hydrogen Energy* **2017**, *42*, 7820–7825. [CrossRef]
28. Ghadimi, A.; Saidur, R.; Metsetaar, H.S.C. A review of nanofluid stability properties and characterization in stationary conditions. *Int. J. Heat Mass Transf.* **2011**, *54*, 4051–4068. [CrossRef]
29. Awais, M.; Ullah, N.; Ahmad, J.; Sikandar, F.; Ehsan, M.M.; Salehin, S. Bhuiyan, A.A. Heat transfer and pressure drop performance of Nanofluid: A state-of-the-art review. *Int. J. Thermofluids* **2021**, *9*, 100065. [CrossRef]
30. Hwang, Y.; Lee, J.K.; Lee, C.H.; Jung, Y.M.; Cheong, S.I.; Lee, C.G.; Ku, B.C.; Jang, S.P. Stability and thermal conductivity characteristics of nanofluids. *Thermochim. Acta* **2007**, *455*, 70–74. [CrossRef]
31. Iacobazzi, F.; Colangelo, G.; Milanese, M.; De Risi, A. Thermal conductivity difference between nanofluids and micro-fluids: Experimental data and theoretical analysis using mass difference scattering. *Therm. Sci.* **2019**, *23*, 3797–3807. [CrossRef]
32. Iacobazzi, F.; Milanese, M.; Colangelo, G.; de Risi, A. A critical analysis of clustering phenomenon in Al_2O_3 nanofluids. *J. Therm. Anal. Calorim.* **2019**, *135*, 371–377. [CrossRef]
33. Bogdan, N.; Vetrone, F.; Ozin, G.A.; Capobianco, J.A. Synthesis of ligand-free colloidally stable water dispersible brightly luminescent lanthanide-doped upconverting nanoparticles. *Nano Lett.* **2011**, *11*, 835–840. [CrossRef] [PubMed]
34. Lee, J.; Kim, M.; Hong, C.K.; Shim, S.E. Measurement of the dispersion stability of pristine and surface-modified multiwalled carbon nanotubes in various nonpolar and polar solvents. *Meas. Sci. Technol.* **2007**, *18*, 3707. [CrossRef]
35. Pantzali, M.N.; Mouza, A.A.; Paras, S.V. Investigating the efficacy of nanofluids as coolants in plate heat exchangers (PHE). *Chem. Eng. Sci.* **2009**, *64*, 3290–3300. [CrossRef]
36. Colangelo, G.; Raho, B.; Milanese, M.; de Risi, A. Numerical evaluation of a HVAC system based on a high-performance heat transfer fluid. *Energies* **2021**, *14*, 3298. [CrossRef]
37. Available online: https://www.clivet.com (accessed on 16 March 2022).
38. Available online: https://www.isprambiente.gov.it/files2020/pubblicazioni/rapporti/Rapporto317_2020.pdf (accessed on 16 March 2022).

Article

Co-Existence of Iron Oxide Nanoparticles and Manganese Oxide Nanorods as Decoration of Hollow Carbon Spheres for Boosting Electrochemical Performance of Li-Ion Battery

Karolina Wenelska *, Martyna Trukawka, Wojciech Kukulka, Xuecheng Chen and Ewa Mijowska

Department of Nanomaterials Physicochemistry, Faculty of Chemical Technology and Engineering, West Pomeranian University of Technology, Szczecin, Piastow Ave. 42, 71-065 Szczecin, Poland; mtrukawka@zut.edu.pl (M.T.); wkukulka@zut.edu.pl (W.K.); xchen@zut.edu.pl (X.C.); emijowska@zut.edu.pl (E.M.)
* Correspondence: kwenelska@zut.edu.pl

Abstract: Here, we report that mesoporous hollow carbon spheres (HCS) can be simultaneously functionalized: (i) endohedrally by iron oxide nanoparticle and (ii) egzohedrally by manganese oxide nanorods ($Fe_xO_y/MnO_2/HCS$). Detailed analysis reveals a high degree of graphitization of HCS structures. The mesoporous nature of carbon is further confirmed by N_2 sorption/desorption and transmission electron microscopy (TEM) studies. The fabricated molecular heterostructure was tested as the anode material of a lithium-ion battery (LIB). For both metal oxides under study, their mixture stored in HCS yielded a significant increase in electrochemical performance. Its electrochemical response was compared to the HCS decorated with a single component of the respective metal oxide applied as a LIB electrode. The discharge capacity of $Fe_xO_y/MnO_2/HCS$ is 1091 mAhg^{-1} at 5 Ag^{-1}, and the corresponding coulombic efficiency (CE) is as high as 98%. Therefore, the addition of MnO_2 in the form of nanorods allows for boosting the nanocomposite electrochemical performance with respect to the spherical nanoparticles due to better reversible capacity and cycling performance. Thus, the structure has great potential application in the LIB field.

Keywords: battery; metal oxide nanoparticles; carbon spheres

1. Introduction

Due to a deteriorating environmental situation (e.g., global warming, and the gradual depletion of oil and hard coal resources), the development of balanced and clean energy resources is extremely important. Clean energy sources such as wind, sun and tidal energy are preferred alternatives to fossil fuels, and they are best utilized with high-efficiency energy storage technologies. Lithium-ion batteries (LIBs)are appealing storage sources due to their unique characteristics, such as their extended life, energy density, low maintenance costs, environmental friendliness and lack of memory effect [1,2]. Although the performance of lithium-ion batteries continues to improve, their energy density, cycle lifetime and productivity remain insufficient for large-scale applications in consumer electronics, and transportation and storage of renewable energy. Much effort has been made to create new electrode materials or to design unique electrode architecture to address the ever-increasing demand for batteries with higher energy density and longer cycle life [3–7]. The electrodes must maintain their integrity across multiple discharge–recharge cycles, which is one of the challenges in their design. Li-alloying agglomeration or the formation of passivation layers, which prohibit the fully reversible injection of Li ions into negative electrodes, reduce the life spans of electrode systems [8,9]. Transition metal oxides (TMOs) have recently found use as electrode material for energy storage devices including LIBs [10]. These materials exhibit a large theoretical specific capacity and high working potential for LIBs (ca. 500–1000 mAhg^{-1}) [11]. This is an advantage of TMO application due to the prevention of lithium dendrite formation, which increases safe use.

TMOs such as SnO_2 [12], Fe_3O_4 [13] and MnO_2 [14] are characterized by a very high theoretical capacity. Furthermore, due to its low cost and low environmental impact, iron oxide is one of the most promising materials. Its disadvantage is that the Li reactivity mechanism of transition metal oxides requires the formation and the decomposition of Li_2O, accompanying the redox reactions of metal nanoparticles [15]. During this conversion reaction, there is usually a large change in volume, which can cause electrode fracture or deterioration of electrochemical efficiency [16]. In order to address these issues, a carbon material can be used, which can prevent both volume change and aggregation of the nanoparticles [17]. In order to obtain high electrochemical performance, the carbon material should possess advantages such as high surface area and high electronic conductivity. A high surface area provides active sites for the pinning on or embedding of nanoparticles on the carbon surface [18].

Mesoporous carbon materials are characterized by a large specific surface area, which can reduce current density per area unit. Another advantage is the thin walls shorten the diffusion paths. In addition, they are an acceptable electrode material due to low cost, high chemical stability and good processing ability [19,20]. Mesoporous hollow carbon nanospheres fully meet these requirements, and therefore, its application as a carrier of metal oxide nanoparticles appears to be reasonable.

Herein, we present a facile synthesis method of hollow carbon nanospheres (HCS) with two stages of functionalization using transition metal oxides (iron oxides in the form of spherical nanoparticles and rod-like manganese dioxide) as advanced anode material for high-performance LIBs. The prepared $Fe_xO_y/MnO_2/HCS$ nanocomposite combines the advantages of empty carbon spheres, such as stability or adaptation to expanding volumes during the cycle, with a high specific capacity of transition metal oxides.

2. Experimental

2.1. Synthesis of Mesoporous Core/Shell Structured Silica Spheres (SiO_2@$mSiO_2$ Spheres)

Core/shell silica spheres were used as a hard template to obtain HCS using our previous reported method [12]. In a typical synthesis, ethanol (100 mL, P.P.H. Stanlab, Lublin, Poland), tetraethyl orthosilicate (TEOS, Sigma Aldrich, Beijing, China) (4 mL) and concentrated ammonia (28 wt.%, 6 mL, CHEMPUR, Piekary Slaskie, Poland) were mixed and stirred for 24 h. Then, solid SiO_2 spheres were dried for further use (SiO_2). To prepare the SiO_2@$mSiO_2$ spheres, 100 mg of SiO_2 spheres were dispersed in water (160 mL), ethanol (80 mL) and ammonia solution (28 wt.%, 0.67 mL). Then, 80 mg of the surfactant cetyltrimethylammonium bromide (CTAB, Sigma Aldrich, Beijing, China) and 0.37 mL of TEOS were added. After stirring for 24 h, the product was dried at 80 °C to get SiO_2@$mSiO_2$ spheres.

2.2. Synthesis of Hollow Carbon Spheres (HCS)

A chemical vapor deposition (CVD) process was applied to prepared HCS. The core-shell SiO_2@$mSiO_2$ was placed in an alumina boat in a tube furnace (Carbolite GERO, Hope, UK) and a CVD process using C_2H_4 as the carbon source occurred for 1 h at 800 °C. After synthesis, SiO_2@$mSiO_2$ covered by carbon was obtained (SiO_2@$mSiO_2$_C). To obtain HCS, SiO_2@$mSiO_2$_C was treated by hydrofluoric acid (CHEMPUR, Piekary Slaskie, Poland) to remove silica and then washed with water and ethanol several times. Finally, the carbon product was obtained by drying the sample in a vacuum at 80 °C for 12 h.

2.3. Functionalization of HCS with Metal Oxides ($Fe_xO_y/MnO_2/HCS$)

To store metal oxide in hollow carbon spheres, iron (III) nitrate nonahydrate (100 mg, CHEMPUR, Piekary Slaskie, Poland) was dissolved in ethanol; added dropwise to the 100 mg HCS, stirring and heating to 50 °C; and then, placed in a furnace for 2 h at 400 °C. Next, Fe_xO_y/HCS (40 mg) was added to a $KMnO_4$ (40 mg, CHEMPUR, Piekary Slaskie, Poland) solution in a round bottom flask. The reaction was carried out for 0.5 h at 70 °C.

Then, the product was collected by filtration, washed two times with water and ethanol and dried at 80 °C in a vacuum for 24 h.

2.4. Characterization

The FEI Tecnai F30 transmission electron microscope (TEM, FEI Corporation, Hillsboro, OR, USA) with a field emission gun operating at 200 kV was used to investigate the morphology of the samples. The elemental mappings were performed via energy-dispersive X-ray spectroscopy (EDX, FEI Corporation, Hillsboro, OR, USA) as the TEM mode. Raman spectra were collected with a Renishaw micro Raman spectrometer (λ = 785 nm, Renishaw, Edinburg, UK). Thermogravimetric analysis (TGA) was carried out on 10 mg samples at a heating rate of 10 °C/min from room temperature to 900 °C under air using a DTA-Q600 SDT (TA Instruments, New Castle, DE, USA). X-ray diffraction (XRD) was conducted on a Philips diffractometer (Malvern, Cambridge, UK) using Cu Kα radiation. The N_2 adsorption/desorption isotherms were measured on a Micromeritics ASAP 2010M instrument (Micrometrics, Tewksbury, UK) at liquid nitrogen temperature (77 K). To compute the specific surface area and pore size distribution, the Brunauer–Emmett–Teller (BET) and Barrett–Joyner–Halenda (BJH) methods were used, respectively.

2.5. Electrochemical Measurements

The as-prepared Fe_xO_y/MnO_2/HCS nanomaterials were used as electrode materials for LIBs. To prepare the working electrode, active materials, acetylene black (C-NERGY™ SUPER C65, Timcal, Congleton, UK) and PVDF (Solvay Plastics, Warszawa, Poland) were mixed in a weight ratio of 85:10:5. Subsequently, N-methyl-pyrrolidone (NMP, CHEMPUR, Piekary Slaskie, Poland) was added to the powder to form a slurry. The working electrodes were fabricated by coating the slurry onto copper foam (Sigma-Aldrich, Beijing, China) and dried in a vacuum at 80 °C overnight. The testing coin cells were assembled with the working electrode, metallic lithium foil (Sigma-Aldrich, Beijing, China) as a counter electrode, NKK TK4350 film as a separator (Sigma-Aldrich, Beijing, China) and 200 µL $LiPF_6$ in 1:1 ethylene carbonate (EC)/dimethyl carbonate (DMC) as the electrolyte (Sigma-Aldrich, Beijing, China). The assembly of the cells was carried out in an argon-filled glovebox (M. Braun Co., Garching, Germany). Electrochemical studies by means of cyclic voltammetry (CV) and galvanostatic cycling with potential limitation (GCPL) were performed. The measurements were executed on a VMP3 multichannel potentiostat (BioLogic, Seyssient-Pariset, France) at room temperature.

3. Results and Discussion

The morphology and architecture of the HCS and the HCS with metal oxides were presented (Scheme 1) and characterized using TEM analysis. Based on the TEM images (Figure 1), it was found that the cores of solid silica spheres were ~250 nm in diameter. The mesoporous shell has a thickness of ~100 nm. Therefore, the obtained HCS have a size diameter of ~450 nm. The iron oxide nanoparticles were evenly distributed throughout the hollow carbon spheres, and their size was ~15 nm. Microscopic analysis revealed that MnO_2 was deposited onto HCS in a flat form of thin rods with an irregular surface (Figure S1). Their diameter was ~20 nm while the size of the iron oxide was ~20 nm (Figure 2). To confirm the elemental composition of the sample, EDS mapping was performed. Figure 3 reveals that the Mn and Fe was distributed homogeneously of the carbon shell.

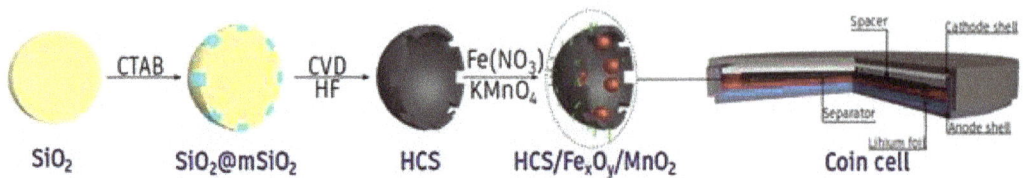

Scheme 1. Scheme presentation of $Fe_xO_y/MnO_2/HCS$ synthesis.

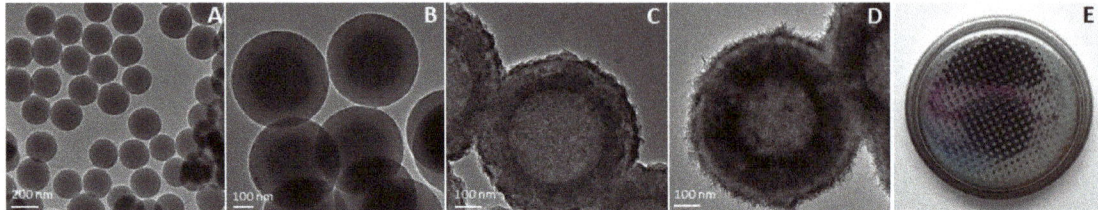

Figure 1. Transmission Electron Microscopy (**A**) images of SiO_2, (**B**) SiO_2@$mSiO_2$, (**C**) hollow carbon spheres, (**D**) $Fe_xO_y/MnO_2/HCS$ and (**E**) the lab-scale lithium cell prototype.

Figure 2. High resolution TEM images of (**A**) MnO_2 and (**B**) Fe_xO_y.

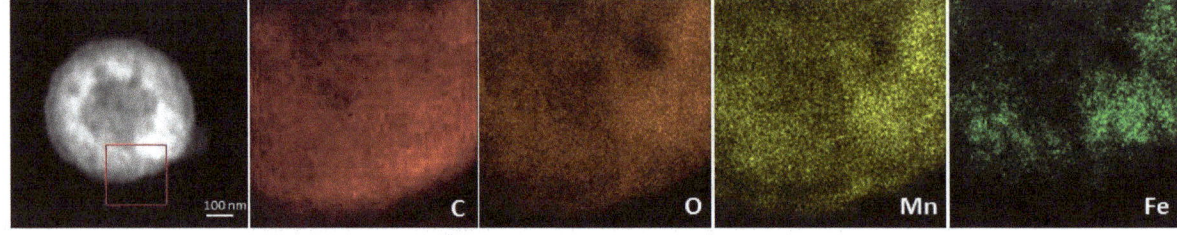

Figure 3. Scanning Transmission Electron Microscope image and high-angle annular dark-field scanning transmission electron microscopy and energy-dispersive X-ray spectroscopy (HAADF-STEM-EDS) mapping images of $Fe_xO_y/MnO_2/HCS$.

In the next step, XRD patterns acquired from HCS, Fe_xO_y/HCS, MnO_2/HCS and the corresponding hybrid material of $Fe_xO_y/MnO_2/HCS$ were depicted (Figure 4). The black line which corresponds to carbon spheres has significant and broad peaks at 2θ = 24.9° and 42° in response to graphitic carbon planes (002) and (100). $Fe_xO_y/MnO_2/HCS$ exhibits further diffraction peaks which reflect the peaks appearing on the patterns of individual components. The peaks at 2θ = 37.9° and 57.8° correspond to the (211) and (600) MnO_2 planes, respectively [21]. Based on the obtained pattern, it was found that the sample contains a mixture of iron oxides: Fe_2O_3 and Fe_3O_4. Diffraction peaks characteristic for Fe_2O_3 were identified at 2θ = 24.4°, 35.7°, 44.2°, 49.6° and 62.9°. They are related to the (211), (110), (024) and (214) planes [22]. Two peaks assigned to Fe_3O_4 were found, at 2θ = 36.6° (311) and 2θ = 57.8° (511) [23].

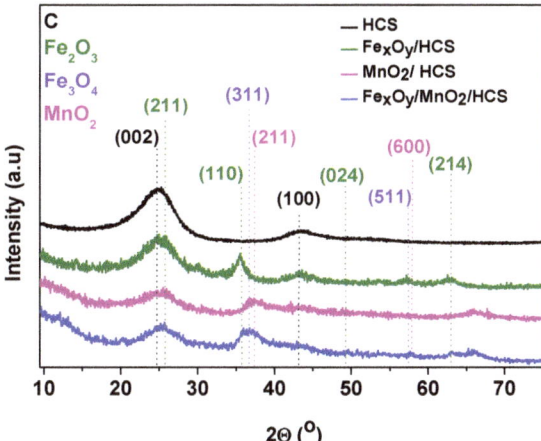

Figure 4. X-ray diffraction patterns of HCS, Fe_xO_y/HCS, MnO_2/HCS and the hybrid material of Fe_xO_y/MnO_2/HCS.

First, the BET method was used to investigate the specific surface area of the HCS and the corresponding hybrid composite. The N_2 adsorption/desorption isotherms are shown in Figure 5A. For HCS, the BET specific surface area is 571 m^2/g. After functionalization with metal oxides, the surface area decreased to 177 m^2/g. A reduction in the specific surface area is related to the fact that nanoparticles of iron oxide are both on the surface and in the pores of nanospheres. Additionally, the presence of MnO_2 on the surface of the nanospheres may cause blockage of the pores [24]. From the adsorption branch, the related mesopore size distribution determined using the BJH approach gives average pore sizes at ~5.53 nm with a predominance of pores of a size at 2.2 nm in the case of the HCS, and 4.97 average pore size and most pores with size 2.57 nm for Fe_xO_y/MnO_2/HCS (Figure 5B). This result suggests the mesoporous nature of the material (Table 1).

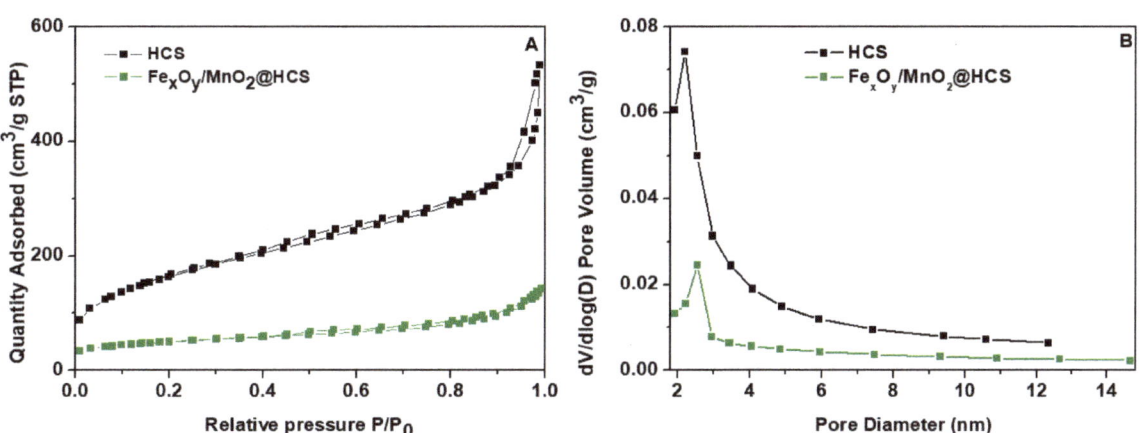

Figure 5. (**A**) N_2 adsorption/desorption isotherms and (**B**) pore size distribution of HCS and Fe_xO_y/MnO_2/HCS.

Table 1. Specific surface area, pore volume and pore sizes of hollow carbon spheres (HCS) and $Fe_xO_y/MnO_2/HCS$.

Sample	S_{BET} (m^2/g)	V_{TOTAL} (cm^3/g)	Average Pore Size (nm)
HCS	571	0.20	5.53
$Fe_xO_y/MnO_2/HCS$	177	0.06	4.97

The purity of HCS and the quantitative analysis of the nanocomposite was verified by thermogravimetric analysis as shown in Figure 6A. At 540 °C, HCS began to decompose in air. The weight loss accelerated as the temperature rose, until all the carbon spheres were depleted at approximately 735 °C. HCS has an ash percentage of 0 wt.% after combustion at 900 °C, indicating that it is of high purity. In comparison to the pristine HCS, the stability of $Fe_xO_y/MnO_2/HCS$ was weaker. During heating, the nanocomposite burned at 230 °C and ended at 430 °C. About 50 wt.% of the sample decomposed; thus, it can be concluded that the metal oxides are half the mass of the sample. Raman spectroscopy is commonly used to characterize all sp^2 carbons. The Raman spectra of HCS and the corresponding nanocomposite shows two prominent peaks at 1320 and 1600 cm^{-1} (Figure 6B). The former, named D-band, reveals the defect of the C atomic lattice, while the latter peak, called G-band, represents the stretching vibration of C atom sp^2 hybrid plane. The relative intensity of D to G provides an indicator for determining the in-plane crystallite size or the amount of disorder in the sample [25,26]. Upon deposition of the metal oxides' nanoparticles, the relation between the D-band and G-band intensities increases; thus, additional defects formed in the HCS structures.

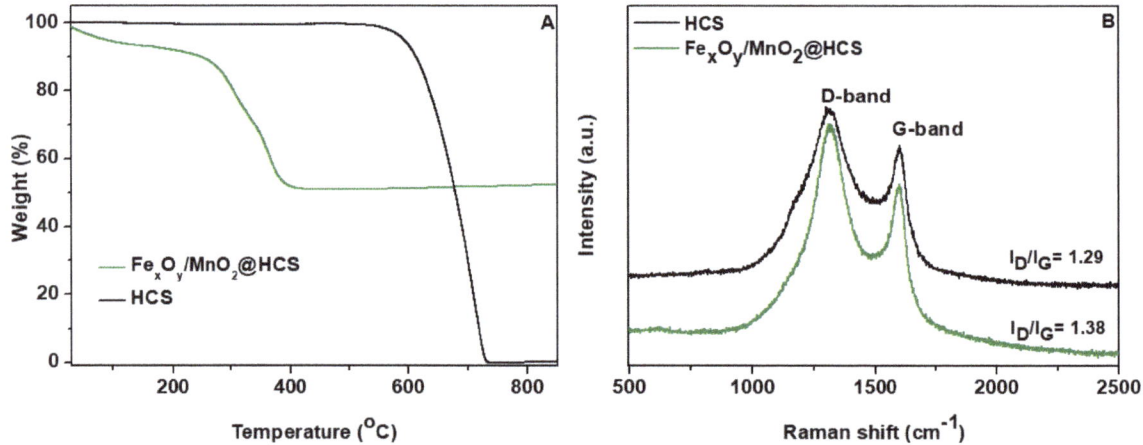

Figure 6. (A) Thermogravimetric analysis profiles, and (B) Raman spectra of HCS and $Fe_xO_y/MnO_2/HCS$.

The prepared materials ($Fe_xO_y/MnO_2/HCS$) were further evaluated as anode material for Li-ion batteries. The CV curves of the $Fe_xO_y/MnO_2/HCS$ electrode recorded at a scan rate of 0.5 mV s^{-1} are shown in Figure 7A. During the first discharge cycle of the $Fe_xO_y/MnO_2/HCS$ electrode, a strong reduction peak in cathodic scan was observed at ~0.5 V, which is in agreement with the reduction of Mn^{2+} and Fe^{3+} to their metallic states due to the formation of Li$_2$O, as illustrated in Equation (1),

$$FeMnO_3 + 6Li^+ + 6e^- \rightarrow Fe + Mn + 3Li_2O \quad (1)$$

accompanied by electrolyte decomposition into a solid electrolyte interphase (SEI) layer. During the first anodic [27] charge process of the $Fe_xO_y/MnO_2/HCS$ electrode, only two anodic peaks (1.19 and 2.08 V) can be attributed to the oxidation of metallic Mn and Fe, which are illustrated in Equations (2) and (3):

$$Fe + xLi_2O - 2xe^- \rightarrow 2\, FeO_x + 2xLi^+ \tag{2}$$

$$Mn + xLi_2O - 2xe^- \rightarrow 2\, MnO_x + 2xLi^+ \tag{3}$$

These two peaks shift to 0.75 and 0.3 V in the following reduction step, indicating improved kinetics. The large peaks at 1.2 and 1.7 V in the charge process are attributable to the two-step oxidation of Mn(0) and Fe(0) to MnO_x and FeO_x, respectively [28]. The two pairs of reduction and oxidation peaks that correspond to the FeO_x/Fe and MnO_x/Mn conversions appear to be well overlapped, indicating that the two-step electrochemical reactions are highly reversible.

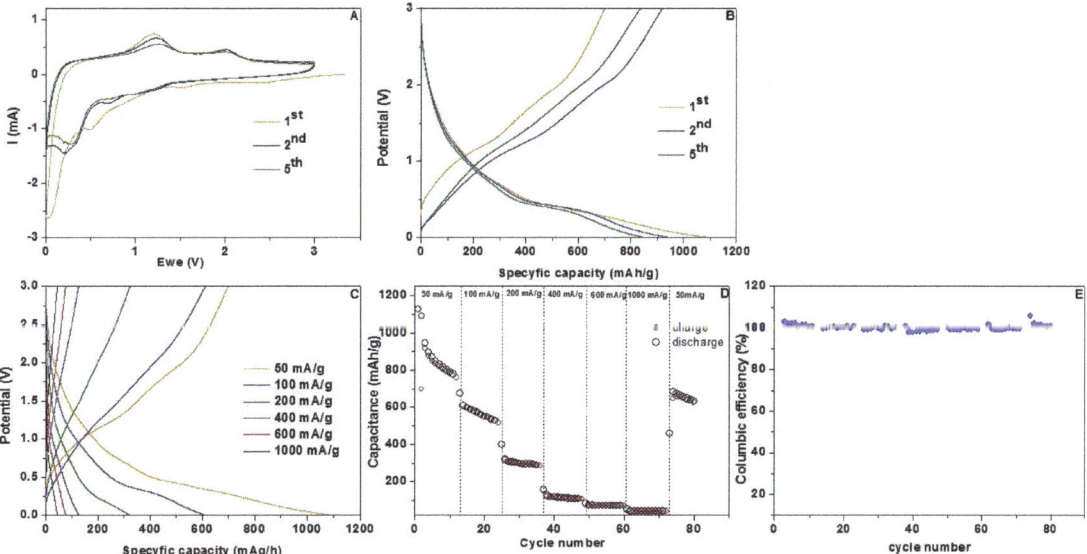

Figure 7. (**A**) cyclic voltammetry curves performed over a potential window from 0.05 to 3 V at a scan rate from 0.5 mVs^{-1}, (**B**) galvanostatic charge/discharge profiles at a current density of 50 mAg^{-1} in the voltage range of 0.05–3.0 V, (**C**) voltage–capacity curves, (**D**) gravimetric specific capacities vs. cycle number and (**E**) Coulombic efficiency of $Fe_xO_y/MnO_2/HCS$.

Figure 7B shows the first, the second and the fifth galvanostatic discharge/charge curves of the $Fe_xO_y/MnO_2/HCS$ between 0.05 and 3.0 V (versus Li/Li$^+$). The orange line is attributed to the first cycle charge and discharge capacity of $Fe_xO_y/MnO_2/HCS$ 625 mAhg^{-1} and 1100 mAhg^{-1}, respectively. It can be assigned to irreversible effects such as the formation of the SEI layer. After cycling, thin SEI form on the $Fe_xO_y/MnO_2/HCS$ electrode, and additional mesopores form in the hollow structure, resulting in the establishment of linked spaces that are conducive to fast Li$^+$ ion and electron transport. The stable SEI layer and hollow space on electrodes can help to stabilize lithiation/de-lithiation and reduce mechanical deterioration caused by discharge volume expansion. In the next step, the charge/discharge profile, with different current densities, was measured. The charge/discharge curves of the $Fe_xO_y/MnO_2/HCS$ composite at various rates are shown in Figure 7C. On both discharge and charge profiles, multiple plateaus can be seen, which are in good agreement with the CV curves. A sequential decay in reversible capacities as the rate increases can be observed. The electrode delivered reversible capacities of 1100, 610, 320, 126, 75 and 42 mAhg^{-1} at current densities from 50 to 1000 mAg^{-1}. As shown in Figure 7D, the new anode material exhibits good Li$^+$-ion storage capacity and cyclic stability at each current density from 50 to 1000 mAg^{-1}. Notably, the fused Fe$_x$O$_y$/MnO$_2$/HCS presented much higher capacities at each stage compared to pristine HCS. Charge-discharge profiles obtained by different applied current densities were used to further study the rate behavior of the Fe$_x$O$_y$/MnO$_2$/HCS electrode. As the rate increased, there was a sequential decrease in reversible capabilities. Figure 7D shows that the electrode delivered reversible capacities of 611, 323, 135, 83 and 46 mAhg^{-1} at current densities from 50 mAg^{-1} to 1000 mAg^{-1}. When the current density was reduced to 50 mAg^{-1}, the capacity immediately returned to 675 mAhg^{-1}. The above results indicate that the hybrid material of the Fe$_x$O$_y$/MnO$_2$/HCS electrode has an excellent rate capability. Additionally, the Fe$_x$O$_y$/MnO$_2$/HCS electrode displays high CE, which in many cycles exceeds 100% (Figure 7E). This can be associated with the irreversible side reaction during the charge process or with an irregular amount of transported Li$^+$ ions during charge–discharge processes. In the first case, the side reaction can suggest more capacity is generated than the amount of Li$^+$ ions released from the active material. Capacity does not reference the actual storage ability of the electrode; it is estimated based on coulomb counting, e.g., to integrate current vs. time until the cut-off potential is reached. Therefore, if there are any side reactions that consume current without affecting the voltage (e.g., the charge is not actually intercalating to a site in the electrode), then this current is integrated, adding to the capacity. If this occurs in the discharge step, then the Coulombic efficiency (CE) result can be >100%. In the second case, if the Li$^+$ ion is less intercalated due to structural interference during the discharge process and the maximum amount of Li-ion is released during the charge process, the CE may exceed 100%. Both scenarios can lead to a gradual degeneration of the structure of the electrode active material and, thus, to a reduced stability (Figure S2).

Hollow carbon spheres decorated by iron and manganese oxides with large surface area and a conductive network enable high accessibility of the active material. Fe$_x$O$_y$/MnO$_2$/HCS composite initially reaches the full theoretical capacity, but degradation effects lead to poor cycle stability. A comparison with the electrochemical performance of other reported HCS composites with Fe$_x$O$_y$, or MnO$_2$, shows that this is a promising approach to optimize the cycling stability of the battery. Graphene-wrapped Fe$_3$O$_4$ synthesized by Zhao et al. [29] shows a better charge/discharge capacity but not satisfying stability. Zhu et al. [30] obtained porous olive-like carbon decorated by Fe$_3$O$_4$, which presented lower dis- and charge capacities. Wu et al. [31] presented a novel foam-like Fe$_3$O$_4$/C composite made with gelatin as the carbon source and ferric nitrate as the iron source using a sol-gel process. As a result, the Fe$_3$O$_4$/C composite electrode demonstrates good rate performance with a reversible capacity of 660 and 580 mAhg^{-1} at 3 and 5 °C, respectively, whereas all composite manganese oxide/carbon presented lower capacity and stability compared to our data [32,33]. Therefore, a combination of two oxides (iron and manganese) significantly improved the capacity of obtained electrodes. The state-of-the-art process provides information about the synergistic effect of such a combination [34]. The synergistic effect of combining such components manifests in improving the reversibility of the electrochemical reaction, buffering large distortions and stresses during discharge-charging processes and preventing aggregation of the active material. This results in high reversible capacity, excellent cycling performance and excellent rate capabilities. The unique MnO$_2$ nanorods morphology has been reported as anode material for lithium-ion batteries [35–37]. This rod-like morphology is reported to enhance electrochemical properties and was proven in our study. These features, along with the high performance of iron oxides, recommend this hybrid structure as promising for boosting the performance of energy storage devices.

4. Conclusions

In summary, the present data demonstrate a facile route for the synthesis of a nanocomposite consisting of HCS and metal oxide hybrid material. The synergistic effect of the components in Fe$_x$O$_y$/MnO$_2$/HCS composite displays the enhancement of electrochemical properties in comparison to pristine HCS. The addition of MnO$_2$ nanorods boosts the reversible capacity and cycling performance. The discharge capacity of $Fe_xO_y/MnO_2/HCS$ is 1091 mAhg^{-1} at 5 Ag^{-1}, and the corresponding CE is as high as 98%. Therefore, the co-existence of two metal oxides stored in HCS resulted in design of composite that has the potential to be used as anode material for lithium-ion batteries with high cycling stability and boosted performance in comparison to the single metal oxide functionalization approach.

Supplementary Materials: The following are available online at https://www.mdpi.com/article/10.3390/ma14226902/s1, Figure S1: TEM images of MnO$_2$ rods on the HCS surface; Figure S2: Cycling stability at 100 mA/g.

Author Contributions: Conceptualization, E.M. and X.C.; methodology, K.W., W.K. and M.T.; writing—original draft preparation, K.W., X.C. and M.T.; writing—review and editing E.M. and W.K. All authors have read and agreed to the published version of the manuscript.

Funding: This research was funded by National Science Centre, Poland, within Beethoven UMO-2016/23/G/ST5/04200.

Institutional Review Board Statement: Not applicable.

Informed Consent Statement: Not applicable.

Data Availability Statement: All datasets generated for this study are included in the article.

Conflicts of Interest: The authors declare no conflict of interest.

References

1. Amjad, S.; Rudramoorthy, R.; Neelakrishnan, S.; Varman, K.S.R.; Arjunan, T.; Shaik, A. Evaluation of energy requirements for all-electric range of plug-in hybrid electric two-wheeler. *Energy* **2011**, *36*, 1623–1629. [CrossRef]
2. Wang, J.; Zhao, H.; Zeng, Z.; Lv, P.; Li, Z.; Zhang, T.; Yang, T. Nano-sized Fe$_3$O$_4$/carbon as anode material for lithium ion battery. *Mater. Chem. Phys.* **2014**, *148*, 699–704. [CrossRef]
3. Kang, B.; Ceder, G. Battery materials for ultrafast charging and discharging. *Nat. Cell Biol.* **2009**, *458*, 190–193. [CrossRef] [PubMed]
4. Armstrong, A.R.; Lyness, C.; Panchmatia, P.M.; Islam, M.S.; Bruce, P.G. The lithium intercalation process in the low-voltage lithium battery anode Li1+xV1−xO$_2$. *Nat. Mater.* **2011**, *10*, 223–229. [CrossRef]
5. Kaskhedikar, N.A.; Maier, J. Lithium Storage in Carbon Nanostructures. *Adv. Mater.* **2009**, *21*, 2664–2680. [CrossRef]
6. Wang, N.; Zhang, X.; Ju, Z.; Yu, X.; Wang, Y.; Du, Y.; Bai, Z.; Dou, S.; Yu, G. Thickness-independent scalable high-performance Li-S batteries with high areal sulfur loading via electron-enriched carbon framework. *Nat. Commun.* **2021**, *12*, 4519. [CrossRef]
7. Peng, C.; Kong, L.; Li, Y.; Fu, C.; Sun, L.; Feng, Y.; Feng, W. Fluorinated graphene nanoribbons from unzipped single-walled carbon nanotubes for ultrahigh energy density lithium-fluorinated carbon batteries. *Sci. China Mater.* **2021**, *64*, 1367–1377. [CrossRef]
8. Courtney, I.A.; McKinnon, W.R.; Dahn, J.R. On the Aggregation of Tin in SnO Composite Glasses Caused by the Reversible Reaction with Lithium. *J. Electrochem. Soc.* **1999**, *146*, 59–68. [CrossRef]
9. Denis, S.; Baudrin, E.; Touboul, M.; Tarascon, J.M. Synthesis and electrochemical properties vs. Li of amorphous/crystallized indium vanadates, Molecular crystals and liquid crystals science and technology. *Sect. A Mol. Cryst. Liq. Cryst.* **1998**, *311*, 351–357. [CrossRef]
10. Li, Q.; Yao, K.; Zhang, G.; Gong, J.; Mijowska, E.; Kierzek, K.; Chen, X.; Zhao, X.; Tang, T. Controllable Synthesis of 3D Hollow-Carbon-Spheres/Graphene-Flake Hybrid Nanostructures from Polymer Nanocomposite by Self-Assembly and Feasibility for Lithium-Ion Batteries. *Part. Part. Syst. Charact.* **2015**, *32*, 874–879. [CrossRef]
11. Zhang, W.M.; Wu, X.L.; Hu, J.S.; Guo, Y.G.; Wan, L.J. Carbon Coated Fe$_3$O$_4$ Nanospindles as a Superior Anode Material for Lithium-Ion Batteries. *Adv. Funct. Mater.* **2008**, *18*, 3941–3946. [CrossRef]
12. Chen, X.; Kierzek, K.; Wilgosz, K.; Machnikowski, J.; Gong, J.; Feng, J.; Tang, T.; Kalenczuk, R.J.; Chen, H.; Chu, P.K.; et al. New easy way preparation of core/shell structured SnO$_2$@carbon spheres and application for lithium-ion batteries. *J. Power Sources* **2012**, *216*, 475–481. [CrossRef]
13. He, Y.; Huang, L.; Cai, J.S.; Zheng, X.M.; Sun, S.G. Structure and electrochemical performance of nanostructured Fe$_3$O$_4$/carbon nanotube composites as anodes for lithium ion batteries. *Electrochim. Acta* **2010**, *55*, 1140–1144. [CrossRef]
14. Zang, J.; Ye, J.; Qian, H.; Lin, Y.; Zhang, X.; Zheng, M.; Dong, Q. Hollow carbon sphere with open pore encapsulated MnO$_2$ nanosheets as high-performance anode materials for lithium ion batteries. *Electrochim. Acta* **2018**, *260*, 783–788. [CrossRef]
15. Poizot, P.; Laruelle, S.; Grugeon, S.; Dupont, L.; Tarascon, J.-M. Nano-sized transition-metal oxides as negative-electrode materials for lithium-ion batteries. *Nat. Cell Biol.* **2000**, *407*, 496–499. [CrossRef] [PubMed]
16. Zhao, K.; Pharr, M.; Vlassak, J.J.; Suo, Z. Fracture of electrodes in lithium-ion batteries caused by fast charging. *J. Appl. Phys.* **2010**, *108*, 073517. [CrossRef]

17. Etacheri, V.; Wang, C.; O'Connell, M.J.; Chan, C.K.; Pol, V.G. Porous carbon sphere anodes for enhanced lithium-ion storage. *J. Mater. Chem. A* **2015**, *3*, 9861–9868. [CrossRef]
18. Li, F.; Zou, Q.; Xia, Y. CoO-loaded graphitable carbon hollow spheres as anode materials for lithium-ion battery. *J. Power Sources* **2008**, *177*, 546–552. [CrossRef]
19. Fang, B.; Kim, M.-S.; Kim, J.H.; Yu, J.-S. Hierarchical Nanostructured Carbons with Meso-Macroporosity: Design, Characterization, and Applications. *Acc. Chem. Res.* **2013**, *46*, 1397–1406. [CrossRef]
20. Chen, X.; Kierzek, K.; Jiang, Z.; Chen, H.; Tang, T.; Wojtoniszak, M.; Kalenczuk, R.J.; Chu, P.K.; Borowiak-Palen, E. Synthesis, Growth Mechanism, and Electrochemical Properties of Hollow Mesoporous Carbon Spheres with Controlled Diameter. *J. Phys. Chem. C* **2011**, *115*, 17717–17724. [CrossRef]
21. Kalubarme, R.S.; Cho, M.-S.; Yun, K.-S.; Kim, T.-S.; Park, C.-J. Catalytic characteristics of MnO_2 nanostructures for the O_2 reduction process. *Nanotechnology* **2011**, *22*, 395402. [CrossRef]
22. Suresh, S.; Karthikeyan, S.; Jayamoorthy, K. Effect of bulk and nano-Fe_2O_3 particles on peanutplant leaves studied by Fourier transform infraredspectral studies. *J. Adv. Res.* **2016**, *7*, 739–747. [CrossRef]
23. Ruíz-Baltazar, A.; Esparza, R.; Rosas, G.; Pérez, R. Effect of the Surfactant on the Growth and Oxidation of Iron Nanoparticles. *J. Nanomater.* **2015**, *2015*, 1–8. [CrossRef]
24. Wenelska, K.; Ottmann, A.; Schneider, P.; Thauer, E.; Klingeler, R.; Mijowska, E. Hollow carbon sphere/metal oxide nanocomposites anodes for lithium-ion batteries. *Energy* **2016**, *103*, 100–106. [CrossRef]
25. Dresselhaus, M.; Jorio, A.; Saito, R. Characterizing Graphene, Graphite, and Carbon Nanotubes by Raman Spectroscopy. *Annu. Rev. Condens. Matter Phys.* **2010**, *1*, 89–108. [CrossRef]
26. Zhang, J.; Wang, B.; Zhou, J.; Xia, R.; Chu, Y.; Huang, J. Preparation of Advanced CuO Nanowires/Functionalized Graphene Composite Anode Material for Lithium Ion Batteries. *Materials* **2017**, *10*, 72. [CrossRef] [PubMed]
27. Cao, K.; Liu, H.; Xu, X.; Wang, Y.; Jiao, L. $FeMnO_3$: A high-performance Li-ion battery anode Material. *Chem. Commun.* **2016**, *52*, 11414–11417. [CrossRef] [PubMed]
28. He, C.; Wu, S.; Zhao, N.; Shi, C.; Liu, E.; Li, J. Carbon-Encapsulated Fe_3O_4 Nanoparticles as a High-Rate Lithium Ion Battery Anode Material. *ACS Nano* **2013**, *7*, 4459–4469. [CrossRef] [PubMed]
29. Zhou, G.; Wang, D.-W.; Li, F.; Zhang, L.; Li, N.; Wu, Z.-S.; Wen, L.; Lu, G.Q.; Cheng, H.-M. Graphene-Wrapped Fe_3O_4 Anode Material with Improved Reversible Capacity and Cyclic Stability for Lithium Ion Batteries. *Chem. Mater.* **2010**, *22*, 5306–5313. [CrossRef]
30. Zhu, J.; Ng, K.Y.S.; Deng, D. Porous Olive-Like Carbon Decorated Fe_3O_4 Based Additive-Free Electrode for Highly Reversible Lithium Storage. *ACS Appl. Mater. Interfaces* **2011**, *3*, 3276. [CrossRef]
31. Wu, F.; Huang, R.; Mu, D.; Wu, B.; Chen, S. New Synthesis of a Foamlike Fe_3O_4/C Composite via a Self-Expanding Process and Its Electrochemical Performance as Anode Material for Lithium-Ion Batteries. *ACS Appl. Mater. Interfaces* **2014**, *12*, 19254–19264. [CrossRef]
32. Cai, Z.; Xu, L.; Yan, M.; Han, C.; He, L.; Hercule, K.M.; Niu, C.; Yuan, Z.; Xu, W.; Qu, L.; et al. Manganese Oxide/Carbon Yolk–Shell Nanorod Anodes for High Capacity Lithium Batteries. *Nano Lett.* **2015**, *15*, 738–744. [CrossRef]
33. Ko, Y.N.; Park, S.B.; Choi, S.H.; Kang, Y.C. One-pot synthesis of manganese oxide-carbon composite microspheres with three dimensional channels for Li-ion batteries. *Sci. Rep.* **2014**, *4*, 5751. [CrossRef] [PubMed]
34. Gu, X.; Chen, L.; Ju, Z.; Xu, H.; Yang, J.; Qian, Y. Controlled Growth of Porous α-Fe_2O_3 Branches on β-MnO_2 Nanorods for Excellent Performance in Lithium-Ion Batteries. *Adv. Funct. Mater.* **2013**, *23*, 4049–4056. [CrossRef]
35. Chen, J.; Wang, Y.; He, X.; Xu, S.; Fang, M.; Zhao, X.; Shang, Y. Electrochemical properties of MnO_2 nanorods as anode materials for lithium ion batteries. *Electrochim. Acta* **2014**, *142*, 152–156. [CrossRef]
36. Ma, Z.; Zhao, T. Freestanding graphene/MnO_2 cathodes for Li-ion batteries. *Electrochim. Acta* **2017**, *201*, 165–171. [CrossRef]
37. Liu, H.; Hu, Z.; Su, Y.; Ruan, H.; Hu, R.; Zhang, L. In Situ Synthesis of MnO_2/Porous Graphitic Carbon Composites as High-Capacity Anode Materials for Lithium-Ion Batteries. *Appl. Surf. Sci.* **2017**, *392*, 777–784. [CrossRef]

Article

Constructing High-Performance Carbon Nanofiber Anodes by the Hierarchical Porous Structure Regulation and Silicon/Nitrogen Co-Doping

Yujia Chen [1], Jiaqi Wang [1], Xiaohu Wang [1,2], Xuelei Li [1,3,*], Jun Liu [1], Jingshun Liu [1,3], Ding Nan [4,5] and Junhui Dong [1,*]

[1] Inner Mongolia Key Laboratory of Graphite and Graphene for Energy Storage and Coating, School of Materials Science and Engineering, Inner Mongolia University of Technology, Hohhot 010051, China; cyj202103@163.com (Y.C.); wjq024466341811@163.com (J.W.); wxh20220208@163.com (X.W.); clxylj@163.com (J.L.); jingshun_liu@163.com (J.L.)
[2] Rising Graphite Applied Technology Research Institute, Chinese Graphite Industrial Park-Xinghe, Ulanqab 013650, China
[3] Collaborative Innovation Center of Non-Ferrous Metal Materials and Processing Technology Co-Constructed by the Province and Ministry, Inner Mongolia Autonomous Region, Inner Mongolia University of Technology, Hohhot 010051, China
[4] Inner Mongolia Enterprise Key Laboratory of High Voltage and Insulation Technology, Inner Mongolia Power Research Institute Branch, Inner Mongolia Power (Group) Co., Ltd., Hohhot 010020, China; nd@imu.edu.cn
[5] College of Chemistry and Chemical Engineering, Inner Mongolia University, West University Street 235, Hohhot 010021, China
* Correspondence: lglixuelei@163.com (X.L.); jhdong@imut.edu.cn (J.D.)

Abstract: Due to the rapid development of bendable electronic products, it is urgent to prepare flexible anode materials with excellent properties, which play a key role in flexible lithium-ion batteries. Although carbon fibers are excellent candidates for preparing flexible anode materials, the low discharge specific capacity prevents their further application. In this paper, a hierarchical porous and silicon (Si)/nitrogen (N) co-doped carbon nanofiber anode was successfully prepared, in which Si doping can improve specific capacity, N doping can improve conductivity, and a fabricated hierarchical porous structure can increase the reactive sites, improve the ion transport rate, and enable the electrolyte to penetrate the inner part of carbon nanofibers to improve the electrolyte/electrode contacting area during the charging–discharging processes. The hierarchical porous and Si/N co-doped carbon nanofiber anode does not require a binder, and is flexible and foldable. Moreover, it exhibits an ultrahigh initial reversible capacity of 1737.2 mAh g^{-1}, stable cycle ability and excellent rate of performance. This work provides a new avenue to develop flexible carbon nanofiber anode materials for lithium-ion batteries with high performance.

Keywords: lithium-ion battery; flexible anode; hierarchical porous structure; Si/N co-doping; high performances

1. Introduction

With the rapid development of bendable electronic products, such as flexible displays, wearable electronics and medical electronics, flexible energy storage systems with higher energy density have become an urgent demand [1,2]. Lithium-ion batteries, as energy storage devices, have attracted enormous attention due to their long cycle life, fast charge–discharge, high energy density, and no memory effect [3–6]. Flexible anode materials are a particularly important factor to obtain lithium-ion batteries with high electrochemical performance [7,8]. Therefore, the development of flexible anode materials with bendable function and excellent electrochemical performance has become one of the research hotspots.

Carbon fibers are one of the most widespread anode materials of flexible lithium-ion batteries [9,10]. However, untreated carbon fibers make it difficult to achieve the high performance demands of lithium-ion batteries due to their low reversible capacity. To obtain high electrochemical performance, it is common to fabricate holes of different sizes in carbon fibers [11–14]. Wang et al. reported that micropores can increase the number of active sites of lithium storage and the ability of ion adsorption [15]. Guo et al. reported that mesopores greatly enhance the capability rate and facilitate the transport of ions [16]. Chen et al. reported that abundant meso/macropores in carbon nanofibers can offer more active sites for Li storage and facilitate electrolyte penetration of the inner part of carbon nanofibers, improving the electrolyte/electrode contacting area [17]. In addition, element doping is another measure of modifying electrode materials. Silicon materials have been widely concerned because of their high theoretical capacity of 4200 mAh g^{-1} [18,19]. However, silicon materials undergo serious volume changes during the charging–discharging process, which leads to particle pulverization and rapid capacity decay [20,21]. Silicon–carbon composites can make use of the structural strength of carbon materials, so that the volume change in silicon materials can be alleviated during the charging–discharging process [22–25]. For example, Jang et al. prepared a pyrolytic carbon-coated silicon nanofiber anode, which has high capacity and excellent cycling performance [26]. Xu et al. reported a flexible 3D Si/C fiber paper anode with capacity of 1600 mAh g^{-1}, which was synthesized by simultaneously electro spraying nano-Si and polyacrylonitrile fibers, followed by carbonization [27]. Traditionally, carbon materials are used to modify silicon materials to obtain silicon-based anode materials with high electrochemical properties. Similarly, it may be effective to use silicon materials to dope carbon fibers to obtain high-capacity carbon fiber anodes. In addition, N doping was also used to improve the conductivity of carbon materials [28–30]. However, it is difficult to fully meet the requirements of high electrochemical properties using a single modification strategy. Therefore, the synergistic effect of various modification methods may be effective for the preparation of high-performance carbon nanofiber anodes.

Based on the above analysis results, we prepared a hierarchical porous and Si/N co-doped carbon nanofiber anode by novel gas–electric co-spinning technology in this work. Si doping can improve the specific capacity, N doping can improve the conductivity, and the fabricated micropores, mesopores and macropores can increase the number of reactive sites, improve the ion transport rate, and enable the electrolyte to penetrate the inner part of carbon nanofibers to improve the electrolyte/electrode contacting area during charging–discharging processes. This modified carbon nanofiber anode does not require a binder, and has the advantages of flexibility and foldability. Moreover, it exhibits a high initial reversible capacity of 1737.2 mAh g^{-1}, good capacity retention and outstanding rate ability. This work provides a new avenue for the development of flexible lithium-ion batteries.

2. Experiment

2.1. Materials

The reagents used in this paper included: polyacrylonitrile (PAN) (solid, molecular weight 150,000, Macklin), graphene (commercially available), N, N-dimethylformamide (DMF) (liquid, analytical purity, Macklin), nano silicon particles (Si) (50 nm) and polymethylmethacrylate (PMMA) (solid, Macklin). These reagents were analytically pure so could be used directly.

2.2. Preparation of Materials

The flow chart of preparation of the hierarchical porous and Si/N co-doped carbon nanofiber anode materials is shown in Figure 1. Firstly, 0.1 g graphene, 2 g polyacrylonitrile (PAN) and a certain amount of silicon were dissolved in 50 mL N,N-dimethylformamide (DMF) and vibrated with ultrasound for 20 min. Then, the pore-making agent polymethylmethacrylate (PMMA) was added to this solution, heated at 70 °C using a water bath and stirred for 10 h. The obtained precursor solution was continuously spun on the flat plate for about 1 h to obtain a fiber cloth by the gas–electric co-spinning method. The feed speed was 4 mL h^{-1}, the voltage was 5 kV, and the air flow was 10 Psi. Subsequently, the fiber cloth was placed in a blast drying oven for pre-oxidation, in which the temperature was raised from room temperature to 280 °C with a heating rate of 2 °C min^{-1} for 6 h. The purpose of pre-oxidation was to make PAN undergo three chemical reactions: cyclization, dehydrogenation and oxidation in this process, so as to make the carbon fiber cloth more stable before carbonization. Finally, the fiber cloth was calcined at 950 °C under the protection of argon for 1 h, and then activated in ammonia for 30 min. High-temperature tubular furnace was used to increase the temperature from room temperature to 200 °C with a heating rate of 2 °C min^{-1}, and then the temperature was reduced to room temperature to obtain the hierarchical porous and Si/N co-doped carbon nanofiber anode materials.

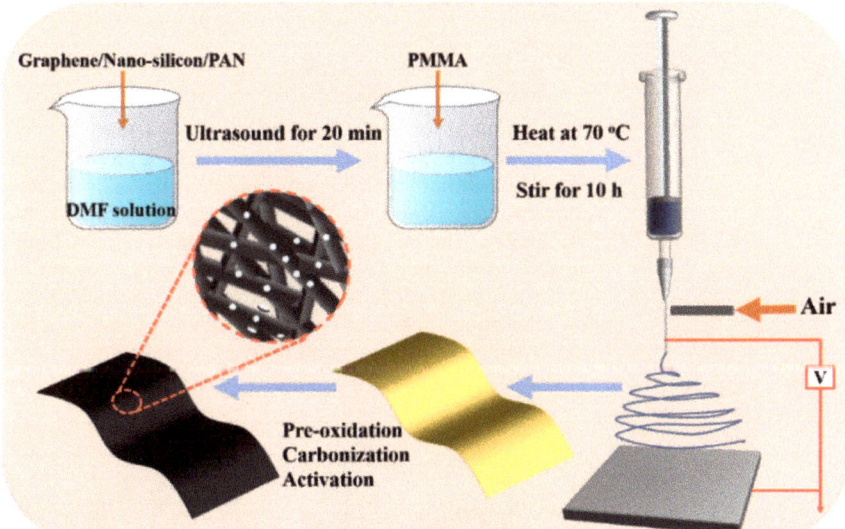

Figure 1. The flow chart of preparation of the hierarchical porous and Si/N co-doped carbon nanofiber anode materials.

In the process of preparing the materials above, by changing the mass ratio of PAN:PMMA = 2:1, 2:2, 2:3 and 2:4, carbon nanofibers with different proportions of PMMA were obtained (named CNFs-PMMA1, CNFs-PMMA2, CNFs-PMMA3 and CNFs-PMMA4, respectively, according to the added PMMA masses of 1 g, 2 g, 3 g and 4 g). After determining the optimal amount of PMMA, according to the added Si mass (0%, 5%, 10% and 15%, the mass ratio of Si/PAN), the hierarchical porous Si/N co-doped carbon nanofiber anode materials were named CNFs-Si0%, CNFs-Si5%, CNFs-Si10% and CNFs-Si15%, respectively.

2.3. Materials Characterization

Microscopic morphology of the samples was observed using scanning electron microscopy (SEM, HITACHI-SU8220, Tokyo, Japan), and the corresponding element mapping on the surface of materials was analyzed by an energy dispersive spectrometer (EDS). The hole size distribution of samples was measured using a nitrogen adsorption–desorption apparatus (BET, BELSORP-miniII, BEL Japan Inc., Osaka, Japan).

2.4. Electrochemical Measurements

The prepared anode sheet of lithium-ion batteries was a flexible carbon fiber, which did not require binders to be added and did not have to be coated on the copper foil, compared with traditional electrodes. The prepared hierarchical porous and Si/N co-doped carbon fiber, after being sliced, can be used directly as a battery anode. The weight of the anode was ~0.76 mg, the diameter of the electrode was 9 mm, and the mass loading was ~1.2 mg cm^{-2}. A metal lithium sheet was used as the counter electrode, the electrolyte was added in the ratio EC:DEC:EMC = 1:1:1 (v/v) to the solvent containing 20% fluoroethylene carbonate (FEC) and lithium hexafluorophosphate (LiPF$_6$), and the separator was porous polypropylene. The electrochemical experiments were carried out by using commercial CR2032 cells at room temperature. The blue electric system (Land CT2001A, Blue Power Company) was set to maintain constant current charge–discharge, and the voltage range was from 0.01 to 3 V. The rate performance of batteries was tested at different current densities (0.05, 0.1, 0.2, 0.5 and 1 C, 1 C = 1000 mAh g^{-1}). Princeton (PMC1000A) electrochemical workstation was used to test the cyclic voltammetry (CV) in the voltage range of 0.01~3 V with a scanning rate of 0.01 mV s^{-1}. The electrochemical impedance spectroscopy (EIS) tests were carried out in a frequency range between 0.1 Hz and 100 kHz with an amplitude of 5 mV.

3. Results and Discussion

Figure 2a shows the digital photo of a hierarchical porous and Si/N co-doped carbon nanofiber cloth, which is flexible and foldable. It can be used directly as a flexible lithium-ion anode without the addition of binders. The microstructure of the carbon nanofiber cloth is further tested, and the SEM images of CNFs-PMMA1 and CNFs-PMMA3 are shown in Figure 2b,c. The diameter of the carbon nanofibers is about 400 nm and the fibers are cross-linked. With the increase in the PMMA content from CNFs-PMMA1 to CNFs-PMMA3, many macropores of different sizes emerge in the fibers. It is speculated that micropores and mesopores may also exist in the fibers. However, it is difficult to judge whether there are smaller holes through SEM images. Therefore, the pore size distribution is further explored by a nitrogen adsorption–desorption test in the follow-up study.

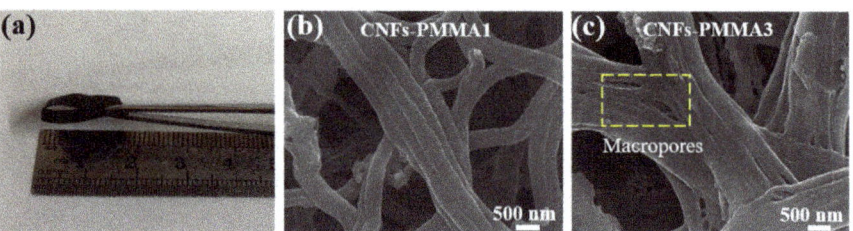

Figure 2. (**a**) A digital photo of the hierarchical porous and Si/N co-doped carbon nanofiber cloth, SEM images of (**b**) CNFs-PMMA1 and (**c**) CNFs-PMMA3.

To obtain the optimal addition content of PMMA, the cycle and rate performances of CNFs-PMMA1, CNFs-PMMA2, CNFs-PMMA3, and CNFs-PMMA4 anodes in lithium-ion batteries are tested, and the results are used as the basis for selecting the content of PMMA. Figure 3a shows cycle performance curves of the four anodes at a current density of 0.05 C, from 0.01 to 3 V. The CNFs-PMMA3 anode has the highest initial capacity of 1437.2 mAh g^{-1} among the four anodes. After 100 cycles, the CNFs-PMMA3 anode can still reach 985.3 mAh g^{-1}. Figure 3b shows the rate performance curves of four anodes. Compared with CNFs-PMMA1, CNFs-PMMA2, and CNFs-PMMA4 anodes, the CNFs-PMMA3 anode shows excellent rate performance. The specific capacity of the CNFs-PMMA3 anode is 1469.9, 1307.6, 1213.1, 956.3 and 686.3 mAh g^{-1} at the current density of 0.05, 0.1, 0.2, 0.5 and 1 C, respectively. Therefore, the amount of PMMA added in CNFs-PMMA3 is selected as the optimal result.

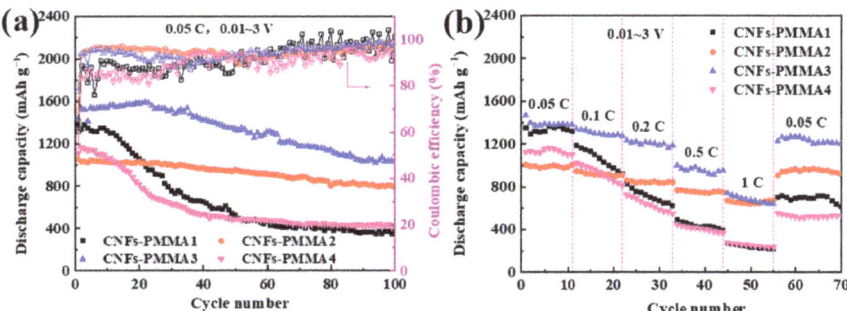

Figure 3. (**a**) Cycle performance and (**b**) rate performance curves of CNFs-PMMA1, CNFs-PMMA2, CNFs-PMMA3, and CNFs-PMMA4 anodes in lithium-ion batteries.

After determining the amount of PMMA, the effect of the silicon content on the hierarchical porous carbon nanofiber anode is further studied. The microscopic morphologies of CNFs-Si0%, CNFs-Si5%, CNFs-Si10% and CNFs-Si15% materials are shown in Figure 4a–d. These carbon nanofibers have a good fiber shape with a diameter of about 1 μm, which are randomly intertwined to form a fiber cloth. With the increase in the Si content from CNFs-Si0%, CNFs-Si5%, CNFs-Si10% to CNFs-Si15%, the white particles increase, due to the agglomeration of nano silicon particles. The SEM image of CNFs-Si10% and the corresponding EDS mapping of elements are shown in Figure 4e. Some holes exist in the fiber, and the nanofiber is cross-linked. In addition, Si, C and N elements are evenly distributed, indicating that Si and N are effectively doped in nanofibers. It is worth noting that the N element is doped through ammonia activation during the experiment. It was reported that N doping can improve the conductivity of carbon materials [28–30], which is not discussed further in this work.

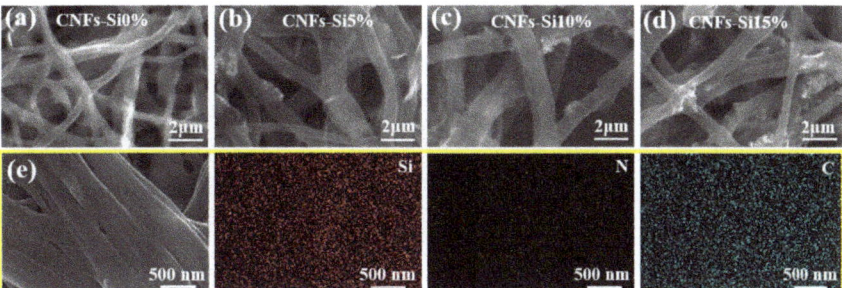

Figure 4. SEM images of (**a**) CNFs-Si0%, (**b**) CNFs-Si5%, (**c**) CNFs-Si10% and (**d**) CNFs-Si15%. (**e**) SEM image and the corresponding EDS mapping of Si, C, and N elements in partial CNFs-Si10%.

Nitrogen adsorption–desorption curves of CNFs-Si0%, CNFs-Si5%, CNFs-Si10% and CNFs-Si15% are shown in Figure 5a. When $P/P_0 > 0.4$, a hysteresis loop appears in the nitrogen adsorption–desorption curves of all samples, which belong to the IV isotherm. The corresponding pore size distribution curves are shown in Figure 5b, which show that all samples have micropores (pore size less than 2 nm) and mesopores (pore size between 2 and 50 nm), which aligns with the experimental expectation. Combined with the SEM images and EDS mapping in Figure 4, this result indicates that a Si/N co-doped carbon nanofiber anode with a hierarchical porous structure has been synthesized.

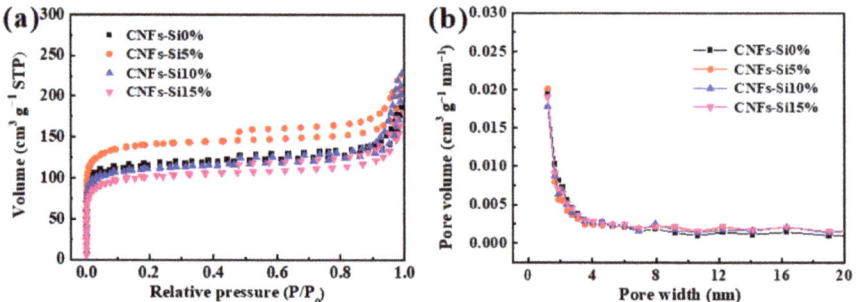

Figure 5. (a) Nitrogen adsorption–desorption curves, and (b) the corresponding pore size distribution curves of CNFs-Si0%, CNFs-Si5%, CNFs-Si10% and CNFs-Si15%.

Figure 6a shows cycle performance curves of CNFs-Si0%, CNFs-Si5%, CNFs-Si10% and CNFs-Si15% anodes at 0.05 C, from 0.01 to 3 V, in lithium-ion batteries. The initial capacity of CNFs-Si10% is 1737.2 mAh g^{-1}, and still maintains a high reversible capacity of 985.3 mAh g^{-1} after 100 cycles, which is the highest capacity among the four anodes. CNFs-Si10% and CNFs-Si15% anodes have a high initial capacity, which can mainly be due to the addition of Si with high capacity, compared with CNFs-Si0%. The capacity of CNFs-Si15% anode decreases significantly after 20 cycles, because the nano-silicon in the materials experiences large volume expansion with the increase in silicon content, destroying its structure. The initial Coulombic efficiency of the CNFs-Si0%, CNFs-Si5%, CNFs-Si10% and CNFs-Si15% anodes is only 49.72%, 50.24%, 77.82% and 50.74%, respectively. The low initial coulombic efficiency is due to the fact that with the addition of the pore-forming agent PMMA, the number of micropores increases, providing many reactive sites, thus forming a high irreversible capacity with the formation of an SEI film with a large area during the first charge–discharge process. After the first charge–discharge, the Coulombic efficiency of each anode is close to stable, basically around 90%, which can be attributed to meso/macropores, which can offer more active sites for Li storage and enable the electrolyte to penetrate the inner part of carbon nanofibers, improving the electrolyte/electrode contacting area.

Figure 6b shows the rate performance curves of CNFs-Si0%, CNFs-Si5%, CNFs-Si10% and CNFs-Si15% anodes in lithium-ion batteries at different current densities. The CNFs-Si10% anode still exhibits excellent specific capacities of 1719.9, 1593.1, 1484.3, 1255.7 and 995.8 mAh g^{-1} at 0.05, 0.1, 0.2, 0.5 and 1 C, respectively. It can be seen that when the current density changes to 0.05 C again, the specific capacity of the CNFs-Si10% anode is greater than 1518.1 mAh g^{-1}, indicating good rate performance. This is because the synergistic effect of N doping and the hierarchical porous structure can improve the conductivity, increase the reactive sites, and improve the Li-ion transport rate. The results of cycling and rate performances indicate that proper silicon/nitrogen co-doping and hierarchical porous structure regulation are important to improve the electrochemical performance of carbon nanofiber anodes. To the best of our knowledge, the outstanding electrochemical performances of the CNFs-Si10% anode are much higher than other reported flexible carbon fiber anodes for lithium-ion batteries (Table 1).

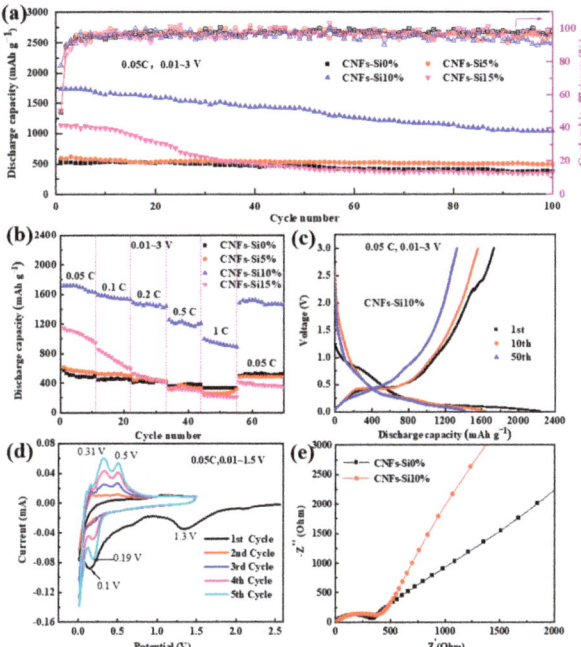

Figure 6. (a) Cycle performance and (b) rate performance curves of CNFs-Si0%, CNFs-Si5%, CNFs-Si10% and CNFs-Si15% anodes in lithium-ion batteries. (c) The charge–discharge curves of the CNFs-Si10% anode, (d) CV curves of the CNFs-Si10% anode, and (e) Nyquist plots of CNFs-Si0% and CNFs-Si10% anodes in lithium-ion batteries.

Table 1. Electrochemical performance of flexible carbon fiber anodes for lithium-ion batteries reported in this work.

Anodes	Mass Loading (mg cm^{-2})	First Reversible Capacity (mAh g^{-1})	Cycle Performance (mAh g^{-1})	Rate Performance (mAh g^{-1})	Reference
V$_2$O$_3$/MCCNFs	1.5–2.5	790.6 (0.1 A g^{-1})	487.7 (5 A g^{-1}, 5000 cycles)	456.8 (5 A g^{-1})	[31]
In$_2$O$_3$@CF	1.4	510 (0.1 A g^{-1})	435 (0.1 A g^{-1}, 500 cycles)	190 (1.5 A g^{-1})	[32]
γ-Fe$_2$O$_3$/C films	1	923.97 (0.2 A g^{-1})	1088 (0.2 A g^{-1}, 300 cycles)	380 (5 A g^{-1})	[33]
am-Fe$_2$O$_3$/rGO/CNFs	1.5–2.0	825 (0.1 A g^{-1})	739 (1 A g^{-1}, 400 cycles)	570 (2 A g^{-1})	[34]
FeCo@NCNFs-600	1.77–2.65	736.3 (0.1 A g^{-1})	566.5 (0.1 A g^{-1}, 100 cycles)	130 (2 A g^{-1})	[35]
Sn@C@CNF	2	891.2 (0.1 A g^{-1})	610.8 (0.2 A g^{-1}, 180 cycles)	305.1 (2 A g^{-1})	[36]
NCNFs	7.64	752.3 (0.05 A g^{-1})	411.9 (0.1 A g^{-1}, 160 cycles)	148.8 (2 A g^{-1})	[16]
γ-Fe$_2$O$_3$@CNFs	2.0	1065 (0.5 A g^{-1})	430 (6 A g^{-1}, 1000 cycles)	222 (60 A g^{-1})	[37]
G/Si@CFs	0.65~1	1036 (0.1 A g^{-1})	896.8 (0.1 A g^{-1}, 200 cycles)	543 (1 A g^{-1})	[38]
C/CuO/rGO	1.30–1.95	550 (0.1 A g^{-1})	400 (1 A g^{-1}, 600 cycles)	300 (2 A g^{-1})	[39]
CNF@SnO$_2$	1.77–3.54	793 (0.5 A g^{-1})	485 (0.1 A g^{-1}, 850 cycles)	359 (4 A g^{-1})	[40]
Fe$_3$O$_4$/NCNFs	1.33	686 (0.1 A g^{-1})	522 (0.1 A g^{-1}, 200 cycles)	407 (5 A g^{-1})	[41]
MoO$_2$/C	85.7	752.5 (0.2 A g^{-1})	450 (2 A g^{-1}, 500 cycles)	432 (2 A g^{-1})	[42]
FCNF-3/4	1.0	775 (0.2 A g^{-1})	630 (0.2 A g^{-1}, 100 cycles)	250 (5 A g^{-1})	[43]
10-SnO$_2$@CNFs/CNT	1.5–2.5	500.9 (0.1 A g^{-1})	460.3 (0.1 A g^{-1}, 200 cycles)	222.2 (3.2 A g^{-1})	[44]
ZnSe@CNFs-2.5	0.8–1.2	737.5 (0.1 A g^{-1})	426.1 (5 A g^{-1}, 3000 cycles)	547.6 (5 A g^{-1})	[45]
CNFs/CNTs	1.27	1500.5 (0.05 A g^{-1})	545 (0.2 A g^{-1}, 400 cycles)	344.8 (2 A g^{-1})	[46]
SnO$_2$/TiO$_2$@CNFs	/	1061.2 (0.1 A g^{-1})	729.6 (0.1 A g^{-1}, 150 cycles)	206.2 (3 A g^{-1})	[47]
CNFs-Si10%	1.2	1737.2 (0.05 A g^{-1})	985.3 (0.05 A g^{-1}, 100 cycles)	995.8 (1 A g^{-1})	This work

Figure 6c shows galvanostatic charge–discharge curves of the CNFs-Si10% anode during the 1st, 10th and 50th cycle at 0.05 C in the voltage range of 0.01–3 V. During the first cycle, the charge–discharge capacity is 2232.5 and 1737.2 mAh g^{-1}, respectively, corresponding to an initial Coulombic efficiency of 77.82%. Irreversible capacity loss is mainly due to the formation of the SEI film and the decomposition of the electrolyte during the initial charge–discharge process [48]. From the 10th to 50th cycle, the discharge capacity dramatically reduces from 1640 to 1421.7 mAh g^{-1}, which may be due to the production of a residual irreversible side reactant after charging and discharging multiple times. It is noteworthy that the Coulombic efficiency value increases to 95.7% and 96.6% during the 10th and 50th cycle, respectively.

To further appraise the lithium storage performance of the CNFs-Si10% anode, CV curves of five initial cycles were obtained from 0.01 to 1.5 V at a scan rate of 0.1 mV s^{-1}, as shown in Figure 6d. During the first lithiation, generation of the SEI film and the decomposition of the electrolyte could lead to a broad irreversible peak at 1.3 V, which disappears in the following cycles. Subsequently, a sharp anodic peak below 0.1 V corresponds to the phase transformation from Si to Li$_x$Si and lithium-ion insertion with carbon nanofiber materials in the CNFs-Si10% anode [48]. Upon delithiation, two cathodic peaks at around 0.31 and 0.5 V are attributed to the dealloying process of Li$_x$Si into Si, which is consistent with the reported results [49]. In the second cycle, a broad anodic peak located at around 0.19 V appears, which is associated with the lithiation of Si. In the following cycles, the cathodic and anodic peak positions are consistent, indicating good concurrency and reversibility. For the anodic peak at 0.19 V and the two cathodic peaks at 0.31 and 0.5 V, the current intensities gradually enlarge in subsequent cycles, corresponding to the activation process as a result of the reaction of more active sites with lithium-ion [50]. This phenomenon is typical for the Si-based anode for lithium-ion batteries [51,52]. The results show that the CNFs-Si10% anode has good stability.

To further explore the effect of Si doping on the performance of lithium-ion batteries, the EIS of CNFs-Si0% and CNFs-Si10% anodes were tested. Figure 6e shows the Nyquist plots of CNFs-Si0% and CNFs-Si10% anodes in lithium-ion batteries. The curve consists of an overlapping semicircle in the high-frequency region and an oblique line in the low-frequency region. The semicircle at high frequency refers to the diffusion and migration process of Li$^+$ in the SEI, and the oblique line at low frequency represents the migration impedance of Li$^+$ in the active substance [53]. The semicircle of the CNFs-Si0% anode is almost the same as that of the CNFs-Si10% anode, meaning that the small amount of silicon doping does not reduce the impedance of CNFs. Compared with the CNFs-Si0% anode, the CNFs-Si10% anode has a larger slope at low frequency. The larger the slope, the lower the Li-ion diffusion resistance. Therefore, the CNFs-Si10% anode has a higher capacity and lower lithium-ion diffusion resistance than the CNFs-Si0% anode. Meanwhile, the excellent cycle and rate capability results of the CNFs-Si10% anode also show that the nano silicon is dispersed in the porous of carbon fibers, and the porous structure can provide a buffer space for the Si volume change during the Li deintercalation process [53,54].

4. Conclusions

In this work, we successfully synthesized hierarchical porous and Si/N co-doped carbon nanofiber anodes. SEM images and nitrogen adsorption–desorption tests confirm that a hierarchical porous structure exists in carbon nanofiber materials, and the EDS mapping confirms that Si and N elements are doped into carbon nanofiber materials. Compared with other anodes, the CNFs-Si10% anode obtained has a high initial reversible specific capacity, high capacity retention, and excellent rate performance in lithium-ion batteries. This excellent electrochemical performance is mainly due to the fact that Si doping can improve specific capacity, N doping can improve conductivity, and the fabricated hierarchical porous structure can increase the number of reactive sites, improve the ion transport rate, and enable the electrolyte to penetrate the inner part of carbon nanofibers to improve the electrolyte/electrode contacting area during charging–discharging processes. This work provides a new avenue for developing flexible anode materials for lithium-ion batteries with high performance.

Author Contributions: Y.C.: conceptualization, methodology, writing—original draft; J.W.: software, data curation; X.W.: cell fabrication and testing; X.L.: supervision, data interpretation, writing—review & editing; J.L. (Jun Liu): conceptualization; J.L. (Jingshun Liu): methodology; D.N. methodology; J.D.: supervision, data interpretation. All authors have read and agreed to the published version of the manuscript.

Funding: This work was financially supported by Natural Science Foundation of Inner Mongolia (no. 2019MS05068), Inner Mongolia Major Science and Technology Project (no. 2020ZD0024), Scientific Research Project of Inner Mongolia University of Technology (no. ZZ202106), the Alashan League's Project of Applied Technology Research and Development Fund (no. AMYY2020-01), the research project of Inner Mongolia Electric Power (Group) Co., Ltd. for post-doctoral studies, Program for Innovative Research Team in Universities of Inner Mongolia Autonomous Region (no. NMGIRT2211), Inner Mongolia University of Technology Key Discipline Team Project of Materials Science (no. ZD202012), Inner Mongolia Natural Science Cultivating Fund for Distinguished Young Scholars (no. 2020JQ05), Science and Technology Planning Project of Inner Mongolia Autonomous Region (no. 2020GG0267), and Local Science and Technology Development Project of the Central Government (no. 2021ZY0006).

Institutional Review Board Statement: Not applicable.

Informed Consent Statement: Not applicable.

Data Availability Statement: Not applicable.

Conflicts of Interest: The authors declare no conflict of interest.

References

1. Wang, X.; Weng, Q.; Yang, Y.; Bando, Y.; Golberg, D. Hybrid two-dimensional materials in rechargeable battery applications and their microscopic mechanisms. *Chem. Soc. Rev.* **2016**, *45*, 4042–4073. [CrossRef] [PubMed]
2. Zhou, Y.; Yang, Y.; Hou, G.; Yi, D.; Zhou, B.; Chen, S.; Lam, T.D.; Yuan, F.; Golberg, D.; Wang, X. Stress-relieving defects enable ultra-stable silicon anode for Li-ion storage. *Nano. Energy* **2020**, *70*, 104568. [CrossRef]
3. Qi, W.; Shapter, J.G.; Wu, Q.; Yin, T.; Gao, G.; Cui, D. Nanostructured anode materials for lithium-ion batteries: Principle, recent progress and future perspectives. *J. Mater. Chem. A* **2017**, *5*, 19521–19540. [CrossRef]
4. Zhou, C.; Liu, J.; Gong, X.; Wang, Z. Optimizing the function of SiO_x in the porous Si/SiO_x network via a controllable magnesiothermic reduction for enhanced lithium storage. *J. Alloys Compd.* **2021**, *874*, 159914. [CrossRef]
5. Vernardou, D. Progress and challenges in industrially promising chemical vapour deposition processes for the synthesis of large-area metal oxide electrode materials designed for aqueous battery systems. *Materials* **2021**, *14*, 4177. [CrossRef]
6. Valvo, M.; Floraki, C.; Paillard, E.; Edstrom, K.; Vernardou, D. Perspectives on iron oxide-based materials with carbon as anodes for Li- and K-ion batteries. *Nanomaterials* **2022**, *12*, 1436. [CrossRef]
7. Fu, J.; Kang, W.; Guo, X.; Wen, H.; Zeng, T.; Yuan, R.; Zhang, C. 3D hierarchically porous NiO/graphene hybrid paper anode for long-life and high rate cycling flexible Li-ion batteries. *J. Energy Chem.* **2020**, *47*, 172–179. [CrossRef]
8. Zeng, L.; Qiu, L.; Cheng, H.M. Towards the practical use of flexible lithium ion batteries. *Energy Storage Mater.* **2019**, *23*, 434–438. [CrossRef]
9. Elizabeth, I.; Singh, B.P.; Gopukumar, S. Electrochemical performance of Sb_2S_3/CNT free-standing flexible anode for Li-ion batteries. *J. Mater. Sci.* **2019**, *54*, 7110–7118. [CrossRef]
10. McCreery, R.L. Advanced carbon electrode materials for molecular electrochemistry. *Chem. Rev.* **2008**, *108*, 2646–2687. [CrossRef]
11. Xu, Z.; Fan, L.; Ni, X.; Han, J.; Guo, R. Sn-encapsulated N-doped porous carbon fibers for enhancing lithium-ion battery performance. *RSC Adv.* **2019**, *9*, 8753–8758. [CrossRef]
12. Nan, D.; Huang, Z.H.; Lv, R.; Yang, L.; Wang, J.G.; Shen, W.; Lin, Y.; Yu, X.; Ye, L.; Sun, H. Nitrogen-enriched electrospun porous carbon nanofiber networks as high-performance free-standing electrode materials. *J. Mater. Chem. A* **2014**, *2*, 19678–19684. [CrossRef]
13. Miao, Y.E.; Huang, Y.; Zhang, L.; Fan, W.; Lai, F.; Liu, T. Electrospun porous carbon nanofiber@ MoS_2 core/sheath fiber membranes as highly flexible and binder-free anodes for lithium-ion batteries. *Nanoscale* **2015**, *7*, 11093–11101. [CrossRef]
14. Zhang, Q.; Zhou, K.; Lei, J.; Hu, W. Nitrogen dual-doped porous carbon fiber: A binder-free and high-performance flexible anode for lithium ion batteries. *Appl. Surf. Sci.* **2019**, *467*, 992–999. [CrossRef]
15. Wang, X.; Sun, N.; Dong, X.; Huang, H.; Qi, M. The porous spongy nest structure compressible anode fabricated by gas forming technique toward high performance lithium ions batteries. *J. Colloid Interf. Sci.* **2022**, *623*, 584–594. [CrossRef]
16. Guo, J.; Liu, J.; Dai, H.; Zhou, R.; Wang, T.; Zhang, C.; Ding, S.; Wang, H.G. Nitrogen doped carbon nanofiber derived from polypyrrole functionalized polyacrylonitrile for applications in lithium-ion batteries and oxygen reduction reaction. *J. Colloid Interf. Sci.* **2017**, *507*, 154–161. [CrossRef]
17. Chen, X.; Gao, G.; Wu, Z.; Xiang, J.; Li, X.; Guan, G.; Zhang, K. Ultrafine MoO_2 nanoparticles encapsulated in a hierarchically porous carbon nanofiber film as a high-performance binder-free anode in lithium ion batteries. *RSC Adv.* **2019**, *9*, 37556–37561. [CrossRef]
18. Yang, Z.; Wu, C.; Li, S.; Qiu, L.; Yang, Z.; Zhong, Y.; Zhong, B.; Song, Y.; Wang, G.; Liu, Y. A unique structure of highly stable interphase and self-consistent stress distribution radial-gradient porous for silicon anode. *Adv. Funct. Mater.* **2021**, *32*, 2107897. [CrossRef]

19. Tian, H.; Tian, H.; Yang, W.; Zhang, F.; Yang, W.; Zhang, Q.; Wang, Y.; Liu, J.; Silva, S.R.P.; Liu, H. Stable hollow-structured silicon suboxide-based anodes toward high-performance lithium-ion batteries. *Adv. Funct. Mater.* **2021**, *31*, 2101796. [CrossRef]
20. Feng, K.; Li, M.; Liu, W.; Kashkooli, A.G.; Xiao, X.; Cai, M.; Chen, Z. Silicon-based anodes for lithium-ion batteries: From fundamentals to practical applications. *Small* **2018**, *14*, 1702737. [CrossRef]
21. Yu, C.; Du, Y.; He, R.; Ma, Y.; Liu, Z.; Li, X.; Luo, W.; Zhou, L.; Mai, L. Hollow SiO_x/C microspheres with semigraphitic carbon coating as the "lithium host" for dendrite-free lithium metal anodes. *ACS Appl. Energy Mater.* **2021**, *4*, 3905–3912. [CrossRef]
22. Terranova, M.L.; Orlanducci, S.; Tamburri, E.; Guglielmotti, V.; Rossi, M. Si/C hybrid nanostructures for Li-ion anodes: An overview. *J. Power Sources* **2014**, *246*, 167–177. [CrossRef]
23. Han, X.; Zhang, Z.; Chen, S.; Yang, Y. Low temperature growth of graphitic carbon on porous silicon for high-capacity lithium energy storage. *J. Power Sources* **2020**, *463*, 228245. [CrossRef]
24. Yang, J.; Wang, Y.X.; Chou, S.L.; Zhang, R.; Xu, Y.; Fan, J.; Zhang, W.X.; Liu, H.K.; Zhao, D.; Dou, S.X. Yolk-shell silicon-mesoporous carbon anode with compact solid electrolyte interphase film for superior lithium-ion batteries. *Nano Energy* **2015**, *18*, 133–142. [CrossRef]
25. Rehman, S.; Guo, S.; Hou, Y. Rational design of Si/SiO_2@ hierarchical porous carbon spheres as efficient polysulfide reservoirs for high-performance Li-S battery. *Adv. Mater.* **2016**, *28*, 3167–3172. [CrossRef]
26. Jang, S.M.; Miyawaki, J.; Tsuji, M.; Mochida, I.; Yoon, S.H. The preparation of a novel Si-CNF composite as an effective anodic material for lithium-ion batteries. *Carbon* **2009**, *47*, 3383–3391. [CrossRef]
27. Xu, Y.; Zhu, Y.; Han, F.; Luo, C.; Wang, C. 3D Si/C fiber paper electrodes fabricated using a combined electrospray/electrospinning technique for Li-ion batteries. *Adv. Energy Mater.* **2015**, *5*, 1400753. [CrossRef]
28. Peng, Y.T.; Lo, C.T. Electrospun porous carbon nanofibers as lithium ion battery anodes. *J. Solid State Electrochem.* **2015**, *19*, 3401–3410. [CrossRef]
29. Yuan, X.L.; Ma, Z.; Jian, S.F.; Ma, H.; Lai, Y.N.; Deng, S.L.; Tian, X.C.; Wong, C.P.; Xia, F.; Dong, Y.F. Mesoporous nitrogen-doped carbon MnO_2 multichannel nanotubes with high performance for Li-ion batteries. *Nano Energy* **2022**, *97*, 107235. [CrossRef]
30. Park, S.W.; Kim, J.C.; Dar, M.A.; Shim, H.W.; Kim, D.W. Superior lithium storage in nitrogen-doped carbon nanofibers with open-channels. *Chem. Eng. J.* **2017**, *315*, 1–9. [CrossRef]
31. Zhang, T.; Zhang, L.; Zhao, L.N.; Huang, X.X.; Li, W.; Li, T.; Shen, T.; Sun, S.N.; Hou, Y.L. Free-standing, foldable V_2O_3/multichannel carbon nanofibers electrode for flexible Li-ion batteries with ultralong lifespan. *Small* **2020**, *16*, 2005302. [CrossRef] [PubMed]
32. Zhao, H.; Yin, H.; Yu, X.X.; Zhang, W.; Li, C.; Zhu, M.Q. In_2O_3 nanoparticles/carbonfiber hybrid mat as free-standing anode for lithium-ion batteries with enhanced electrochemical performance. *J. Alloy. Compd.* **2018**, *735*, 319–326. [CrossRef]
33. Chen, Y.J.; Zhao, X.H.; Liu, Y.; Razzaq, A.A.; Haridas, A.K.; Cho, K.K.; Peng, Y.; Deng, Z.; Ahn, J.H. γ-Fe_2O_3 nanoparticles aligned in porous carbon nanofibers towards long life-span lithium ion batteries. *Electrochim. Acta* **2018**, *289*, 264–271. [CrossRef]
34. Zhao, Q.S.; Liu, J.L.; Li, X.X.; Xia, Z.Z.; Zhang, Q.X.; Zhou, M.; Tian, W.; Wang, M.; Hu, H.; Li, Z.T.; et al. Graphene oxide-induced synthesis of button-shaped amorphous Fe_2O_3/rGO/CNFs films as flexible anode for high-performance lithium-ion batteries. *Chem. Eng. J.* **2019**, *369*, 215–222. [CrossRef]
35. Li, X.Q.; Xiang, J.; Zhang, X.K.; Li, H.B.; Yang, J.N.; Zhang, Y.M.; Zhang, K.Y.; Chu, Y.Q. Electrospun FeCo nanoparticles encapsulated in N-doped carbon nanofibers as self-supporting flexible anodes for lithium-ion batteries. *J. Alloy. Compd.* **2021**, *873*, 159703. [CrossRef]
36. Zhu, S.Q.; Huang, A.M.; Wang, Q.; Xu, Y. MOF-derived porous carbon nanofibers wrapping Sn nanoparticles as flexible anodes for lithium/sodium ion batteries. *Nanotechnology* **2021**, *32*, 165401. [CrossRef]
37. Su, Y.; Fu, B.; Yuan, G.L.; Ma, M.; Jin, H.Y.; Xie, S.H.; Li, J.Y. Three-dimensional mesoporous γ-Fe_2O_3@carbon nanofiber network as high performance anode material for lithium- and sodium-ion batteries. *Nanotechnology* **2020**, *31*, 155401. [CrossRef]
38. Ma, X.X.; Hou, G.M.; Ai, Q.; Zhang, L.; Si, P.C.; Feng, J.K.; Ci, L.J. A heart-coronary arteries structure of carbon nanofibers/graphene/silicon composite anode for high performance lithium ion batteries. *Sci. Rep.* **2017**, *7*, 9642. [CrossRef]
39. Wu, S.H.; Han, Y.D.; Wen, K.C.; Wei, Z.H.; Chen, D.J.; Lv, W.Q.; Lei, T.Y.; Xiong, J.; Gu, M.; He, W.D. Composite nanofibers through in-situ reduction with abundant active sites as flexible and stable anode for lithium ion batteries. *Compos. Part B Eng.* **2019**, *161*, 369–375. [CrossRef]
40. Abe, J.; Takahashi, K.; Kawase, K.; Kobayashi, Y.; Shiratori, S. Self-standing carbon nanofiber and SnO_2 nanorod composite as a high-capacity and high-rate-capability anode for lithium-ion batteries. *ACS Appl. Nano Mater.* **2018**, *1*, 2982–2989. [CrossRef]
41. Guo, L.G.; Sun, H.; Qin, C.Q.; Li, W.; Wang, F.; Song, W.L.; Du, J.; Zhong, F.; Ding, Y. Flexible Fe_3O_4 nanoparticles/N-doped carbon nanofibers hybrid film as binder-free anode materials for lithium-ion batteries. *Appl. Surf. Sci.* **2018**, *459*, 263–270. [CrossRef]
42. Zhang, X.Y.; Gao, M.Z.; Wang, W.; Liu, B.; Li, X.B. Encapsulating MoO_2 nanocrystals into flexible carbon nanofibers via electrospinning for high-performance lithium storage. *Polymers* **2021**, *13*, 22. [CrossRef]
43. Chen, R.Z.; Hu, Y.; Shen, Z.; Pan, P.; He, X.; Wu, K.S.; Zhang, X.W.; Cheng, Z.L. Facile fabrication of foldable electrospun polyacrylonitrile-based carbon nanofibers for flexible lithium-ion batteries. *J. Mater. Chem. A* **2017**, *5*, 12914–12921. [CrossRef]
44. Zhang, S.G.; Yue, L.C.; Wang, M.; Feng, Y.; Li, Z.; Mi, J. SnO_2 nanoparticles confined by N-doped and CNTs-modified carbon fibers as superior anode material for sodium-ion battery. *Solid State Ionics* **2018**, *323*, 105–111. [CrossRef]

45. Zhang, T.; Qiu, D.P.; Hou, Y.L. Free-standing and consecutive ZnSe@carbon nanofibers architectures as ultra-long lifespan anode for flexible lithium-ion batteries. *Nano Energy* **2022**, *94*, 106909. [CrossRef]
46. Huang, L.; Guan, Q.; Cheng, J.L.; Li, C.; Ni, W.; Wang, Z.P.; Zhang, Y.; Wang, B. Free-standing N-doped carbon nanofibers/carbon nanotubes hybrid film for flexible, robust half and full lithium-ion batteries. *Chem. Eng. J.* **2018**, *334*, 682–690. [CrossRef]
47. Mou, H.Y.; Chen, S.X.; Xiao, W.; Miao, C.; Li, R.; Xu, G.L.; Xin, Y.; Nie, S.Q. Encapsulating homogenous ultra-fine SnO_2/TiO_2 particles into carbon nanofibers through electrospinning as high-performance anodes for lithium-ion batteries. *Ceram. Int.* **2021**, *47*, 19945–19954. [CrossRef]
48. Nie, P.; Le, Z.; Chen, G.; Liu, D.; Liu, X.; Wu, H.B.; Xu, P.; Li, X.; Liu, F.; Chang, L. Graphene caging silicon particles for high-performance lithium-ion batteries. *Small* **2018**, *14*, 1800635. [CrossRef]
49. Zhang, J.; Zuo, S.; Wang, Y.; Yin, H.; Wang, Z.; Wang, J. Scalable synthesis of interconnected hollow Si/C nanospheres enabled by carbon dioxide in magnesiothermic reduction for high-performance lithium energy storage. *J. Power Sources* **2021**, *495*, 229803. [CrossRef]
50. Chan, C.K.; Peng, H.; Liu, G.; McIlwrath, K.; Zhang, X.F.; Huggins, R.A.; Cui, Y. High-performance lithium battery anodes using silicon nanowires. *Nat. Nanotechnol.* **2008**, *3*, 31–35. [CrossRef]
51. Zuo, X.; Wang, X.; Xia, Y.; Yin, S.; Ji, Q.; Yang, Z.; Wang, M.; Zheng, X.; Qiu, B.; Liu, Z. Silicon/carbon lithium-ion battery anode with 3D hierarchical macro-/mesoporous silicon network: Self-templating synthesis via magnesiothermic reduction of silica/carbon composite. *J. Power Sources* **2019**, *412*, 93–104. [CrossRef]
52. He, J.; Zhao, H.; Li, X.; Su, D.; Zhang, F.; Ji, H.; Liu, R. Superelastic and superhydrophobic bacterial cellulose/silica aerogels with hierarchical cellular structure for oil absorption and recovery. *J. Hazard. Mater.* **2018**, *346*, 199–207. [CrossRef] [PubMed]
53. Liu, H.; Meng, X.; Chen, Y.; Zhao, Y.; Guo, X.; Ma, T. Synthesis and surface engineering of composite anodes by coating thin-layer silicon on carbon cloth for lithium storage with high stability and performance. *ACS Appl. Energy Mater.* **2021**, *4*, 6982–6990. [CrossRef]
54. Li, Y.; Liu, X.; Zhang, J.; Yu, H.; Zhang, J. Carbon-coated Si/N-doped porous carbon nanofibre derived from metal-organic frameworks for Li-ion battery anodes. *J. Alloy. Compd.* **2022**, *902*, 163635. [CrossRef]

Article

Flexible Porous Silicon/Carbon Fiber Anode for High−Performance Lithium−Ion Batteries

Gang Liu [1,†], Xiaoyi Zhu [1,†], Xiaohua Li [2], Dongchen Jia [1], Dong Li [1], Zhaoli Ma [3] and Jianjiang Li [1,*]

[1] School of Environmental Science and Engineering, Qingdao University, No. 308, Ningxia Road, Qingdao 266071, China; 2019025785@qdu.edu.cn (G.L.); xyzhu@qdu.edu.cn (X.Z.); 2020025847@qdu.edu.cn (D.J.); 2019205603@qdu.edu.cn (D.L.)

[2] School of Material Science and Engineering, Qingdao University, No. 308, Ningxia Road, Qingdao 266071, China; 2019020442@qdu.edu.cn

[3] School of Chemical Experimental Teaching Center, Qingdao University, No. 308, Ningxia Road, Qingdao 266071, China; zlma@qdu.edu.cn

[*] Correspondence: ljjqdu@163.com

[†] These authors contributed equally to this work.

Abstract: We demonstrate a cross−linked, 3D conductive network structure, porous silicon@carbon nanofiber (P−Si@CNF) anode by magnesium thermal reduction (MR) and the electrospinning methods. The P−Si thermally reduced from silica (SiO_2) preserved the monodisperse spheric morphology which can effectively achieve good dispersion in the carbon matrix. The mesoporous structure of P−Si and internal nanopores can effectively relieve the volume expansion to ensure the structure integrity, and its high specific surface area enhances the multi−position electrical contact with the carbon material to improve the conductivity. Additionally, the electrospun CNFs exhibited 3D conductive frameworks that provide pathways for rapid electron/ion diffusion. Through the structural design, key basic scientific problems such as electron/ion transport and the process of lithiation/delithiation can be solved to enhance the cyclic stability. As expected, the P−Si@CNFs showed a high capacity of 907.3 mAh g^{-1} after 100 cycles at a current density of 100 mA g^{-1} and excellent cycling performance, with 625.6 mAh g^{-1} maintained even after 300 cycles. This work develops an alternative approach to solve the key problem of Si nanoparticles' uneven dispersion in a carbon matrix.

Keywords: porous silicon; flexible; binder−free; anode; Li−ion batteries

1. Introduction

Li−ion batteries (LIBs) are widely used in different fields as a new energy storage technology, including portable electronic devices, electrical transportation, and even energy storage systems, owing to their high energy density, long cycle life, and environmental friendliness [1]. Among all candidate anode materials, silicon (Si) is the most promising anode for the next generation of high−capacity LIBs because of its high capacity (4200 mAh g^{-1}), low platform working potential (~0.4 V vs. Li/Li$^+$), and rich natural resources [2,3]. However, the volume change (~400%) during the process of lithiation/delithiation [4] and the low electrical conductivity severely limit its commercial development. To mitigate or solve these critical problems, developing nanostructured Si (nanotubes, nanowires, nanoparticles) and Si/C composites represents two effective strategies [5,6]. The combination of high−capacity nanostructured Si with stable conductivity carbon can form a stable SEI and improve the electrode conductivity to obtain a good rate performance and cycling stability [7]. However, nanostructured Si is easily agglomerated, and an incomplete coating of carbon on Si leads to a poor Si–C interface and dispersion and further influences the electrode conductivity and cycling stability [8]. Therefore, achieving a good dispersion of nanostructured Si in a carbon matrix is the key problem in Si/C composite research [9].

Porous silicon (P−Si) nanomaterials have attracted increasing attention owing to their highly porous structure and high specific surface area which are of great significance for enhancing the Li−ion transferring rate and cycling stability [10,11]. Developing porous Si/C composites can exert synergy between the porous structure and carbon matrix [12]. The pores inside the silicon material can relieve the volume expansion and mechanical stress during lithium storage. By dispersing P−Si into the carbon matrix, the overall electronic conductivity of the material can be improved, and the irreversible capacity loss caused by the side reaction can be reduced due to the less direct contact between the porous silicon and electrolyte [13]. Guo et al. [14] fabricated a Si/C hybrid with P−Si nanoparticles loaded in void carbon spheres. The nanoporous structure and special void space are beneficial for the structure integrity and provide channels for the fast transport of electrons/ions during the cycling. The obtained P−Si/C hybrid displayed an improved electrochemical performance. Lu's team reported a unique yolk–shell structure with a graphene cage encapsulating mesoporous Si spheres. As excepted, the as−synthesized P−Si/C composite displayed extraordinary lithium storage performance in terms of a high specific capacity, a high rate capability, and excellent cycling performance [11].

In recent years, electrostatic spinning technology has been widely used to prepare flexible, binder−free Si/C anode composites [15]. CNFs prepared by electrospinning exhibit 3D carbon frameworks with good mechanical strength, which can greatly accommodate the volume change problem of Si, and can also build a conductive network to further improve the conductivity. Most importantly, this flexible material can be cut into any shape and used as an independent electrode directly without conductive additives, copper foil, and polymer binders, which can greatly improve the energy density [16]. Chen et al. prepared pyrolytic carbon−coated Si/C nanofibers (Si−C/CNFs) via electrospinning, carbonization, and secondary thermal treatment. The Si/C−CNF composite exhibited a more stable cycle performance due to the coating material and nanofiber structure. Si NPs were distributed along the fibers, but the agglomeration was still evident [17]. Therefore, achieving a good dispersion of Si NPs in a carbon matrix is the key problem of the Si/C structure.

Herein, we report a cross−linked, 3D conductive network structure, porous silicon@carbon nanofiber (P−Si@CNF) anode synthesized by magnesium thermal reduction (MR) and the electrospinning method (Scheme 1). Firstly, P−Si NPs were prepared through the MR process from SiO_2 obtained by the well−established Stöber method. P−Si NPs can preserve the original monodisperse spheric morphology of SiO_2 which is beneficial to the homogeneous dispersion of P−Si NPs in PAN fibers during the subsequent electrospinning process. Meanwhile, the internal nanopores of P−Si can alleviate the volume change, ensuring a uniform stress distribution and the integrity of the structure. Secondly, P−Si NPs were dispersed in PAN to obtain P−Si@PAN fibers via a typical electrospinning process. Thirdly, through a "pre−oxidation−slicing−carbonization" process, the final flexible, binder−free P−Si@CNFs were obtained. In P−Si@CNFs, the higher specific surface area of P−Si enhances the multi−position electrical contact with the carbon material to enhance the overall conductivity. In addition, CNFs, serving as a 3D conductive framework, can not only accommodate the volume expansion of Si but also accelerate the electron/ion transfer. Additionally, P−Si@CNFs can be used as an independent electrode to obtain a high capacity density.

This work develops an alternative approach to improve the uniform dispersion of Si in a carbon matrix which is the key problem of traditional Si/C composites. The porous structure of Si and its high specific surface area allow multi−position electrical contact between Si and the carbon material to effectively realize enhanced electrical conductivity.

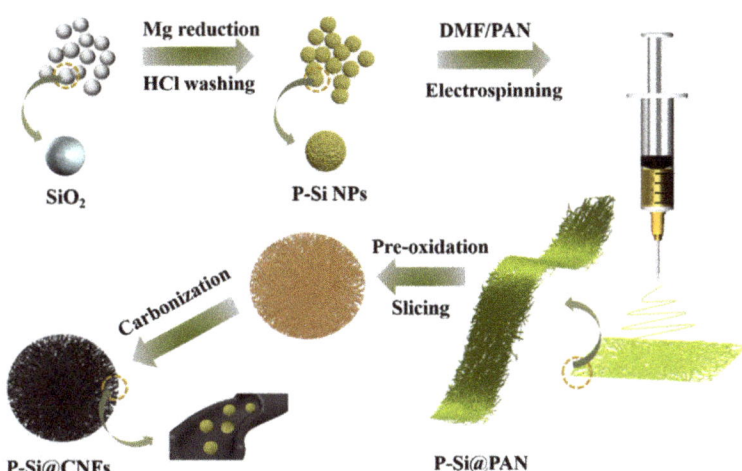

Scheme 1. Schematic diagram of the synthesis mechanism of the P−Si@CNF anode.

2. Materials and Methods

2.1. Synthesis of SiO$_2$ Microspheres

The chemical reagents used in this experiment were obtained from Sinopharm Group Co., Ltd. (Shanghai, China). SiO$_2$ microspheres were prepared using the well−established Stöber method [18]. First, 200 mL ethanol–water mixture ($V_E:V_W$ = 4:1) was prepared, and ultrasonic treatment was performed for 3 h to form a uniformly dispersed solution. Then, 5 mL ammonia and 3 mL tetraethylorthosilicate (TEOS) were added separately into the above solution and reacted for 24 h. SiO$_2$ microspheres were collected by centrifugation, washed with ethanol–water, and vacuum dried for 12 h.

2.2. Synthesis of P−Si NPs

In the typical MR process, magnesium powders (Mg 99%) and the obtained SiO$_2$ microspheres (Mg:SiO$_2$ = 1:1) were uniformly mixed well and placed on the end of a graphite boat. Sodium chloride (NaCl: SiO$_2$: = 9:1) was placed on the other side. The graphite boat was placed in the center of the quartz tube of the tube furnace (Kejing, Shenzhen, China), heated to 750 °C at a rate of 5 °C min^{-1} in an Ar/H$_2$ (5 % H$_2$) atmosphere, and maintained for 6 h. After cooling to room temperature, 1 mol L^{-1} hydrochloric acid (HCl) was used to remove NaCl, magnesium oxide (MgO), and other impurities. Finally, the P−Si powders were collected by centrifuging three times with ethanol and vacuum drying for 12 h.

2.3. Synthesis of P−Si@CNFs

P−Si@CNFs were synthesized via a typical electrospinning process. Firstly, the P−Si obtained above and 0.5 g polyacrylonitrile (PAN, Mw = 150,000) were dispersed in 4.5 g N, N−DimethylFormamide (DMF) solvent and stirred for 12 h. Then, electrospinning was operated at a voltage of 17.5 kV with a flow rate of 0.75 mL h^{-1}. The collected film was named P−Si@PAN. Secondly, the film was heated to 250 °C for 2 h for pre−oxidation at the rate of 2 °C min^{-1} in air. Then, the film was sliced into wafers with a diameter of 1 cm. Finally, the wafers were heated to 850 °C for carbonization for 2 h in an Ar atmosphere. The denoted samples and corresponding synthesis conditions are listed in Table S1.

3. Results and Discussions

The X−ray diffraction (XRD) patterns of SiO$_2$ by the well−established Stöber method and the reduced P−Si NPs via the MR process are shown in Figure S1. The broad peak located at about 25° relates to amorphous SiO$_2$. After the MR process, three distinct diffraction peaks located at the 2θ values of 28.4, 47.2, and 56.1° were assigned to the crystalline Si phase of the (111), (220), and (311) planes, respectively (JCPDSNO.27−1402) [19,20], suggesting that the spherical amorphous SiO$_2$ synthesized by the Stöber method was reduced [21]. The respective XRD patterns of P−Si@CNFs−100, 150, and 200 are shown in Figure 1a. Obviously, besides the typical peaks of Si, the broad peak at 25° related to the (002) plane was observed in the three samples due to the amorphous carbon produced by the pyrolytic carbonization of PAN [22].

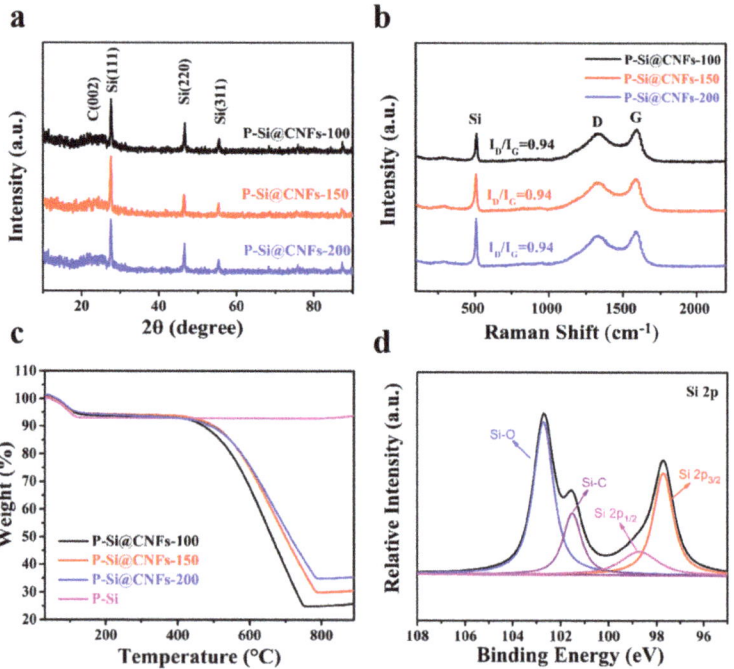

Figure 1. (**a**) XRD pattern, (**b**) Raman spectra, and (**c**) TGA curves of P−Si@CNFs−100, 150, and 200 in air, and (**d**) high−resolution spectrum of Si 2p of P−Si@CNFs−150.

The Raman spectra of P−Si@CNFs−100, 150, and 200 are exhibited in Figure 1b. The peak at around 500 cm^{-1} relates to Si. The peak at around 1357 (D band) is due to structure defect− and disorder−induced features in the graphene layers of the carbon, and the peak at 1584 cm^{-1} is attributed to the high−frequency E$_{2g}$ first−order graphitic crystallites of the carbon [23]. The calculated I$_D$/I$_G$ results of P−Si@CNFs−100, 150, and 200 were all calculated as 0.94. In P−Si@CNFs, the D band is attributed to amorphous carbon derived from the PAN pyrolytic carbonization and corresponds to the broad diffraction peak at 25° in XRD.

To further clarify the proportion of Si content in the composite, thermogravimetric analysis (TGA) was operated under oxygen at the rate of 10 °C min^{-1}. Figure 1c shows the TGA curves of P−Si@CNFs−100, 150, and 200. The substantial weight loss between 500 and 700 °C was due to the combustion of amorphous carbon coming from PAN. After 800 °C, the weight increased slightly due to the oxidization of P−Si. Thus, the C content coming from PAN in the P−Si@CNFs−100, P−Si@CNFs−150, and P−Si@CNFs−200 composites was calculated as 74.15, 68.76, and 63.53%, respectively. The corresponding Si content was 24.81, 29.99, and 35.01%, respectively. The TGA curve of the comparative sample Si@CNFs−150 is shown in Figure S2. This sample had about 28.12% Si content, which is similar to the above curves.

The elemental composition of P−Si@CNFs−150 was determined by XPS. Figure S3 shows the spectra, confirming the presence of C 1s, O 1s, and Si 2p. Figure 1d reveals the spectrum of Si 2p. It is subdivided into Si−Si including Si2p$_{1/2}$ (98.5 eV) and Si2p$_{3/2}$ (97.6 eV), Si−C (101.5 eV), and Si−O (102.7 eV) [24]. The spectra of C 1s and O 1s are shown in Figure S4. In Figure S4a, the spectrum of C 1s is divided into C−Si (283.1 eV), C−C (283.6 eV), and C−O (284.3 eV) [25,26]. In Figure S4b, the spectrum of O 1s is divided into O−Si (531.8 eV) and O−C (532.7 eV) [3]. Si−O may come from the residual silica of the MR process and the surface oxidization of the reduced Si [11].

The scanning electron microscopy (SEM) images of the monodisperse spheric SiO$_2$ by the Stöber method with a diameter of around 200 nm are shown in Figure S5a. The spheric shape was preserved after the MR process (Figure S5b). The images of P−Si@CNFs−100, 150, and 200 are displayed in Figure 2. Only few P−Si NPs are distributed in the CNFs in Figure 2a,b. As the content of P−Si increased to 150 mg (Figure 2c,d), P−Si NPs were completely dispersed in the CNFs with little agglomeration. As the P−Si content increased to 200 mg, P−Si NPs presented a distinct agglomeration phenomenon (Figure 2e,f). It can be concluded that P−Si@CNFs−150 with particles uniformly scattered along the CNFs is expected to exhibit good electrochemical performance with the optimum P−Si content. When using commercial Si NPs instead of P−Si, severe particle agglomeration along the CNFs was presented in the comparative sample Si@CNFs−150 (Figure S6). Compared with the commercial Si NPs (80 nm), P−Si with around a 200 nm diameter maintained the original monodisperse spherical morphology, which is beneficial to the uniform dispersion of Si in CNFs.

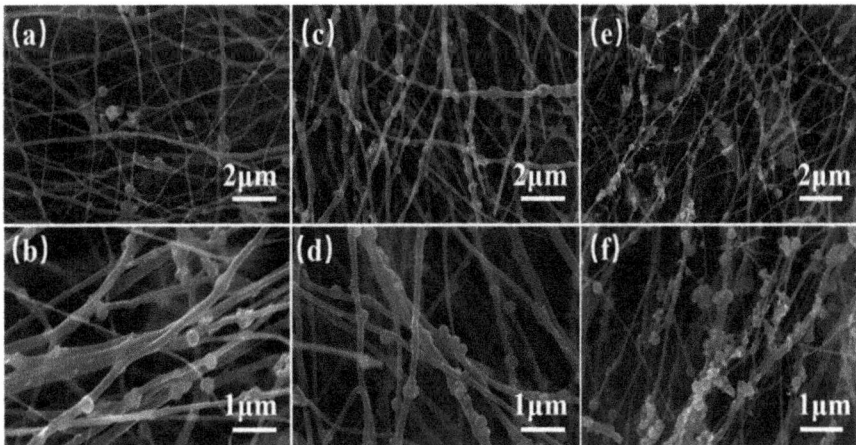

Figure 2. SEM images of (**a**,**b**) P−Si@CNFs−100, (**c**,**d**) 150, and (**e**,**f**) 200.

The transmission electron microscope (TEM) images of SiO$_2$ and the corresponding P−Si are displayed in Figure 3a–c. Figure 3b,c reveal that P−Si presented a spheric morphology and was composed of nanosized Si particles with diameters of 5–10 nm. The clearly porous structure can shorten the Li−ion diffusion paths, buffer the Si volume expansion, and facilitate the multi−site contact between the Si and conductive carbon. The Si NPs in Si@CNFs−150 are agglomerated together in Figure S7. Figure 3d shows a typical fiber of P−Si@CNFs−150. It can be found that P−Si NPs were fully embedded and dispersed in the CNFs. The edge of the particles in the CNFs was unsmooth, which confirms the porous structure of P−Si compared with the SiO$_2$ nanospheres directly dispersed in the CNFs (Figure S8). In Figure 3e, the high−resolution transmission electron microscopy (HRTEM) image further illustrates that the D−spacing of 0.31 nm was related to Si (111) [27–29]. The photo of P−Si@CNFs−150 is shown in Figure 3f. This sample proved to be a flexible electrode due to its good flexible characteristics withstanding multiple bending events.

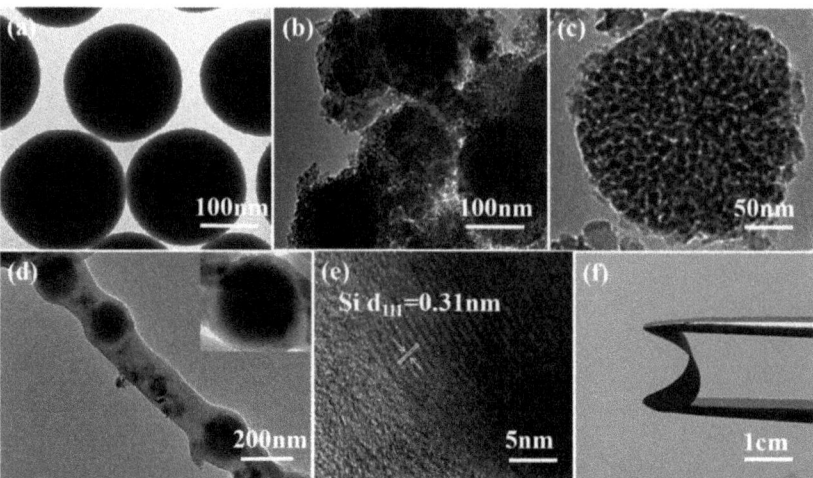

Figure 3. TEM images of (**a**) SiO$_2$, (**b**,**c**) P−Si via the MR process, and (**d**) P−Si@CNFs−150, (**e**) HRTEM image of P−Si@CNFs−150, and (**f**) photo of the flexible P−Si@CNFs.

Figure 4a exhibits the cyclic voltammetry (CV) curve of P−Si@CNFs−150 for the first three cycles. The broad peak at about 0.7~0.85 V in the initial cathodic cycle (lithiation) was attributed to the formation of the irreversible SEI layer and disappeared in the next cycle [30]. The peak at about 0.18 V was due to the lithiation of Si [31]. Correspondingly, the oxidative peak at approximately 0.5 V was due to the dealloying of Li$_{4.4}$Si [32]. The peaks in the reductive and oxidative cycle became stronger, ascribed to the Li ion diffusion and the electrical conductivity of the electrode.

Figure 4b shows the initial charge/discharge results of P−Si@CNFs at a current density of 100 mA g^{-1}. According to the first discharging and charging capacity process, P−Si@CNFs−100, 150, and 200 displayed an outstanding capacity of 1100.9 and 808.8 mAh g^{-1}, 1552.9 and 1241.5 mAh g^{-1}, and 1728.7 and 1374.9 mAh g^{-1}, with the corresponding initial coulomb efficiencies (ICE) of 73.46, 79.94, and 79.53%, respectively. From the second cycle, the voltage platform between 0.5 and 0.8 V disappeared, which is consistent with the result of the CV curve, confirming that the generated SEI film was stable [33]. The discharge/charge curve of P−Si@CNFs−150 for the first five cycles is shown in Figure S9. The CE reached 95.89% after the second cycle and then increased to 98.02% in the fifth cycle. In addition, the voltage stretch was maintained well in the cycle, which indicates that the electrochemical efficiency was greatly improved.

The cycling property of P−Si@CNFs−100, 150, and 200 at 100 mA g^{-1} is shown in Figure 4c. After 100 cycles, they displayed a reversible capacity of 680.8, 907.3, and 691.7 mAh g^{-1}, respectively. In contrast, P−Si@CNFS−150 displayed a relatively high reversible capacity and better capacity retention owing to the optimal Si content. Si@CNFs−150 delivered poor cycling properties due to the serious aggregation of Si NPs. Figure S10 shows the long−term cycle performance of P−Si@CNFS−150 at 100 mA g^{-1}, where it maintained a special reversible capacity of 625.6 mAh g^{-1} after 300 cycles, which is consistent with the SEM analysis discussed before. P−Si@CNFS−150 exhibited better dispersion along the CNFs, and the homogeneous mesoporous structure can effectively accommodate Si volume expansion. We disassembled the battery after cycling and show its photo and SEM images in Figure S11. No obvious cracks can be seen, and the electrode microstructure maintained its integrity even after long cycling.

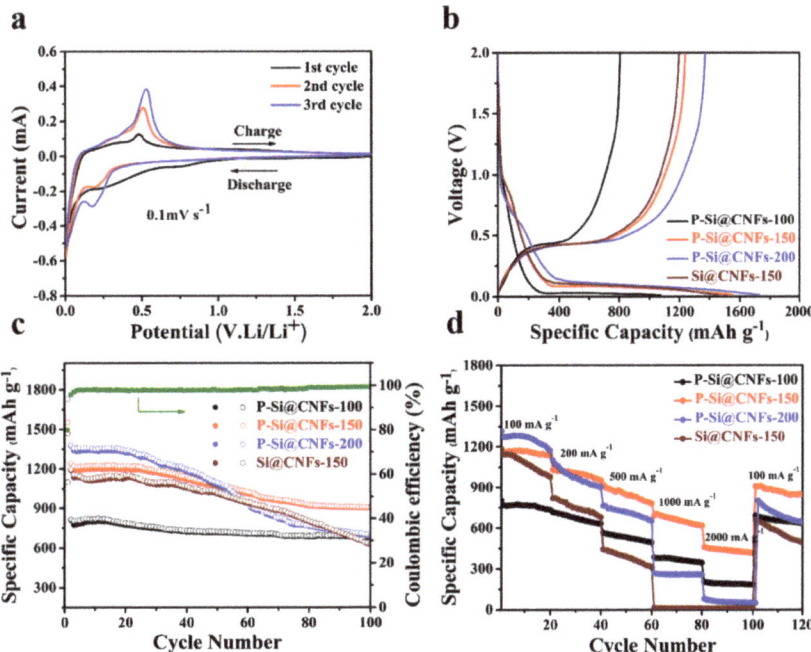

Figure 4. Electrochemical properties: CV profiles of P−Si@CNFs−150 (**a**), discharge/charge curves (**b**), cycle life of P−Si@CNF samples and Si@CNFs−150 at a current density of 100 mA g^{-1} and Coulombic efficiency of P−Si@CNFs−150 (**c**), and rate performances of P−Si@CNF samples and Si@CNFs−150 (**d**).

Moreover, the rate performance of P−Si@CNFs−100, 150, and 200 carried out at different current densities is displayed in Figure 4d. Obviously, P−Si@CNFs−150 exhibited the best rate performance. Even after a high current density of 2 A g^{-1}, the average reversible charge capacity still reached 420.3 mAh g^{-1}. When P−Si@CNFs−150 was cycled at 100 mA g^{-1} again, it still had a high reversible capacity (906.7 mAh g^{-1}), confirming the structure stability. P−Si@CNFs−200 displayed a higher initial reversible specific capacity but showed a worse rate performance due to the much higher P−Si content. The excellent rate performance of P−Si@CNFs−150 was attributed to the high specific surface area of P−Si which enhances the multi−site electrical contact with the CNFs, resulting in improved electrical conductivity. Table S2 shows the Brunauer−Emmett−Teller (BET) surface area, pore volume, and average pore size of P−Si@CNFs−150 and Si@CNFs−150. The surface areas of the two samples were 261.67 and 229.35, respectively. Obviously, P−Si@CNFs−150 with multiple pores can facilitate the penetration of the electrolyte into the composite structure and shorten the Li ion diffusion paths. Figure S12 shows the schematic diagram of the mechanism of lithiation and delithiation. The P−Si@CNF composite presented a cross−linked 3D conductive framework structure. The original monodisperse spheric morphology realized P−Si uniformly dispersed along the CNFs during the electrospinning process. Meanwhile, the internal nanopores of P−Si effectively alleviated the volume expansion and maintained the structure integrity; thus, the cycling performance was obtained. On the other hand, the Li ion and electron transfer was effectively accelerated owing to the CNF conductive framework and thus improved the electrode rate performance. Additionally, the flexible electrode film obtained by electrospinning can be easily sliced for use as an independent electrode and makes a high capacity density possible. The comparison with other silicon carbon fiber anodes is shown in Figure S13. P−Si@CNFs−150 exhibited a superior capacity and cycling stability, which proves its good structural design.

4. Conclusions

In conclusion, a flexible P−Si@CNF electrode with a cross−linked 3D conductive framework microstructure was prepared by the process of magnesium thermal reduction and electrostatic spinning. The optimized P−Si@CNFs with 29.99% Si displayed an excellent capacity (907.3 mAh g^{-1}) after 100 cycles at 100 mA g^{-1}. Even after 300 cycles, they also maintained a unique capacity of 625.6 mAh g^{-1}. The outstanding cyclic stability was attributed to their unique microstructure and flexible characteristics. The reduced P−Si from the MR process exhibited a mesoporous structure which can not only relieve the stress and strain caused by volume expansion but also increase the multi−site electronic contact site between the Si and C, effectively enhancing the overall electrical conductivity. More importantly, P−Si obtained from the reduction of SiO_2 preserved the single−sphere dispersion in CNFs which is the key factor in structural stability. Overall, the structural design in this work effectively improves the key homogeneous dispersion problem of Si/C composites and has a great prospect in the future application of Si/C anodes in LIBs.

Supplementary Materials: The following supporting information can be downloaded at: https://www.mdpi.com/article/10.3390/ma15093190/s1, Figure S1. The X−ray diffraction (XRD) patterns of SiO_2 and P−Si. Figure S2. TGA curves of Si@CNFs−150. Figure S3. The survey XPS spectra of P−Si@CNFs−150. Figure S4. XPS spectrum of C 1s (a) and O 1s (b) of P−Si@CNFs−150. Figure S5. The SEM images of SiO_2(a) and P−Si(b). Figure S6. The SEM images of Si@CNFs−150. Figure S7. The TEM images of Si@CNFs−150. Figure S8. The TEM images of SiO_2@CNFs. Figure S9. The charge and discharge of P−Si@CNFs−150 from the first to the fifth cycle. Figure S10. Cycle performance of P−Si@CNFs−150 at 100 mA g^{-1} after 300 cycles. Figure S11. Electrode photo (a) and SEM images (b) after 300 cycling. Figure S12. The schematic illustration of lithiation and delithiation processes. Figure S13. Comparison of flexible binder−free Si/C anodes reported for LIBs. Table S1. Synthesis conditions and nomenclature of the samples. Table S2. The BET surface area, pore volume and average pore size of P−Si@CNFs−150 and Si@CNFs−150.

Author Contributions: Conceptualization, J.L.; methodology, G.L. and X.Z.; software, G.L. and J.L.; validation, G.L., X.L. and D.J.; formal analysis, G.L.; investigation, J.L.; resources, Z.M.; data curation, G.L. and D.L.; writing—original draft preparation, G.L.; writing—review and editing, X.Z. and J.L.; visualization, X.Z.; supervision, G.L.; project administration, X.Z. and J.L.; funding acquisition, X.Z. All authors have read and agreed to the published version of the manuscript.

Funding: This research received no external funding.

Institutional Review Board Statement: Not applicable.

Informed Consent Statement: Not applicable.

Data Availability Statement: The data presented in this study are available on request from the corresponding author.

Conflicts of Interest: The authors declare no conflict of interest.

References

1. Almehmadi, F.A.; Alqaed, S.; Mustafa, J.; Jamil, B.; Sharifpur, M.; Cheraghian, G. Combining an active method and a passive method in cooling lithium−ion batteries and using the generated heat in heating a residential unit. *J. Energy Storage* **2022**, *49*, 104181. [CrossRef]
2. Xu, Q.; Sun, J.-K.; Li, J.-Y.; Yin, Y.-X.; Guo, Y.-G. Scalable synthesis of spherical Si/C granules with 3D conducting networks as ultrahigh loading anodes in lithium−ion batteries. *Energy Storage Mater.* **2018**, *12*, 54–60. [CrossRef]
3. Yan, M.-Y.; Li, G.; Zhang, J.; Tian, Y.-F.; Yin, Y.-X.; Zhang, C.-J.; Jiang, K.-C.; Xu, Q.; Li, H.-L.; Guo, Y.-G. Enabling SiO_x/C Anode with High Initial Coulombic Efficiency through a Chemical Pre−Lithiation Strategy for High−Energy−Density Lithium−Ion Batteries. *ACS Appl. Mater. Interfaces* **2020**, *12*, 27202–27209. [CrossRef] [PubMed]
4. Kim, N.; Chae, S.; Ma, J.; Ko, M.; Cho, J. Fast−charging high−energy lithium−ion batteries via implantation of amorphous silicon nanolayer in edge−plane activated graphite anodes. *Nat. Commun.* **2017**, *8*, 812. [CrossRef] [PubMed]
5. Zhu, X.; Yang, D.; Li, J.; Su, F. Nanostructured Si−Based Anodes for Lithium−Ion Batteries. *J. Nanosci. Nanotechnol.* **2015**, *15*, 15–30. [CrossRef] [PubMed]

6. Alqaed, S.; Almehmadi, F.A.; Mustafa, J.; Husain, S.; Cheraghian, G. Effect of nano phase change materials on the cooling process of a triangular lithium battery pack. *J. Energy Storage* **2022**, *51*, 104326. [CrossRef]
7. Xu, S.; Hou, X.; Wang, D.; Zuin, L.; Zhou, J.; Hou, Y.; Mann, M. Insights into the Effect of Heat Treatment and Carbon Coating on the Electrochemical Behaviors of SiO Anodes for Li−Ion Batteries. *Adv. Energy Mater.* **2022**. [CrossRef]
8. Zhu, X.; Chen, H.; Wang, Y.; Xia, L.; Tan, Q.; Li, H.; Zhong, Z.; Su, F.; Zhao, X.S. Growth of silicon/carbon microrods on graphite microspheres as improved anodes for lithium−ion batteries. *J. Mater. Chem. A* **2013**, *1*, 4483–4489. [CrossRef]
9. Wang, J.; Gao, C.; Yang, Z.; Zhang, M.; Li, Z.; Zhao, H. Carbon−coated mesoporous silicon shell−encapsulated silicon nano−grains for high performance lithium−ion batteries anode. *Carbon* **2022**, *192*, 277–284. [CrossRef]
10. Zhang, Y.; Du, N.; Zhu, S.; Chen, Y.; Lin, Y.; Wu, S.; Yang, D. Porous silicon in carbon cages as high−performance lithium−ion battery anode Materials. *Electrochim. Acta* **2017**, *252*, 438–445. [CrossRef]
11. Nie, P.; Le, Z.; Chen, G.; Liu, D.; Liu, X.; Wu, H.B.; Xu, P.; Li, X.; Liu, F.; Chang, L.; et al. Graphene Caging Silicon Particles for High−Performance Lithium−Ion Batteries. *Small* **2018**, *14*, 1800635. [CrossRef] [PubMed]
12. Ren, Y.; Zhou, X.; Tang, J.; Ding, J.; Chen, S.; Zhang, J.; Hu, T.; Yang, X.-S.; Wang, X.; Yang, J. Boron−Doped Spherical Hollow−Porous Silicon Local Lattice Expansion toward a High−Performance Lithium−Ion−Battery Anode. *Inorg. Chem.* **2019**, *58*, 4592–4599. [CrossRef] [PubMed]
13. Zhang, Z.; Xi, F.; Ma, Q.; Wan, X.; Li, S.; Ma, W.; Chen, X.; Chen, Z.; Deng, R.; Ji, J.; et al. A nanosilver−actuated high−performance porous silicon anode from recycling of silicon waste. *Mater. Today Nano* **2022**, *17*, 100162. [CrossRef]
14. Guo, S.; Hu, X.; Hou, Y.; Wen, Z. Tunable Synthesis of Yolk–Shell Porous Silicon@Carbon for Optimizing Si/C−Based Anode of Lithium−Ion Batteries. *ACS Appl. Mater. Interfaces* **2017**, *9*, 42084–42092. [CrossRef]
15. Shao, F.; Li, H.; Yao, L.; Xu, S.; Li, G.; Li, B.; Zou, C.; Yang, Z.; Su, Y.; Hu, N.; et al. Binder−Free, Flexible, and Self−Standing Non−Woven Fabric Anodes Based on Graphene/Si Hybrid Fibers for High−Performance Li−Ion Batteries. *ACS Appl. Mater. Interfaces* **2021**, *13*, 27270–27277. [CrossRef] [PubMed]
16. Li, X.; Wang, X.; Li, J.; Liu, G.; Jia, D.; Ma, Z.; Zhang, L.; Peng, Z.; Zhu, X. High−performance, flexible, binder−free silicon–carbon anode for lithium storage applications. *Electrochem. Commun.* **2022**, *137*, 107257. [CrossRef]
17. Chen, Y.; Hu, Y.; Shao, J.; Shen, Z.; Chen, R.; Zhang, X.; He, X.; Song, Y.; Xing, X. Pyrolytic carbon−coated silicon/carbon nanofiber composite anodes for high−performance lithium−ion batteries. *J. Power Sources* **2015**, *298*, 130–137. [CrossRef]
18. Liu, N.; Lu, Z.; Zhao, J.; McDowell, M.T.; Lee, H.-W.; Zhao, W.; Cui, Y. A pomegranate−inspired nanoscale design for large−volume−change lithium battery anodes. *Nat. Nanotechnol.* **2014**, *9*, 187–192. [CrossRef] [PubMed]
19. Ghanooni Ahmadabadi, V.; Shirvanimoghaddam, K.; Kerr, R.; Showkath, N.; Naebe, M. Structure−rate performance relationship in Si nanoparticles−carbon nanofiber composite as flexible anode for lithium−ion batteries. *Electrochim. Acta* **2020**, *330*, 135232. [CrossRef]
20. Wang, C.; Zhang, L.; Al-Mamun, M.; Dou, Y.; Liu, P.; Su, D.; Wang, G.; Zhang, S.; Wang, D.; Zhao, H. A Hollow−Shell Structured V_2O_5 Electrode−Based Symmetric Full Li−Ion Battery with Highest Capacity. *Adv. Energy Mater.* **2019**, *9*, 1900909. [CrossRef]
21. Zhang, L.; Huang, Q.; Liao, X.; Dou, Y.; Liu, P.; Al-Mamun, M.; Wang, Y.; Zhang, S.; Zhao, S.; Wang, D.; et al. Scalable and controllable fabrication of CNTs improved yolk−shelled Si anodes with advanced in operando mechanical quantification. *Energy Environ. Sci.* **2021**, *14*, 3502–3509. [CrossRef]
22. Mu, G.; Ding, Z.; Mu, D.; Wu, B.; Bi, J.; Zhang, L.; Yang, H.; Wu, H.; Wu, F. Hierarchical void structured Si/PANi/C hybrid anode material for high−performance lithium−ion batteries. *Electrochim. Acta* **2019**, *300*, 341–348. [CrossRef]
23. Ji, L.; Zhang, X. Electrospun carbon nanofibers containing silicon particles as an energy−storage medium. *Carbon* **2009**, *47*, 3219–3226. [CrossRef]
24. Liu, H.; Chen, Y.; Jiang, B.; Zhao, Y.; Guo, X.; Ma, T. Hollow−structure engineering of a silicon–carbon anode for ultra−stable lithium−ion batteries. *Dalton Trans.* **2020**, *49*, 5669–5676. [CrossRef] [PubMed]
25. Xie, Q.; Qu, S.; Zhao, P. A facile fabrication of micro/nano−sized silicon/carbon composite with a honeycomb structure as high−stability anodes for lithium−ion batteries. *J. Electroanal. Chem.* **2021**, *884*, 115074. [CrossRef]
26. Zhou, X.; Liu, Y.; Ren, Y.; Mu, T.; Yin, X.; Du, C.; Huo, H.; Cheng, X.; Zuo, P.; Yin, G. Engineering Molecular Polymerization for Template−Free SiO_X/C Hollow Spheres as Ultrastable Anodes in Lithium−Ion Batteries. *Adv. Funct. Mater.* **2021**, *31*, 2101145. [CrossRef]
27. He, Y.; Xiang, K.; Zhou, W.; Zhu, Y.; Chen, X.; Chen, H. Folded−hand silicon/carbon three−dimensional networks as a binder−free advanced anode for high−performance lithium−ion batteries. *Chem. Eng. J.* **2018**, *353*, 666–678. [CrossRef]
28. Zhang, L.; Zhang, M.; Wang, Y.; Zhang, Z.; Kan, G.; Wang, C.; Zhong, Z.; Su, F. Graphitized porous carbon microspheres assembled with carbon black nanoparticles as improved anode materials in Li−ion batteries. *J. Mater. Chem. A* **2014**, *2*, 10161–10168. [CrossRef]
29. Guan, P.; Li, J.; Lu, T.; Guan, T.; Ma, Z.; Peng, Z.; Zhu, X.; Zhang, L. Facile and Scalable Approach To Fabricate Granadilla−like Porous−Structured Silicon−Based Anode for Lithium Ion Batteries. *ACS Appl. Mater. Interfaces* **2018**, *10*, 34283–34290. [CrossRef]
30. Jia, H.; Zheng, J.; Song, J.; Luo, L.; Yi, R.; Estevez, L.; Zhao, W.; Patel, R.; Li, X.; Zhang, J.-G. A novel approach to synthesize micrometer−sized porous silicon as a high performance anode for lithium−ion batteries. *Nano Energy* **2018**, *50*, 589–597. [CrossRef]

31. Yoon, N.; Young, C.; Kang, D.; Park, H.; Lee, J.K. High−conversion reduction synthesis of porous silicon for advanced lithium battery anodes. *Electrochim. Acta* **2021**, *391*, 138967. [CrossRef]
32. Cao, M.; Liao, F.; Wang, Q.; Luo, W.; Ma, Y.; Zheng, X.; Wang, Y.; Zhang, L. Rational design of ZnS/CoS heterostructures in three dimensional N−doped CNTs for superior lithium storage. *J. Alloys Compd.* **2021**, *859*, 157867. [CrossRef]
33. Zhu, G.; Zhang, F.; Li, X.; Luo, W.; Li, L.; Zhang, H.; Wang, L.; Wang, Y.; Jiang, W.; Liu, H.K.; et al. Engineering the Distribution of Carbon in Silicon Oxide Nanospheres at the Atomic Level for Highly Stable Anodes. *Angew. Chem. Int. Ed.* **2019**, *58*, 6669–6673. [CrossRef] [PubMed]

Article

Facile Synthesis of Unsupported Pd Aerogel for High Performance Formic Acid Microfluidic Fuel Cell

Alejandra Martínez-Lázaro [1], Luis A. Ramírez-Montoya [2], Janet Ledesma-García [1], Miguel A. Montes-Morán [3], Mayra P. Gurrola [4], J. Angel Menéndez [3], Ana Arenillas [3,*] and Luis G. Arriaga [5,*]

[1] División de Investigación y Posgrado, Facultad de Ingeniería, Universidad Autónoma de Querétaro, Santiago de Querétaro 76010, Mexico; AleeM.Lazaro@live.com (A.M.-L.); janet.ledesma@uaq.mx (J.L.-G.)
[2] Laboratory for Research on Advanced Processes for Water Treatment, Engineering Institute, Universidad Nacional Autónoma de México (UNAM), Blvd. Juriquilla 3001, Santiago de Querétaro 76230, Mexico; lar-75@hotmail.com
[3] Instituto de Ciencia y Tecnología del Carbono, INCAR-CSIC. Francisco Pintado Fe, 33011 Oviedo 26., Spain; miguel@incar.csic.es (M.A.M.-M.); angelmd@incar.csic.es (J.A.M.)
[4] CONACYT-Tecnológico Nacional de México/Instituto Tecnológico de Chetumal. Av. Insurgentes 330, David Gustavo Gutiérrez, Chetumal 77013, Mexico; mayra.pg@chetumal.tecnm.mx
[5] Centro de Investigación y Desarrollo Tecnológico en Electroquímica, Santiago de Querétaro 76703, Mexico
* Correspondence: aapuente@incar.csic.es (A.A.); larriaga@cideteq.mx (L.G.A.)

Abstract: In this work, unsupported Pd aerogel catalysts were synthesized for the very first time by using microwaves as a heating source followed by a lyophilization drying process and used towards formic acid electro-oxidation in a microfluidic fuel cell. Aerogels were also made by heating in a conventional oven to evaluate the microwave effect during the synthesis process of the unsupported Pd aerogels. The performance of the catalysts obtained by means of microwave heating favored the formic acid electro-oxidation with H_2SO_4 as the electrolyte. The aerogels' performance as anodic catalysts was carried out in a microfluidic fuel cell, giving power densities of up to 14 mW cm^{-2} when using mass loads of only 0.1 mg on a 0.019 cm^2 electrode surface. The power densities of the aerogels obtained by microwave heating gave a performance superior to the resultant aerogel prepared using conventional heating and even better than a commercial Pd/C catalyst.

Keywords: unsupported aerogel; microwave heating; microfluidic fuel cell; electro-oxidation

Citation: Martínez-Lázaro, A.; Ramírez Montoya, L.A.; Ledesma-García, J.; Montes-Morán, M.A.; Gurrola, M.P.; Menéndez, J.A.; Arenillas, A.; Arriaga, L.G. Facile Synthesis of Unsupported Pd Aerogel for High Performance Formic Acid Microfluidic Fuel Cell. *Materials* **2022**, *15*, 1422. https://doi.org/10.3390/ma15041422

Academic Editor: Alessandro Dell'Era

Received: 6 January 2022
Accepted: 10 February 2022
Published: 15 February 2022

Publisher's Note: MDPI stays neutral with regard to jurisdictional claims in published maps and institutional affiliations.

Copyright: © 2022 by the authors. Licensee MDPI, Basel, Switzerland. This article is an open access article distributed under the terms and conditions of the Creative Commons Attribution (CC BY) license (https://creativecommons.org/licenses/by/4.0/).

1. Introduction

Noble metals have been widely used as electrocatalysts for energy conversion devices, including fuel cells, water electrolysis, and metal–air batteries [1]. Fuel cells supply energy in a similar way as batteries, although they do not require charging and operate as long as fuel is provided [2,3]. Fuel cells fed with formic acid supply electricity and heat, based on the electrochemical oxidation of fuels in the anode and the reduction of oxygen in the cathode, where H_2 is formed [4,5]. Pd nanostructures have been used specially in direct formic acid fuel cells that are considered green energy sources for portable electronics and hybrid vehicles due to their high open-circuit voltage, safety and reliability, and low fuel crossover effect, including Pd nanoparticles supported on graphene [6], since nanostructures have simple morphology [7] until hybrid variation of Pd–Cu [8] and Pd–Co [9]. Although these supported materials have great performance towards formic acid oxidation (FAO), the carbonaceous support is still a problem when working with devices that work at high voltages.

Conventional formic acid fuel cells usually employ a physical barrier for the separation of electrodes, which presents many limitations such as membrane fouling and clogging [10]. Therefore, fuel cells working with microfluids would take advantage of laminar flow as a fluid separator and avoid the use of membranes and their drawbacks [11]. Microfluidic fuel cells have the advantage of being portable and of carrying out small-scale processes offering high efficiency of energy conversion. The development of this type of device allows to incorporate the electro-oxidation of formic acid for diverse technological applications on a small scale [12]. On the other hand, miniaturizing the cell may reduce fabrication costs. A study has recently presented a novel microfluidic fuel cell (MFC) that incorporates the innovation of using a laminar flow instead of the conventional solid membrane to separate the fuel and oxidant [13]. Hence, the membrane-related issues are eliminated in this new MFC, which also offers savings in the manufacturing costs. However, it is still necessary to develop effective materials for these types of devices.

The use of catalysts for generating hydrogen from formic acid could minimize the dependence on lithium batteries in a large number of mobile devices [7]. Formic acid can be decomposed catalytically according to the following reactions:

$$HCOOH_{(l)} \rightarrow H_{2\,(g)} + CO_{2\,(g)} \tag{1}$$

$$HCOOH_{(l)} \rightarrow H_2O_{(l)} + CO_{(g)} \tag{2}$$

Noble metals have been extensively studied as catalysts due to their high efficiency, non-toxicity, and stability. Particularly, Pd is widely used in anodes for FAO [9,10]. The synergistic effect of the metallic phase and oxyphilic properties of the Pd surface provides active sites for adsorption and dissociation of formic acid besides providing promoters of oxygen-containing species at low potentials [14,15]. There are a great number of Pd-based catalysts in the bibliography for acid formic oxidation, however, improving their activity is still a requirement in order to be implemented in fuel cells.

Mesoporous materials with low density and a greater number of active sites such as aerogels would allow the use of less mass of the catalyst and, at the same time, to provide a high catalytic activity [12–14]. Noble metal aerogels have approximately 90% air and very low contents of the active metal which reduces the cost of the catalysts [15,16]. These materials are commonly obtained by a sol-gel process and supercritical drying [16,17], however, other techniques such as lyophilization allow promising aerogel qualities [18]. Lyophilization, like supercritical drying, shows high efficiency in the formation of metallic aerogels [19–21]; this has been demonstrated in works such as Cu(II) cryogels [22] and Pd/CeO$_2$-ZrO$_2$ alloy aerogels that have been used in the reduction of CO [23] poly(3-sulfopropylmethacrylate) (p(SPM)) cold gels for H$_2$ production and mostly organic aerogels as supports for other catalytic materials [24–26].

The aerogel synthesis in this work was implemented under microwave radiation. This heating technology allowed, not only the saving of processing time that is usually associated with this type of heating but also to obtain materials with homogeneous and controlled low particles. This fact is quite advantageous, as the particles are usually obtained with more heterogeneous and big particle sizes when the materials are prepared under long conventional heating. The microwave heating allows heating of the bulk precursor solution without gradients, which favors a very good dispersion of nucleation points for the reaction occurring and therefore a better control of the final particle size. Furthermore, the use of microwaves may also influence some chemical reactions and the final products may present some chemical differences in comparison with the ones obtained by conventional heating. In this work, the use of microwave heating provides many benefits such as facile synthesis of homogeneous and low particle size synthesis of Pd aerogels, with high activity towards the oxidation of formic acid in an MFC due to their unique physicochemical characteristics.

2. Experimental

2.1. Pd Aerogels Chemical Synthesis

The synthesis of Pd aerogels was carried out by adding 10 mL of a 2 mg/mL solution of PdCl$_2$ (99%, Sigma-Aldrich ReagentPlus®, anhydrous powder, St. Louis, MO, USA) in deionized water into a solution of 240 mg of sodium carbonate (≥99.5%, J.T. Baker®) and 40 mg of glyoxylic acid monohydrate (98% Sigma-Aldrich) (ratio 6:1) in 40 mL of deionized water at 67.5 °C (Figure 1) for two hours in all cases. Two different devices were used as a heating source for this initial step of the reaction: conventional heating in a lab oven (CON) and microwave heating (MW). In the case of microwave heating, the reaction temperature was controlled by a thermocouple introduced in the precursor mixture and connected to a proportional–integral–derivative (PID) temperature controller installed in the microwave oven.

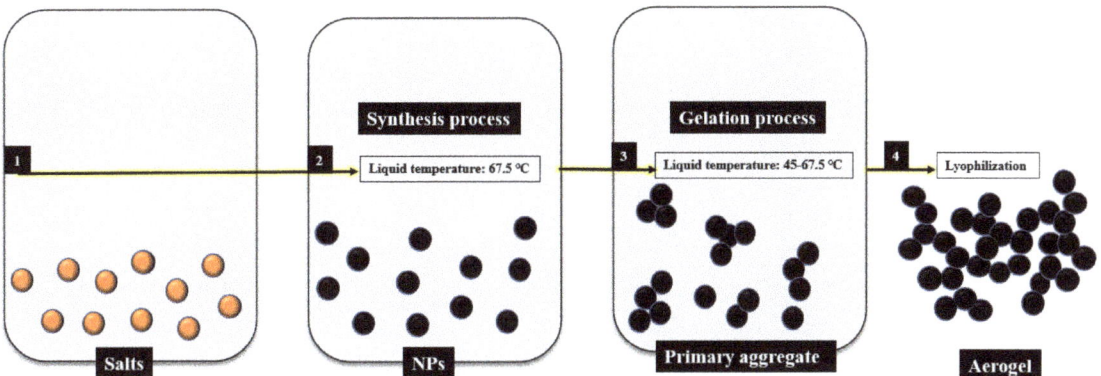

Figure 1. Procedure for obtaining aerogels using the sol-gel methodology in this work.

Once the first reactions took place, the reduction reaction and gelation process were also carried out by heating the mixture either by conventional heating (CON) or microwave heating (MW) under different operating conditions (i.e., 45 °C or 67 °C and 7 h or 24 h), as described in Table 1. After the gelation process, hydrogels were washed several times using deionized water and ethanol to remove the organic residues in the aqueous solution. Before submitting the clean samples to the lyophilization process, they were frozen with Liquid N_2, with a volume of 3 mL of deionized water and finally they were dried in a lyophilizer (LYO). A conventional drying in a lab oven (CON) was also performed for comparative proposes.

Table 1. Synthesis procedures and operating conditions to obtain the aerogels of this work.

Sample	Synthesis at 67.5 °C for 2 h	Heating Reduction Device	Reduction Conditions	Drying Device
PdA-CC	CON	CON	24 h/45 °C	LYO
PdA-MC	MW	CON	24 h/45 °C	LYO
PdA-MM	MW	MW	7 h/45 °C	LYO
PdA-MMT	MW	MW	7 h/67 °C	LYO
PdA-CON	CON	CON	24 h/45 °C	CON

2.2. Physicochemical Characterization

The morphology of the Pd aerogels was characterized using a JEOL JEM-2100F high-resolution transmission electron microscope (HR-TEM) with spherical aberration correction and a scanning electron microscope (SEM, Quanta FEG 650 microscopes from FEI). The crystal structures were measured by X-ray diffraction (XRD; D8-advance diffractometer Bruker) equipped with a CuKα X-ray source (λ = 0.1541 nm, 40 kV, 40 mA), using a step size of 0.02° 2θ and a scan step time of 5 s. The specific surface area and the pore size distribution were determined by nitrogen adsorption–desorption isotherms at -196 °C (Micromeritics ASAP 2020), after an overnight outgassing at 120 °C. The electronic structure of elements was measured by X-ray photoelectron spectroscopy (XPS; K-Alpha+ spectrometer equipped with the Avantage Data System from Thermo Scientific™, Waltham, MA, USA).

2.3. Electrochemical Measurements

2.3.1. Electrocatalytic Activity in Half-Cell Configuration

The electrochemical evaluation of the Pd aerogels was carried out in a Biologic VMP3 Potentiostat/Galvanostat using a conventional three-electrode electrochemical cell in acid media at a scan rate of 20 mV s^{-1}. A glass-carbon electrode (3 mm) was used as the working electrode, Hg/Hg$_2$SO$_4$ electrode as the reference electrode, and Pt wire as the counter electrode. The electrocatalyst ink was prepared using each aerogel sample in a mixture of 500 µL of deionized water and 50 µL of Nafion® (5%) per milligram of catalyst. The ink was sonicated for one hour and then 5 µL were deposited over the electrode surface. A similar ink was prepared using commercial Pd/C (20%, Sigma Aldrich, St. Louis, MI, USA) as catalyst and it was used for comparison. The electrolyte was bubbled with N$_2$ for 30 min before the electrochemical measurement.

The electrochemical profile for each sample was obtained in cyclic voltammetry (CV) experiments in 0.5 M H$_2$SO$_4$ within a potential range of 0–1.4 V vs. RHE, where the faradaic processes were visible in a current (i.e., J) that was tested by mg of catalyst.

2.3.2. FAO Performance

The electrocatalytic activity of the Pd aerogels towards FAO were tested by cyclic voltammetry (CV) in a 0.5 M of HCOOH in 0.5 M H$_2$SO$_4$ electrolyte. As for the electrocatalytic activity in the half-cell configuration, at potential range between 0 and 1.4 V vs. RHE was explored. The results were also compared with the FAO electrocatalytic activity of a commercial reference (Pd/C 20%, Sigma Aldrich).

2.3.3. Stability Performance

The stability performance was carried out by a chronoamperometry (CA) technique at 0.3 V vs. RHE for 24 h at nitrogen atmosphere.

2.3.4. Evaluation of the Microfluidic Fuel Cell System

The description of the MFC used for these experiments has been previously reported [27]. Both the anode and cathode were Pd aerogel samples deposited on Toray carbon paper-060 (TCP) with a transversal area of 0.02 cm^2. The electrocatalyst loading was 0.1 mg for both electrodes. Linear sweep voltammetry (LSV) was performed by injecting 0.5 M HCOOH with H$_2$SO$_4$ as the electrolyte in the anode with the evaluated catalyst; and 0.5 M H$_2$SO$_4$ in the cathode with commercial Pt/C as previous studies reported its best performance [19]. A flow rate of 200 µL min^{-1} was used in the test, i.e., Figure 2.

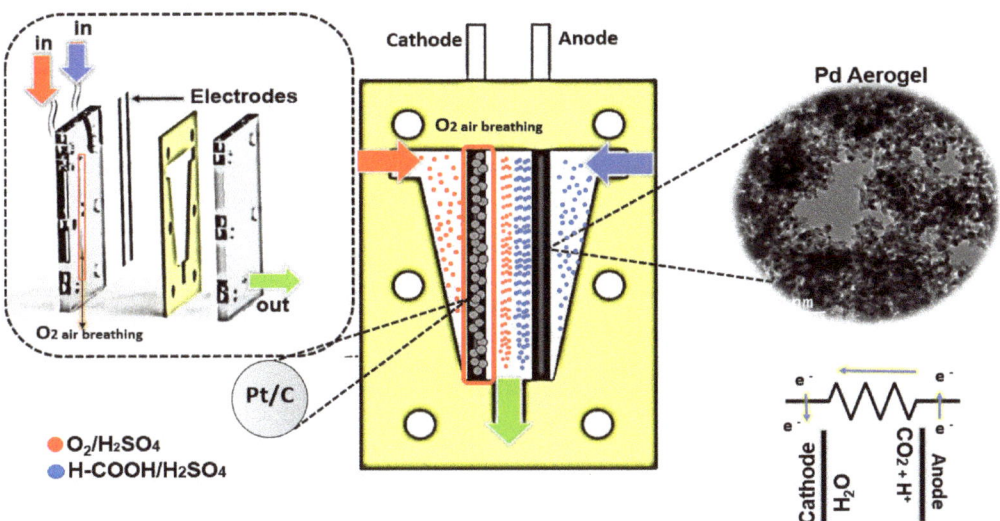

Figure 2. Microfluidic system for formic acid electro-oxidation.

3. Results and Discussion

3.1. Physicochemical Characterization

The route selected in this paper, i.e., in situ reduction and subsequent fusion, is one of the two common strategies found in the literature for the preparation of Pd aerogels [28–31]. This route is essentially a one-pot synthesis that significantly shortens and simplifies the metal nanoparticles aerogels. The aerogels are thus synthetized by mixing noble metal salts ($PdCl_2$ in our case) with a strong reducing agent, typically $NaBH_4$, $LiAlH_4$, hydrazine, sodium citrate, tannic acid or, in our case, glyoxylic acid monohydrate (combined with a base). The two reactants lead to the Pd^{2+} to Pd^0 reduction at moderate temperatures, typically around 60 °C. Under these conditions, the metal nanoparticles in the sol state tend to aggregate until the stability of the solution is compromised and turned into a gel state (Figure 1). The hydrogel obtained after the Pd aggregation is finally washed several times using deionized water by carefully exchanging the supernatant to ensure the integrity of the formed hydrogel but minimizing the presence of impurities (remnant salts and organic matter) and dried either using supercritical CO_2 extraction or, in our case, lyophilization.

Glyoxylic acid monohydrate (combined with a base) is a popular reducing agent in the electroless copper plating [32], and it has been used before in the synthesis of metal aerogels [31]. Although the mechanism of the Pd reduction is not fully understood, it is plausible that the contribution of the glyoxylic acid is twofold according to the following equations. First, the disproportionation of the glyoxylic acid in a basic medium occurs:

$$2OCHCOOH + H_2O \rightarrow HOCH_2COOH + HOOC\text{-}COOH \qquad (3)$$

The oxalic acid would then react with the $PdCl_2$:

$$HOOC\text{-}COOH + PdCl_2 + Na_2CO_3 \rightarrow Pd(COO)_2 + 2NaCl + H_2CO_3 \qquad (4)$$

Finally, the palladium oxalate would be reduced by the glyoxylic acid [32]:

$$Pd(COO)_2 + OCHCOOH + 2OH^- \rightarrow Pd + 2C_2O_4^{2-} + H_2O + 2H^+ \qquad (5)$$

In this work, the yield of the Pd aerogel synthesis was 67%.

The XPS analysis was performed to evaluate the composition and electronic structure of the four Pd aerogels obtained. Figure 3 displays their core-level binding energy Pd 3d5/2 (336.8–337.0 eV) and Pd 3d3/2 (340.0–341.0 eV) XP spectra [21,33]. For all samples, each peak can be deconvoluted into two contributions: 335.0 and 340.2 eV for metallic Pd; and 337.0 and 342.4 eV for Pd^{2+} [34]. XPS data revealed that Pd^0 is the main species on the Pd aerogels surface for PdA-MMT. The high-resolution Pd XPS profiles of samples PdA-CC and PdA-MM are very similar pointing out that the microwave heating produces an analogous reduction process but with a remarkable saving of time (i.e., 7 h vs. 24 h, see Table 1). In addition, it can be also observed that the Pd metallic phase on PdA-MMT aerogel is higher than that on the other samples. This analysis shows that the ratio between Pd^0 and Pd^{2+} favors Pd^0 when the aerogels were prepared by microwave heating, and this would probably be one of the main reasons for high stability and performance of that catalyst, as it will be shown below. Therefore, increasing the temperature from 45 °C to 67 °C during the reduction stage would be preferred because it brings about a greater quantity of metallic Pd^0.

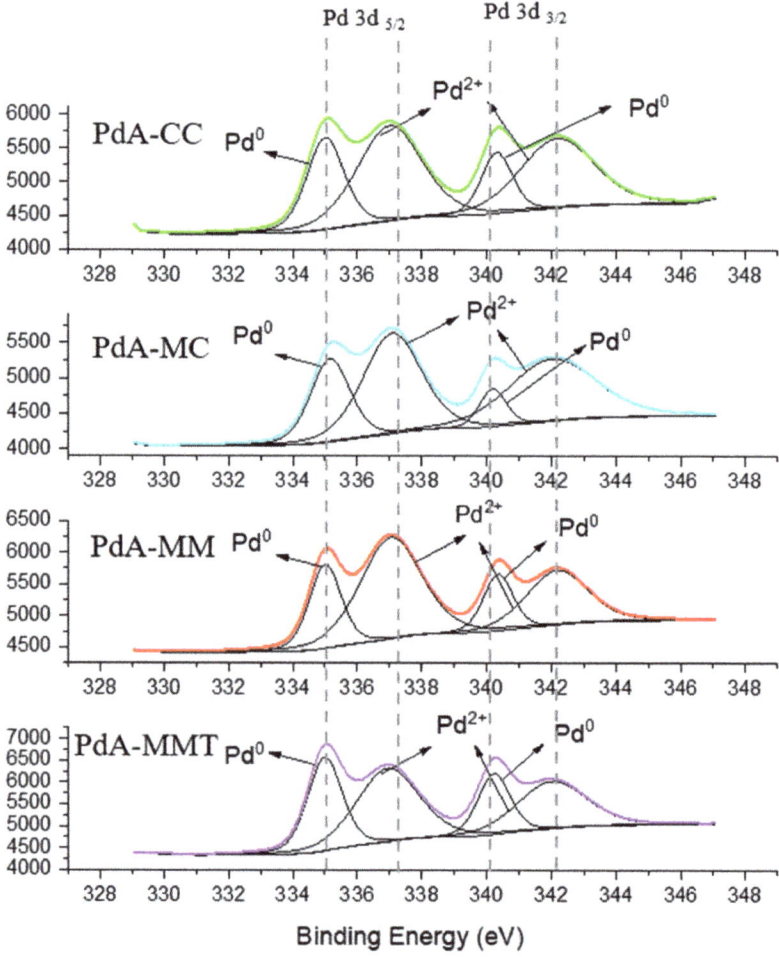

Figure 3. XPS of the Pd aerogels obtained.

Powder X-ray diffraction (XRD) measurements were performed to evaluate the crystallinity of Pd aerogels samples. Figure 4 illustrates the XRD patterns of the synthesized catalysts. In all XRD patterns, two major diffraction peaks appear at about 40.1° 2θ and 46.6° 2θ, which are ascribed to the (111) and (200) reflection planes of metallic Pd, respectively [35]. These peaks agree with a face-centered cubic crystal structure of Pd (JCPDS# 46-1043) [36]. In addition, the crystallite size was calculated using the Scherrer equation:

$$d_{111} = \frac{K\lambda}{\beta_{111} \cos \theta}$$

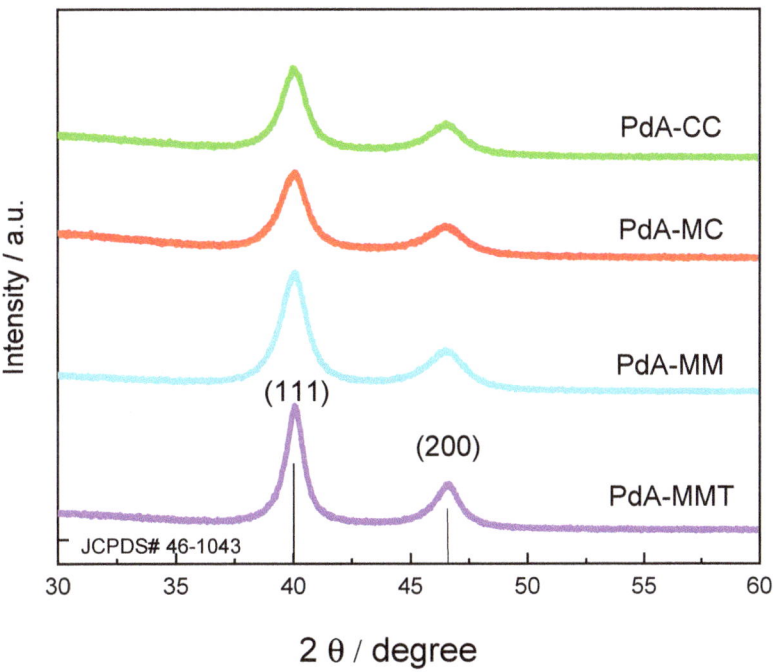

Figure 4. XRD of Pd aerogels samples.

The crystallite sizes for the PdA-CC, PdA-MC, PdA-MM, and PdA-MMT samples were 10.3, 7.2, 6.9, and 7.6 nm, respectively.

The use of microwaves reduces the size of the Pd crystallites, the minimum size being obtained when microwave was used as the heating method in the two stages of the synthesis. The size of the crystal is significantly smaller when using microwaves as a heating method because this heating process is volumetric and heat gradients are minimized. Thus, the reaction occurs uniformly in the precursor mixture. This means that under microwave heating, there are multiple crystallization spots in the precursor solution, whereas in conventional heating, the temperature gradient produces less crystallization spots that grow to form larger particles.

The morphologies of the Pd aerogels were characterized by SEM and the images are presented in Figure 5c–f.

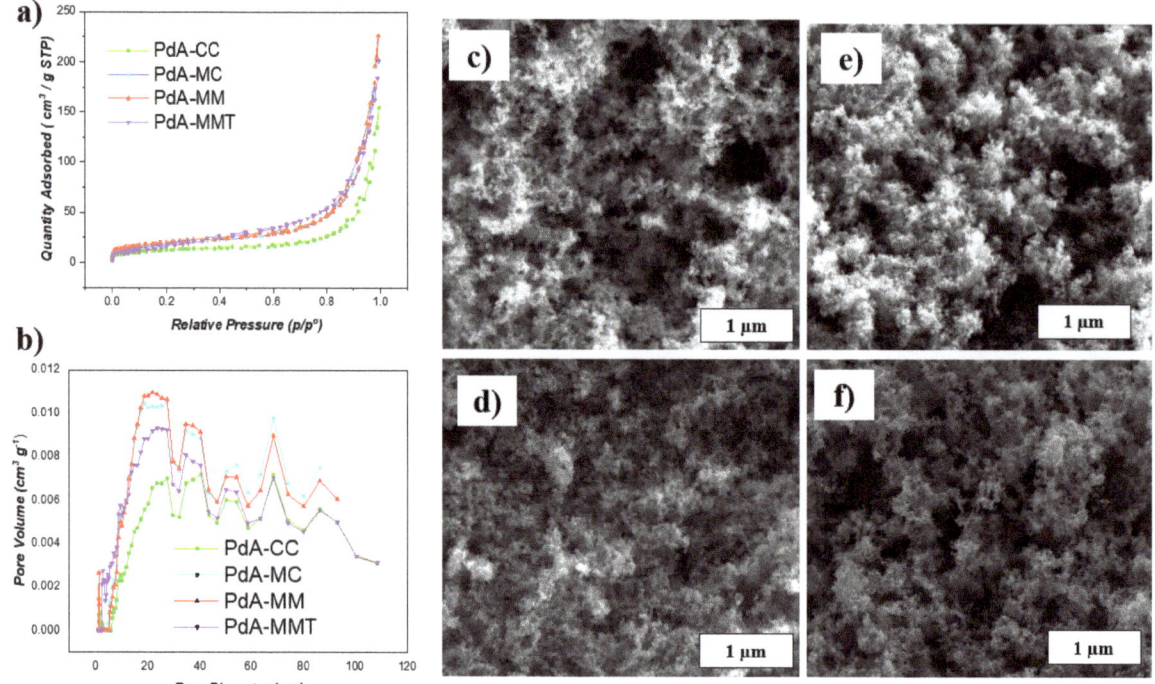

Figure 5. Physicochemical analysis. BET analysis of Pd aerogels: (**a**) N$_2$ adsorption isotherms; and (**b**) pore size distribution. SEM of Pd aerogels: (**c**) PdA-CC; (**d**) PdA-MC; (**e**) PdA-MM; and (**f**) PdA-MMT.

The aerogels obtained present a three-dimensional porous network anchored with nanochains which are extremely thin and make the material look like a sponge with a wide pore size distribution in the range of mesopores and macropores, i.e., Figure 5a,b.

All samples present a similar morphology, although sample PdA-MM (i.e., Figure 5c) seems to present the most open 3D structure. Slight differences in morphology between the different metallic Pd aerogels are attributed, again, to the heating technology applied in their synthesis. Thus, a finer distribution of particles appears in the material synthesized and reduced by microwave heating. This can be explained in Figure 6f where it is observed that the distribution of heat within the solution, by the radiation of microwaves, allows the reaction to occur homogeneously. Instead, larger particles are formed by conventional heating due to the fact that the heat distribution begins at the edges and causes the reaction to occur in an inhomogeneous way, resulting in larger particles that decrease the BET surface area in the aerogels.

TEM images at different magnifications show chains of particles surrounding pores of different sizes. The spherical nanoparticles in those chain structures present different lengths depending on the treatment performed during the synthesis of the aerogels. Particle diameters of up to 30 nm for the aerogels were synthesized in the conventional oven, i.e., Figure 6c, whereas particles less than 9 nm in diameter are characteristic of samples synthesized in MW (Figure 6d–f). To understand the effect of heating on the particle size, the normal distribution of the widths was analyzed, the trend of the means for each sample analyzed is observed in the histograms, being 15.3 nm for PdA-CC, 6.2 nm for PdA-MC, and 5 nm for PdA-MM and Pd-MMT. Therefore, a smaller particle size is attributed to the effect of microwave heating.

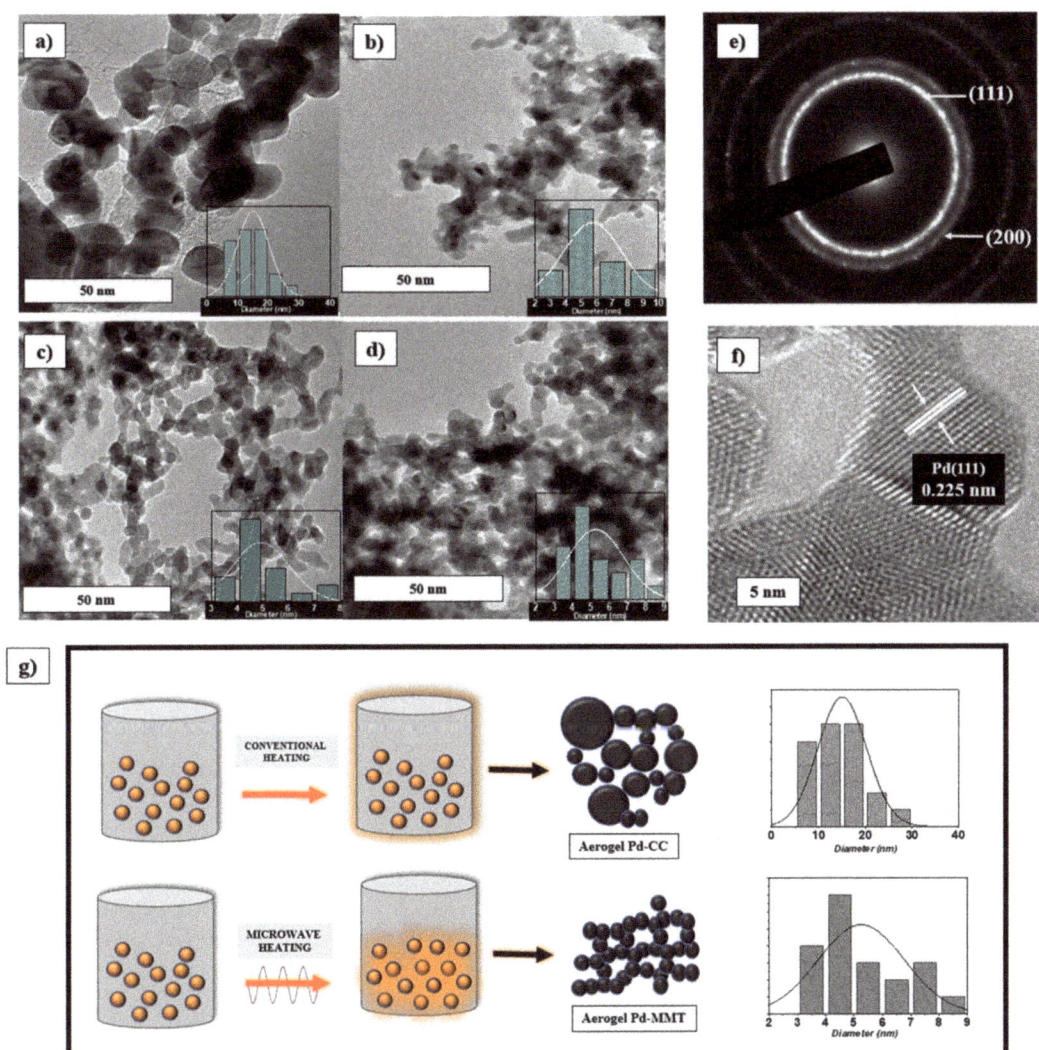

Figure 6. TEM micrographs of Pd aerogels. (**a**) PdA-CC, (**b**) PdA-MC, (**c**) PdA-MM, (**d**) PdA-MMT, (**e**,**f**) crystallographic patterns observed in PdA-MMT aerogel, and (**g**) heating effect on the particle size of Pd-CC and Pd-MMT aerogels.

According to the different structures observed by SEM and TEM, different porosity and therefore availability of reactive surface area of the aerogels studied was expected, which could be relevant for their further use in electrochemistry. Porous properties of samples were investigated by nitrogen adsorption–desorption isotherm at −196 °C (see Figure 5a). The isotherms of the aerogels are of type II according to the IUPAC classification, which are characteristic of meso–macroporous materials [36] according to the low volume of adsorption at low relative pressures, the sharp increase in the adsorption at high relative pressures, and the absence of a hysteresis loop. Furthermore, the pore size distribution reveals a high volume of mesopores (i.e., 2–50 nm) and macropores (i.e., >50 nm) as can be seen in Figure 5b. The surface area of the samples was determined using the BET equation, giving relatively low values due to the lack of microporosity (i.e., pores < 2 nm) in these samples: 45, 65, 75, and 77 m^{2}g^{-1} for PdA-CC, PdA-MC, PdA-MM, and PdA-MMT, respectively. Nevertheless, a trend to increase the BET surface area is observed if microwave heating is used in the different steps during the synthesis.

3.2. Electrocatalytic Performances

The electrocatalytic activity for aerogels samples were evaluated using cyclic voltammetry (CV). First, electrochemical profiles were obtained in a 0.5 M H_2SO_4 aqueous solution at ambient conditions with a sweep rate of 20 mVs^{-1} (Figure 7a). The peaks detected are attributed to (i) the hydrogen desorption in the 0.1–0.25 V range, (ii) hydrogen adsorption at 0.23 V, (iii) reduction of Pd (II) oxide at 0.65–0.75 V, and (iv) formation of Pd (II) oxide at 1–1.2 V. All these phenomena are present in the cyclic voltammograms of all Pd aerogels. However, the use of microwave radiation during any of the synthesis steps clearly improves the electrochemical activity of the materials.

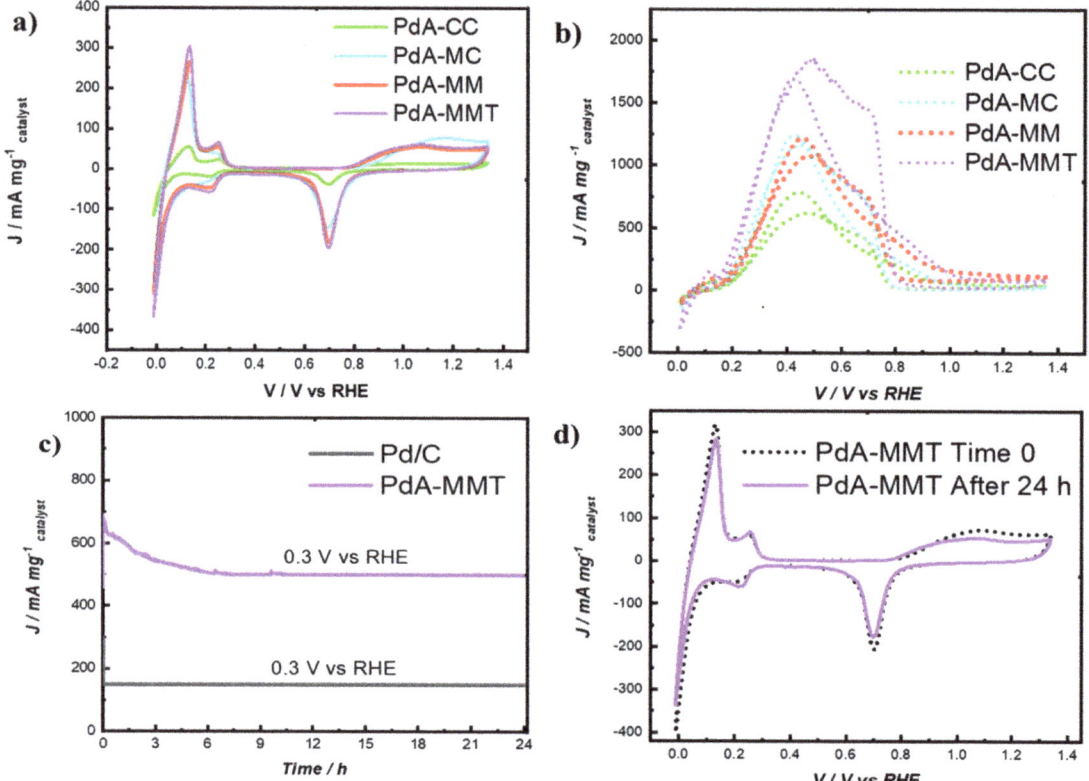

Figure 7. *Cont.*

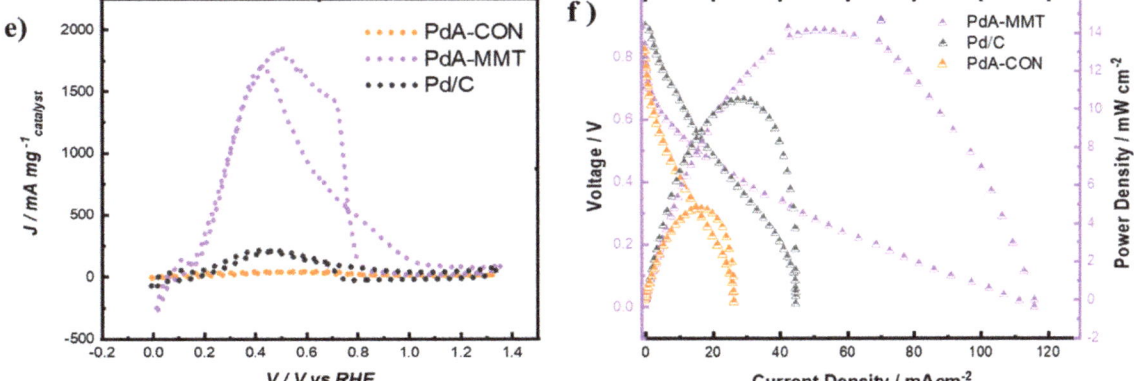

Figure 7. Electrochemical performance. (**a**) Pd aerogels profiles in 0.5 M H_2SO_4; (**b**) comparison between the aerogels in FAO; (**c**) stability performance on Pd-MMT vs. Pd/C; (**d**) PdA-MMT CV before and after the CA for 24 h; (**e**) FAO performance between PdA-CON, PdA-MMT, and Pd/C; and (**f**) MFC performance for the best aerogels obtained (PdA-MMT), a commercial catalyst (Pd/C) and an in-lab catalyst obtained by conventional procedure (PdA-CON).

To quantify this improvement, the electrochemical active surface area (*ECSA*) was evaluated on the electrode surface of each catalyst. The values of *ECSA* for the samples studied in this work were estimated from the cyclic voltammograms (i.e., Figure 7a) by using the reduction charge of Pd (II) oxide according to the following Equation (2):

$$ECSA = \frac{Q_m}{m_{Pd} ed_m}$$

where, Q_m denotes coulombic charge (Q per $\mu C cm^{-2}$) for the reduction of Pd (II) oxide achieved by integrating the charges related to the reduction of Pd (II) oxide for the different samples; m_{Pd} is the mass amount of Pd loaded (g cm^{-2}) on the GC electrode surface and ed_m is a constant (424 $\mu C\ cm^{-2}$), which corresponds to the reduction of a Pd (II) oxide monolayer [37].

The *ECSA* values of the Pd aerogels samples depicted in Figure 7a are shown in Table 2. *ECSA* values for PdA-CON and the commercial Pd/C catalyst are also included for comparative purposes. As expected from the cyclic voltammograms, there is a great increase in the electrochemical active surface area for the samples prepared using MW. The lower particle size detected by TEM, the lower size of the crystals detected by XRD, and the higher content of the Pd^0 evaluated by XPS, in samples obtained using microwave heating show clearly that this process has a huge impact on the electrochemical performance of the resulting Pd aerogels. In other words, by means of microwave heating instead of conventional heating for the synthesis of Pd aerogels, the innovation presented in this work, not only the is the processing time reduced but also the electrochemical behavior of these materials is notably enhanced.

Table 2. Electrochemical active surface area (*ECSA*) values for the samples studied.

Sample	ECSA (m²/g)
PdA-CON	1.3
PdA-CC	22.1
PdA_MC	22.1
PdA-MM	22.8
PdA-MMT	28.5
Pd/C	16

This is further corroborated when performing the CV experiments in an electrolyte containing formic acid (i.e., Figure 7b). The evaluation of FAO was carried out in the same range of potential as the CV tests (0.0–1.4 V vs. RHE, see Figure 7a). Comparison of the electrochemical profiles with FAO curves recorded by GC electrodes clearly demonstrated that Pd aerogels offered strong peaks for the electro-oxidation at room temperature conditions [38]. The maximum current values during FAO occur at 0.4 V vs. RHE. Again, the effect of using microwave heating for the synthesis of the Pd aerogels clearly increases their activity in the electro-oxidation of formic acid. Thus, a maximum mass current (J) of 1750 mA mg^{-1} was for the PdA-MMT sample; in the case of PdA-MM and PdA-MC materials, values of 1200 and 1190 mA mg^{-1} were respectively obtained, being the lowest performance for a PdA-CC sample with 700 mA mg^{-1}. On the other hand, it seems clear that increasing the temperature of the reduction step from 45 °C to 67 °C makes a difference in the formic acid oxidation activity (see Figure 7b).

As for the formic acid electro-oxidation mechanism with these Pd aerogels, the reaction occurs following two parallel paths, one giving rise to CO_2 at reasonably low overpotentials through the so-called active intermediate and a chemical dehydration step leading to adsorbed CO, which will be oxidized to CO_2 at higher potentials [39]. The peaks between 0.2 and 0.6 V for all Pd catalysts (Figure 7d) represent a direct pathway (HCOOH → COOHads/HCOOads + H$^+$ + e$^-$ → CO_2 + 2H$^+$ + 2e$^-$), while the peaks ranging from 0.7 to 0.9 V represent an indirect pathway (HCOOH → COads + H_2O → CO_2 + 2H$^+$ + 2e$^-$) [40]. Since the maximum electrochemical activity of all Pd aerogels studied was measured at ca. 0.4 V, the direct pathway is clearly favored in this case.

The best sample, PdA-MMT, obtained in this work was also compared with the Pd/C commercial catalyst (20 wt%) and PdA-CON in order to show the benefits of using the novel synthesis presented in this work (i.e., microwave heating during the synthesis and lyophilization as the drying procedure). To analyze the catalytic activity of PdA-MMT versus Pd/C, the maximum current intensity shown in Figure 7e should be compared. It can be seen that a current density near to 1900 mA mg^{-1} was observed for the aerogel against almost 300 mA mg^{-1} for Pd/C. Whilst in the case of the behavior of the PdA-MMT vs. the PdA-CON sample, a significant increase in electrochemical activity was detected for the PdA-MMT sample (see Figure 7e).

In order to characterize not only the activity but also the stability of the PdA-MMT sample, a test was carried out for 24 h on this sample and the commercial catalyst Pd/C (Figure 7c,d). The results reveal that the great activity of the sample obtained in this work (PdA-MMT) is totally maintained and stable with time.

3.3. Microfluidic Fuel Cell Performance

A microfluidic fuel cell was used for testing samples to verify their activity under real operating conditions, using 0.5 M formic acid with 0.5 M H_2SO_4 as the electrolyte for the anodic reaction. Three materials have been tested, namely, PdA-MMT, PdA-CON, and the commercial Pd/C catalysts used as reference materials. Linear sweep voltammetry is a specific electrochemical protocol to discriminate the catalytic activity of the anode materials. The microcell works by pumping 200 µL min^{-1} of the electrolyte fuel (0.5 M HCOOH in 0.5 M H_2SO_4) into the anode, and 200 µL min^{-1} of only 0.5 M H_2SO_4 into the Pt/C cathode. Both electrodes were normalized to a mass charge of 0.1 mg of each material over an area of 0.019 cm^2. The polarization curves were obtained for stable open circuit voltage (OCV) and values registered for the PdA-MMT, Pd/C, and PdA-CON samples were 0.88, 0.9, and 0.84 V, respectively. The power density obtained in Figure 7f shows favorable results for the aerogel obtained by MW heating. Small bubbles could be detected on the anode during the experiment, corresponding to the generation of hydrogen. The mass current of the PdA-MMT aerogel is almost three times higher than the commercial Pd/C catalyst (i.e., Figure 7e), corroborating, again, the superior performance of the aerogels obtained by using microwave heating during the synthesis. The superior catalytic activity of PdA-MMT may be attributed to the most active Pd surface. The use of microwave heating during the synthesis of the aerogels leads not only to a better morphology of the aerogels but also to a higher content of metallic Pd^0. Previous studies reveal that Pd^{2+} species are catalytically inactive for FAO [41]. Therefore, the use of microwave heating has a determinant influence on the production of effective Pd aerogel catalysts.

Table 3 shows the performance of other microfluidic devices from the bibliography in comparison with the one used in this work with the Pd catalyst obtained by the innovative method presented. It may be observed that the configuration used in this work using PdA-MMT as anodic material allows to obtain extremely higher yield per mass of catalyst in terms of density power.

Table 3. Anodic catalyst comparison.

Anodic Catalyst	Formic Acid Concentration	Mass Loading/mg cm^{-2}	OCV/V	J/mA cm^{-2}	W Max/mW cm^2	Reference
Pd/C	0.5 M	0.7	0.9	7.4	2.9	[42]
Pd/C	0.5 M	0.1	0.9	43	10.5	This work
Pd/MWCNT	0.5 M	1.3	0.9	9.1	2.3	[43]
Pd50Co50/MWCNT	0.5 M	1.2	0.9	5.9	1.75	[38]
PdA-MMT	0.5 M	0.1	0.88	118.3	14	This work
Pd/graphene	0.5 M	2	0.7	30	15.2	[26]
Pt/CN$_x$	0.5 M	1	1.1	9.79	3.43	[44]
Pt-Ru	3 M	3	0.47	1.2	12.5	[45]
Au-Pt	1 M	-	1.2	28	12	[46]
Pd	0.5 M	10	0.95	125	26	[47]
Pt	0.5 M	-	1.1	8	2.2	[48]

4. Conclusions

The use of microwave heating during the synthesis, followed by lyophilization as the drying step, renders Pd aerogels with a particle size smaller than 5 nm in anchored chains. Furthermore, Pd^0 species are generated in a greater proportion when microwave heating is used during the synthesis. These facts demonstrate that the sample obtained by microwave heating presents a much higher electrochemical active surface area, thus favoring the formic acid electro-oxidation reaction. The PdA-MMT sample showed a three-times superior performance in a microfluidic fuel cell than a commercial Pd/C catalyst, reaching current densities of up to 118.3 mA cm^{-2}.

This result shows that the innovative synthesis route presented in this work, using microwave heating for the synthesis process, leads to the production of very competitive aerogels to be used as electrocatalysts.

Author Contributions: Conceptualization, L.G.A. and J.L.-G.; methodology, A.A. and L.A.R.-M.; formal analysis, A.M.-L.; resources, L.G.A., J.A.M. and A.A.; writing—original draft preparation, A.M.-L. and L.A.R.-M.; writing—review and editing, L.G.A., M.A.M.-M.; J.L.-G., M.P.G. and A.A.; supervision, L.G.A. and A.A. All authors have read and agreed to the published version of the manuscript.

Funding: Mexican Council for Science and Technology (CONACYT) through project Ciencia de Frontera Grant no. 845132 and estancia Sabatica Conacyt 2021-2. Grant PID2020-113001RB-I00 funded by MCIN/AEI/ 10.13039/501100011033 and by the European Union NextGenerationEU/PRTR; Grant PCI2020-112039 funded by MCIN/AEI/ 10.13039/501100011033 and by the European Union NextGenerationEU/PRTR.

Conflicts of Interest: The authors declare no conflict of interest.

References

1. Zhang, B.; Ren, L.; Xu, Z.; Cheng, N.; Lai, W.; Zhang, L.; Hao, W.; Chu, S.; Wang, Y.; Du, Y.; et al. Atomic Structural Evolution of Single-Layer Pt Clusters as Efficient Electrocatalysts. *Small* **2021**, 1–8, 2100732. [CrossRef]
2. Hidayah, N.; Irwan, M. Three-dimensional CFD modeling of a direct formic acid fuel cell. *Int. J. Hydrogen Energy* **2019**, *44*, 30627–30635.

3. Fan, H.; Cheng, M.; Wang, L.; Song, Y.; Cui, Y.; Wang, R. Extraordinary electrocatalytic performance for formic acid oxidation by the synergistic effect of Pt and Au on carbon black. *Nano Energy* **2018**, *48*, 1–9. [CrossRef]
4. Ying, L.; Ouyang, Y.; Wang, S.; Gong, Y.; Jiang, M. Ultrafast synthesis of uniform 4–5 atoms-thin layered tremella-like Pd nanostructure with extremely large electrochemically active surface area for formic acid oxidation. *J. Power Sources* **2020**, *447*, 227248.
5. Choi, M.; Ahn, C.; Lee, H.; Kwan, J.; Oh, S.; Hwang, W.; Yang, S.; Kim, J.; Kim, O.; Choi, I. Bi-modified Pt supported on carbon black as electro-oxidation catalyst for 300 W formic acid fuel cell stack. *Appl. Catal. B Environ.* **2019**, *253*, 187–195. [CrossRef]
6. Yang, Q.; Lin, H.; Wang, X.; Ying, L.; Jing, M.; Yuan, W. Dynamically self-assembled adenine-mediated synthesis of pristine graphene-supported clean Pd nanoparticles with superior electrocatalytic performance toward formic acid oxidation. *J. Colloid Interface Sci.* **2022**, *613*, 515–523. [CrossRef]
7. Wang, H.; Chen, H.; Wang, H.Q.; Ou, C.R.; Li, R.; Liu, H.B. Synthesis of ultrafine low loading Pd–Cu alloy catalysts supported on graphene with excellent electrocatalytic performance for formic acid oxidation. *Int. J. Hydrogen Energy* **2020**, *45*, 10735–10744. [CrossRef]
8. Douk, A.S.; Saravani, H.; Noroozifar, M. A fast method to prepare Pd-Co nanostructures decorated on graphene as excellent electrocatalyst toward formic acid oxidation. *J. Colloid Interface Sci.* **2018**, *739*, 882–891.
9. Hu, S.; Che, F.; Khorasani, B.; Jeon, M.; Won, C. Improving the electrochemical oxidation of formic acid by tuning the electronic properties of Pd-based bimetallic nanoparticles. *Appl. Catal. B Environ.* **2019**, *254*, 685–692. [CrossRef]
10. Kwon, T.; Choi, K.; Han, J. Separation of Ultra-High-Density Cell Suspension via Elasto-Inertial Microfluidics. *Small* **2021**, *17*, 2101880. [CrossRef]
11. Hao, R.; Yu, Z.; Du, J.; Hu, S.; Yuan, C.; Guo, H.; Zhang, Y.; Yang, H. A High-Throughput Nanofluidic Device for Exosome Nanoporation to Develop Cargo Delivery Vehicles. *Small* **2021**, *12*, 2102150. [CrossRef] [PubMed]
12. Moreno-zuria, A.; Ortiz-ortega, E.; Gurrola, M.P. Evolution of microfluidic fuel stack design as an innovative alternative to energy production. *Int. J. Hydrogen Energy* **2017**, *42*, 27929–27939. [CrossRef]
13. Ouyang, T.; Chen, J.; Huang, G.; Lu, J.; Mo, C.; Chen, N. A novel two-phase model for predicting the bubble formation and performance in microfluidic fuel cells. *J. Power Sources* **2020**, *457*, 228018. [CrossRef]
14. Gharib, A.; Arab, A. Electrodeposited Pd, Pdsingle bondCd, and Pdsingle bondBi nanostructures: Preparation, characterization, corrosion behavior, and their electrocatalytic activities for formic acid oxidation. *J. Electroanal. Chem.* **2020**, *866*, 114166. [CrossRef]
15. Bao, Y.; Liu, H.; Liu, Z.; Wang, F.; Feng, L. Pd/FeP catalyst engineering via thermal annealing for improved formic acid electrochemical oxidation. *Appl. Catal. B Environ.* **2020**, *274*, 119106. [CrossRef]
16. Rupa, P.; Harivignesh, R.; Sung, Y.; Kalai, R. Polyol assisted formaldehyde reduction of bi-metallic Pt-Pd supported agro-waste derived carbon spheres as an efficient electrocatalyst for formic acid and ethylene glycol oxidation. *J. Colloid Interface Sci.* **2020**, *561*, 358–371. [CrossRef]
17. Bi, Y.; Liu, M.; Ren, H.; Shi, X.; He, S.; Zhang, Y.; Zhang, L. Investigation on gelation process and microstructure for copper-based aerogel prepared via sol–gel method. *J. Non-Cryst. Solids* **2015**, *425*, 195–198. [CrossRef]
18. Suenaga, S.; Osada, M. Preparation of β-chitin nanofiber aerogels by lyophilization. *Int. J. Biol. Macromol.* **2019**, *126*, 1145–1149. [CrossRef]
19. Herrmann, A.K.; Formanek, P.; Borchardt, L.; Klose, M.; Giebeler, L.; Eckert, J.; Kaskel, S.; Gaponik, N.; Eychmüller, A. Multimetallic Aerogels by Template-Free Self-Assembly of Au, Ag, Pt, and Pd Nanoparticles. *Chem. Mater.* **2014**, *26*, 1074–1083. [CrossRef]
20. Wang, H.; Fang, Q.; Gu, W.; Du, D.; Lin, Y.; Zhu, C. Noble Metal Aerogels. *ACS Appl. Mater. Interfaces* **2020**, *12*, 52234–52250. [CrossRef]
21. Du, R.; Fan, X.; Jin, X.J. Emerging Noble Metal Aerogels: State of the Art and a Look Forward. *Matter* **2019**, *1*, 39–56. [CrossRef]
22. Erdem, A.; Ngwabebhoh, F.A.; Yildiz, U. Novel macroporous cryogels with enhanced adsorption capability for the removal of Cu(II) ions from aqueous phase: Modelling, kinetics and recovery studies. *J. Environ. Chem. Eng.* **2017**, *5*, 1269–1280. [CrossRef]
23. Osaki, T. Factors controlling catalytic CO oxidation activity on Pd/CeO2-ZrO2-Al2O3 cryogels. *Mater. Res. Bull.* **2019**, *118*, 110498. [CrossRef]
24. Yildiz, S.; Aktas, N.; Sahiner, N. Metal nanoparticle-embedded super porous poly(3-sulfopropyl methacrylate) cryogel for H2 production from chemical hydride hydrolysis. *Int. J. Hydrogen Energy* **2014**, *39*, 14690–14900. [CrossRef]
25. Baimenov, A.; Berillo, A.D.; Poulopoulos, G. A review of cryogels synthesis, characterization and applications on the removal of heavy metals from aqueous solutions. *Adv. Coll. Int. Sci.* **2020**, *276*, 102088. [CrossRef]
26. Li, D.; Xu, H.; Zhang, L.; Leung, D.Y.; Vilela, F.; Wang, H. Boosting the performance of formic acid microfluidic fuel cell: Oxygen annealing enhanced Pd@graphene electrocatalyst. *Int. J. Hydrogen Energy* **2016**, *41*, 10249–10254. [CrossRef]
27. Ortiz-Ortega, E.; Goulet, M.A.; Lee, J.W.; Guerra-Balcázar, M.; Arjona, N.; Kjeang, E.; Ledesma-García, J.; Arriaga, L.G. A nanofluidic direct formic acid fuel cell with a combined flow-through and air-breathing electrode for high performance. *Lab Chip* **2014**, *14*, 4596–4598. [CrossRef] [PubMed]
28. Zhu, C.; Du, D.; Eychmüller, A.; Lin, Y. Engineering Ordered and Nonordered Porous Noble Metal Nanostructures: Synthesis, Assembly, and Their Applications in Electrochemistry. *Chem. Rev.* **2015**, *115*, 8896–8943. [CrossRef]
29. Liu, W.; Herrmann, A.K.; Geiger, D.; Borchardt, L.; Simon, F.; Kaskel, S.; Gaponik, N.; Eychmüller, A. High-performance electrocatalysis on palladium aerogels. *Angew. Chem. Int. Ed. Engl.* **2012**, *51*, 5743–5747. [CrossRef]

30. Zhu, C.; Shi, Q.; Fu, S.; Song, J.; Xia, H.; Du, D.; Lin, Y. Efficient Synthesis of MCu (M = Pd, Pt, and Au) Aerogels with Accelerated Gelation Kinetics and their High Electrocatalytic Activity. *Adv. Mater.* **2016**, *28*, 8779–8783. [CrossRef]
31. Douk, A.S.; Saravani, H.; Abad, M.Z.Y.; Noroozifar, M. Three-Dimensional Engineering of Nanoparticles To Fabricate a Pd–Au Aerogel as an Advanced Supportless Electrocatalyst for Low-Temperature Direct Ethanol Fuel Cells. *ACS Appl. Energy Mater.* **2020**, *3*, 7527–7534. [CrossRef]
32. Honma1, H.; Kobayashi, T.J. Electroless Copper Deposition Process Using Glyoxylic Acid as a Reducing Agent. *Electrochem. Soc.* **1994**, *141*, 730. [CrossRef]
33. Maleki, H.; Hüsing, N. Current status, opportunities and challenges in catalytic and photocatalytic applications of aerogels: Environmental protection aspects. *Appl. Catal. B Environ.* **2018**, *221*, 530–555. [CrossRef]
34. Liu, X.; Bu, Y.; Cheng, T.; Gao, W.; Jiang, Q. Flower-like carbon supported Pd–Ni bimetal nanoparticles catalyst for formic acid electrooxidation. *Electrochim. Act.* **2019**, *324*, 134816. [CrossRef]
35. Caglar, A.; Selim, M. Effective carbon nanotube supported metal (M = Au, Ag, Co, Mn, Ni, V, Zn) core Pd shell bimetallic anode catalysts for formic acid fuel cells. *Renew. Energy* **2020**, *150*, 78–90. [CrossRef]
36. Doustkhah, E.; Mohtasham, H.; Farajzadeh, M.; Rostamnia, S.; Wang, Y.; Arandiyan, H. Organosiloxane tunability in mesoporous organosilica and punctuated Pd nanoparticles growth; theory and experiment. *Microporous Mesoporous Mater.* **2020**, *293*, 109832. [CrossRef]
37. Gao, N.; Ma, R.; Wang, X.; Jin, Z. Activating the Pd-Based catalysts via tailoring reaction interface towards formic acid dehydrogenation. *Int. J. Hydrogen Energy* **2020**, *45*, 17575–17582. [CrossRef]
38. Saravani, H.; Farsadrooh, M.; Share, M. Two-dimensional engineering of Pd nanosheets as advanced electrocatalysts toward formic acid oxidation. *Int. J. Hydrogen Energy* **2020**, *45*, 21232–21240. [CrossRef]
39. Herrero, E.; Feliu, J.M. Understanding formic acid oxidation mechanism on platinum single crystal electrodes. *Curr. Opin. Electrochem.* **2018**, *9*, 145–150. [CrossRef]
40. Douk, A.S.; Saravani, H.; Noroozifar, M. Three-dimensional assembly of building blocks for the fabrication of Pd aerogel as a high performance electrocatalyst toward ethanol oxidation. *Electrochim. Act.* **2018**, *275*, 182–191. [CrossRef]
41. Yang, L.; Wang, X.; Liu, D.; Cui, G.; Dou, B.; Wang, J. Efficient anchoring of nanoscale Pd on three-dimensional carbon hybrid as highly active and stable catalyst for electro-oxidation of formic acid. *Appl. Catal. B Environ.* **2020**, *263*, 118304. [CrossRef]
42. Morales-Acosta, D.; G, H.R.; Godinez, L.A.; Arriaga, L.G. Performance increase of microfluidic formic acid fuel cell using Pd/MWCNTs as catalyst. *J. Power Sources* **2010**, *195*, 1862–1865. [CrossRef]
43. Morales-Acosta, D.; Morales-Acosta, M.D.; Godinez, L.A.; Álvarez-Contreras, L.; Duron-Torres, S.M.; Ledesma-García, J.; Arriaga, L.G. PdCo supported on multiwalled carbon nanotubes as an anode catalyst in a microfluidic formic acid fuel cell. *J. Power Sources* **2011**, *196*, 9270–9275. [CrossRef]
44. Jindal, A.; Basu, S.; Chauhan, N.; Ukai, T.; Kumar, D.; Samudhyatha, K.T. Application of electrospun CNx nanofibers as cathode in microfluidic fuel cell. *J. Power Sources* **2017**, *342*, 165–174. [CrossRef]
45. Zhang, H.; Xuan, J.; Leung, D.; Wang, H.; Xu, H.; Zhang, L. Advanced gas-emission anode design for microfluidic fuel cell eliminating bubble accumulation. *J. Micromech. Microeng.* **2017**, *27*, 105016. [CrossRef]
46. Park, H.; Lee, K.; Sung, H.J. Performance of H-shaped membraneless micro fuel cells. *J. Power Sources* **2013**, *226*, 266–271. [CrossRef]
47. Brushett, F.; Jayashree, R.; Zhou, W.; Kenis, P. Investigation of fuel and media flexible laminar flow-based fuel cells. *Electrochim. Act.* **2009**, *54*, 7099–7105. [CrossRef]
48. Choban, E.; Markoski, L.; Wieckowski, A.; Kenis, P. Microfluidic fuel cell based on laminar flow. *J. Power Sources* **2004**, *128*, 54–60. [CrossRef]

Review

Applications of 2D MXenes for Electrochemical Energy Conversion and Storage

Chenchen Ji [1,2,†], Haonan Cui [1,†], Hongyu Mi [1,*] and Shengchun Yang [3,*]

1 School of Chemical Engineering and Technology, Xinjiang University, Urumqi 830046, China; jichenchen@xju.edu.cn (C.J.); chn954512087@stu.xju.edu.cn (H.C.)
2 State Key Laboratory of Fine Chemicals, Dalian University of Technology, Dalian 116024, China
3 MOE Key Laboratory for Non-Equilibrium Synthesis and Modulation of Condensed Matter, Shaanxi Province Key Laboratory of Advanced Functional Materials and Mesoscopic Physics, School of Physics, Xi'an Jiaotong University, Xi'an 710049, China
* Correspondence: mmihongyu@xju.edu.cn (H.M.); ysch1209@mail.xjtu.edu.cn (S.Y.)
† These authors contributed equally.

Citation: Ji, C.; Cui, H.; Mi, H.; Yang, S. Applications of 2D MXenes for Electrochemical Energy Conversion and Storage. *Energies* **2021**, *14*, 8183. https://doi.org/10.3390/en14238183

Academic Editor: Lyes Bennamoun

Received: 12 September 2021
Accepted: 2 December 2021
Published: 6 December 2021

Publisher's Note: MDPI stays neutral with regard to jurisdictional claims in published maps and institutional affiliations.

Copyright: © 2021 by the authors. Licensee MDPI, Basel, Switzerland. This article is an open access article distributed under the terms and conditions of the Creative Commons Attribution (CC BY) license (https://creativecommons.org/licenses/by/4.0/).

Abstract: As newly emerged 2D layered transition metal carbides or carbonitrides, MXenes have attracted growing attention in energy conversion and storage applications due to their exceptional high electronic conductivity, ample functional groups (e.g., -OH, -F, -O), desirable hydrophilicity, and superior dispersibility in aqueous solutions. The significant advantages of MXenes enable them to be intriguing structural units to engineer advanced MXene-based nanocomposites for electrochemical storage devices with remarkable performances. Herein, this review summarizes the current advances of MXene-based materials for energy storage (e.g., supercapacitors, lithium ion batteries, and zinc ion storage devices), in which the fabrication routes and the special functions of MXenes for electrode materials, conductive matrix, surface modification, heteroatom doping, crumpling, and protective layer to prevent dendrite growth are highlighted. Additionally, given that MXene are versatile for self assembling into specific configuration with geometric flexibility, great efforts about methodologies (e.g., vacuum filtration, mask-assisted filtration, screen printing, extrusion printing technique, and directly writing) of patterned MXene-based composite film or MXene-based conductive ink for fabricating more types of energy storage device were also discussed. Finally, the existing challenges and prospects of MXene-based materials and growing trend for further energy storage devices are also presented.

Keywords: MXenes; energy conversion and storage; flexibility device methodologies

1. Introduction

The increasingly prominent climate changes and limited availability of fossil fuel issues have stimulated a tremendous amount of research interest in highly efficient and clean energy storage and conversion devices [1–5]. The development of new classes of advanced two-dimensional (2D) layered materials, including graphene, MoS_2, phosphorene, have promoted tremendous technological progress in those energy resources (e.g., supercapacitor, different-type metal ion batteries (MIBs)), which are attributed to their extraordinary properties [2,6–17]. The enhancement in the performance of these devices by incorporating layered materials indicated that the 2D materials commonly possess the following two unique characteristics:

(i) The large interlayer spaces within the layered structures could provide abundant ion transport pathways, promote the ions' intercalation and diffusion, and limit the volume change during the charge/discharge process [18].

(ii) The layered structures could provide high carrier mobility [19], which ensures high electrical conductivities. Thus, 2D materials represent the most intensively and successfully investigated materials for energy storage devices.

As a novel family of transition metal carbides, carbonitrides, or nitrides, (e.g., Ti_2CT_x, Ti_3CNT_x, V_2CT_x), MXenes have attracted particular research enthusiasm since their discovery in 2011 [20]. The delaminated MXenes have elevated research on the novel 2D materials to a new era in the energy related fields due to their prominent and attractive properties, including adjustable composition, hydrophilic, conductivity, thermal conductivity, tunable band gap, and excellent mechanical strength [21–26]. However, MXenes themselves are easily to spontaneously stack and aggregate into multilayer structures due to strong van der Waals forces between the layers, which lead to the decrease in the interlayer space and sluggish kinetics for redox activities or intercalation/deintercalation behaviors [27]. Thus, the capabilities to enlarge the space between the interlayer and elaborately modulate pathways or active sites for MXenes draw lots of attention. As experienced by other 2D materials, several strategies have been proposed for successfully solving the challenges mentioned above, which includes the surface modification, heteroatom doping, or crumpling [28–33]. Recent investigation also demonstrated that the MXene are versatile for self-assembling into specific configuration with geometric flexibility [34]. In this sense, substantial efforts have also been made in methodologies of patterned MXene-based composite film or MXene-based conductive ink for fabricating more types of energy storage devices, which encompasses vacuum filtration, mask-assisted filtration, screen printing, extrusion printing technique, and directly writing, etc. In general, the materials and electrodes for various energy storage devices need to fulfill the requirements of stable physicochemical properties, superb energy storage performance, and high electron/ion conduction [35]. By now, available and optional strategies for enhancing the ion and electron transfer ability of MXene-based materials and the methodologies to design satisfactory patterns are still limited due to the primary development stage of MXenes [36–38]. It is highly needed to get intensive and systematical understanding about the progress of elaborately constructing advanced MXene-based electrode materials and versatile MXene-based flexible electrodes with favorable configurations. In this regard, the efforts on the synthetic method of MXene, the enhancement in electrochemical performance of MXene materials, and the research strategies and technologies to fabricate flexible electrodes for energy storage are briefly summarized. Meanwhile, the challenges and perspectives of MXene-based materials and technologies for the future energy storage applications are presented.

2. Synthetic Methods

The MXene flakes are commonly synthesized by using hydrofluoric acid or a mixed solution of lithium fluoride and hydrochloric acid to selectively remove the "A" layer from the ternary MAX phase, where the M represents early transition metals (e.g., Ti, Mo, V), the A represents IIIA or IVA group elements (e.g., Al, Ga, Si, Ge), and the X represents C and/or N. The MXenes usually possess accordion-like hexagonal lattices which result from the original metallic backbone, weak M-A bond, and strong M-X bond [39]. Meanwhile, some surface terminal groups (e.g., -OH, -O, -F) are formed during the bond breaking and binding processes, which provides more active sites as well as suitable interlayer spacing (Figure 1). Till now, the routinely used method for synthesizing the MXene and its derivatives is hydrofluoric acid (HF) etching. Early works revealed that HF is the most effective etchant to selectively react with the "A" layer atoms and continual out-diffusion to exfoliate the MAX phase (such as Ti_3AlC_2). Those works also proved that the MAX phases are inert in the conventional acids (e.g., H_2SO_4, HCl, and HNO_3), alkaline liquors (e.g., NaOH), and salt solutions (e.g., NaCl and Na_2SO_4). Nevertheless, HF reagent is a hazardous poison with a highly corrosive property to the human body. The direct use of HF raises the potential for causing considerable safety and environmental problems [40]. Thus, many mild etchants have been developed (e.g., NH_4HF_2, or a mixture of LiF/HCl). It was noted that the diverse routes will bring different surface functional groups such as fluoride (-F), hydroxyl (-OH), and/or oxygen (-O) on MXenes and give rise to different levels of delamination, and hence, the preparation process will directly determine the properties of the final MXenes, e.g., metallic, semi-metallic, and semiconducting types. To understand the formation fundamentals from the MAX phase to the MXene sheets, three basic approaches will be discussed in the sections (e.g., wet-chemical etching in hydrofluoric acid, in situ HF-forming method, and freeze-and-thaw-assisted (FAT) method).

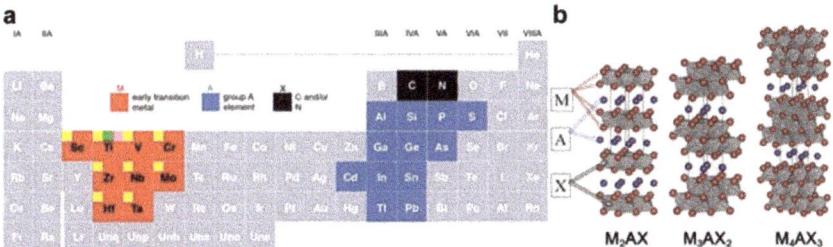

Figure 1. (a) The elements of the periodic table for MAX phases. (b) Structures of M_2AX, M_3AX_2, and M_4AX_3 phases [41].

2.1. Wet-Chemical Etching

The first MXene of Ti_3C_2 was obtained from the MAX phase of Ti_3AlC_2 in the HF solution after successful and selective removal of Al atoms. Subsequently, various kinds of MXene members have been developed by using the wet-chemical etching in hydrofluoric acid. Interestingly, after the etching, the deficiencies of the Al atoms change the structure from undamaged cube to accordion-like cube due to the stable bond between M-X and the weak bonding to the A-M layers. In addition, the as-formed Al deficiencies able to expose more Ti atoms to the atmosphere and further facilitate the exposed terminations to combine with other functional groups. Hence, the extraction processes help to vary the surface chemistries and change their band gap, which results in a higher surface chemical activity of MXene in contrast to MAX phase. The reaction formula for the selective etching process can be ascribed as following [20]:

$$Ti_3AlC_2\ (s) + HF(aq) = AlF_3(aq) + 3/2H_2(g) + Ti_3C_2\ (s) \tag{1}$$

$$Ti_3C_2(s) + 2H_2O(aq) = Ti_3C_2(OH)_2(s) + H_2(g) \tag{2}$$

$$Ti_3C_2(s) + 2HF(aq) = Ti_3C_2F_2(s) + H_2(g) \tag{3}$$

where Equation (1) represents the selective reaction of A layers with HF etchant while Equations (2) and (3) always react together and represent the growth mechanism of the function groups on the exposed Ti atoms [42].

2.2. In Situ HF-Forming Method

The in situ HF-forming strategy commonly employs fluoride-containing acidic solutions (e.g., lithium fluoride (LiF), ammonium hydrogen fluoride (NH_4HF_2), sodium bifluoride ($NaHF_2$), and potassium hydrogen fluoride (KHF_2)) [43–50] to indirectly produce HF etchants for stripping Al atoms, the reaction for the etching process can be described as the following:

$$LiF(aq) + HCl(aq) = HF(aq) + LiCl(aq) \tag{4}$$

The synthetic reaction usually occurs at 40 °C and the obtained samples need to be rinsed to remove the by-products, such as AlF_3, LiCl. However, the lower concentration of hydrogen fluoride and fluoride salts compared with HF etchants and long etching time lead to poor etching effect and the accordion-like morphology is hard to obtain in the reaction system.

2.3. Other Synthesis Method

Given that most of the commonly used synthetic methods have low exfoliation yields, some alternative and effective strategies were developed for boosting the yield of MXenes. Recently, Wu and co-workers reported a gentle water FAT method to prepare the MXene nanosheets by inserting water molecules into the multilayer of MXene and using the volume expansion process to raise the exfoliation efficiency. The yield of these FAT-MXene can reach to 39% and can further increase to 81.4% by the sonication treatment [51]. In addition, due to the water-freezing expansion force, the method can prevent the restacking and expand the space between the multilayer during the reaction process. The FAT-MXene also possesses larger flake size compared with MXenes obtained from other methods (Figure 2). Therefore, films or current-collector-free electrodes assembled from the FAT-MXene exhibit a high level of layer alignment and a low flake-to-flake contact resistance, which enables the FAT-MXene-based on-chip micro-supercapacitor to display a high areal capacitance of 23.6 mF cm^{-2} and a high volumetric capacitance of 591 F cm^{-3}.

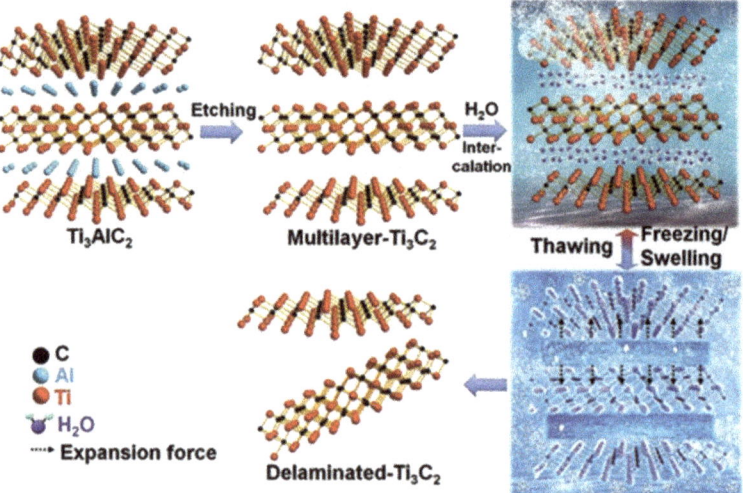

Figure 2. The synthesis process of FAT assisted method [51].

3. Electrochemical Energy Storage Applications

The electrochemical energy storage systems implement energy conversion through electrophysical process (e.g., electrosorption of ions) or electrochemical redox reactions coupled with ions and electrons migration within electrolytes and electrodes. In these systems, 2D materials, such as graphene, black phosphorene, and MoS_2 attract extensive concern due to their favorable morphological and electrical properties [52,53]. Additionally, recent advances in MXenes have enhanced the performance for storage devices (e.g., supercapacitors and metal-ion batteries) and exhibit extraordinary physical and chemical properties, such as mechanical flexibility, high tap density (4.0 g cm^{-3}), high electrical conductance (3.1 × 10^3 S m^{-1} for MXene and 2.0 × 10^3 S m^{-1} for graphene) [42,54,55], outstanding volumetric capacity (up to 1500 F cm^{-3}), and favorable hydrophilicity [56,57].

3.1. Supercapacitors (SCs)

Supercapacitors (SCs) are kinds of energy storage devices that contain negative electrodes, positive electrodes, separators, and electrolytes [58]. Normally, SCs possess significant advantages of fast charge/discharge rates, high power density, and long cycle stability due to their special energy storage mechanism. In general, SCs can be classified as electrical double-layer capacitance (EDLC), pseudocapacitors, and hybrid supercapacitors [59]. The EDLC mechanism is an electro-physical storage process that involves the charge accumulation and reversible adsorption/desorption of electrolyte ions on the interface between electrodes and electrolytes [60]. The commonly used electrode materials for EDLC are mainly porous carbonaceous materials. Pseudocapacitors, also called redox capacitors, store charges via the electrochemical processes of surface redox reactions with electron gain/loss or pseudocapacitive intercalation [61]. Hybrid supercapacitors (HSC) integrate EDLC-type and redox-type electrodes in the device and display an electrophysical and electrochemical mixed mechanism [62]. The schematic illustrations of different SCs are shown in Figure 3.

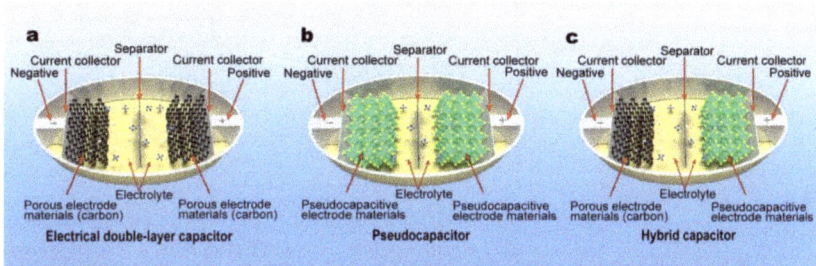

Figure 3. Schematic illustrations of (**a**) a EDLC with porous carbon materials for each electrode, (**b**) a pseudocapacitor with transition metal-related material for each electrode, and (**c**) asymmetric hybrid capacitors utilizing a porous carbon negative electrode and a transition metal oxide positive electrode.

The charge storage mechanism for MXene (e.g., $Ti_3C_2T_x$) is commonly identified as a Faraday redox processes, which involves transitions between various oxidation states of titanium during the intercalation/deintercalation processes of ions in the electrolytes [63]. Yury Gogotsi and co-workers demonstrated that the intercalation/deintercalation of cations (K^+ and NH^{4+}) from alkaline electrolyte into Ti_3C_2 MXene layers in a range of −0.6–0 V are responsible for the electrochemical performance and capacitances of these materials, and the corresponding electrochemical reaction can be ascribed as follows [55,64]:

$$Ti_3C_2T_x + \delta K^+ + \delta e^- \rightleftharpoons K_\delta Ti_3C_2T_x \quad (5)$$

The Faraday processes in a potential window of 0–0.6 V for the MXenes (e.g., Ti_3C_2) as the positive SC electrodes in alkaline electrolyte may involve the following reactions [65].

$$Ti_3C_2O_x(OH)_yF_z + \delta OH^- \rightleftharpoons Ti_3C_2O_x + \delta(OH)_y^- \delta F_z + \delta H_2O + \delta e^- \quad (6)$$

$$Ti_3C_2O_x + \delta OH^- \rightleftharpoons Ti_3C_2O_x(OH)_\delta + \delta e^- \quad (7)$$

MXene-based composites. In order to solve the restacking problem mentioned above and further enhance the charge storage performance of MXenes, Sun and co-workers adopted the Ti_3C_2 MXene and 1T-MoS_2 to construct a 3D interpenetrated architecture of 2D/2D 1T-MoS_2/Ti_3C_2 heterostructure by the magneto-hydrothermal method. Enhanced electrochemical performances are observed from this material, which result from the synergistically interplayed effect of the unique 3D interpenetrated heterogeneous networks (e.g., enlarged ion storage space between Ti_3C_2 and 1T-MoS_2 for storing more electrolyte ions, boosted electron conductivity provided by MXene, and more active sites for redox reaction offered from 1T-MoS_2 phase). As a result, the heterogeneous material exhibits a specific capacitance of 386.7 F g^{-1} at 1 A g^{-1} in a potential window of −0.3–0.2 V with outstanding rate performances (a high capacitance of 207.3 F g^{-1} at 50 A g^{-1}). They also reported that the total capacitance of this material comes from three parts, containing contributions originating from 1T-MoS_2, MXene, and extra H^+ storage in the interpenetrated space within the 3D heterostructure (Figure 4), while the 1T-MoS_2/MXene sample without heterostructure lacks the extra H^+ storage capacitance, indicating the existence of a strong coupled effect between composition and structure for the 3D heterostructure. Furthermore, a symmetric SC based on the as-design material displayed a high areal capacitance of 347 mF cm^{-2} at 2 mA cm^{-2} with excellent cycle stability (91.1% capacitance retention after 20,000 cycles) (Figure 4) [66].

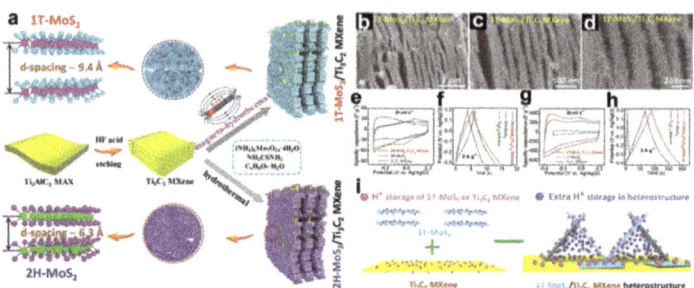

Figure 4. (**a**) Schematic illustrating the preparation process for the two types of the MoS$_2$/Ti$_3$C$_2$ MXene 3D heterostructures. (**b–d**) SEM images of the prepared 1T-MoS$_2$/Ti$_3$C$_2$ MXene sample. (**c**) Electrochemical performance of the prepared heterogeneous electrodes. (**e–h**) Electrochemical performance of the prepared samples. (**i**) Schematic diagram of the storage mechanism for the heterostructure [66].

A Ti$_3$C$_2$/CuS composite as the positive electrode in alkaline electrolyte (within a potential window of 0–0.6 V) delivers a capacity of 169.5 C g^{-1} at 1 A g^{-1} with a capacity retention of 90.5% at 5 A g^{-1} after 5000 long-term cycles. An asymmetric SC fabricated by the Ti$_3$C$_2$/CuS positive electrode and Ti$_3$C$_2$ negative electrode at a voltage range of 0–1.1 V reveals a specific capacitance of 49.3 F g^{-1} to 0–1.5 V with a maximum energy density of 15.4 Wh kg^{-1} and a capacitance retention of 82.4% after 5000 cycles [65]. A MXene/nickel-aluminum layered double hydroxide (MXene/LDH) composite in 6 M KOH aqueous electrolyte within a range of 0–0.6 V displays a specific capacitance of 1,061 F g^{-1} at 1 A g^{-1} with a retention of 70% after 4000 charge/discharge cycles [67]. A nickel sulfide/Ti$_3$C$_2$ (Ni–S/Ti$_3$C$_2$) nanohybrid in 6.0 M KOH aqueous electrolyte within a window of 0–0.6 V exhibits a capacity of 840.4 C g^{-1} with enhanced rate performance (a capacity retention of 64.3% at 30 A g^{-1}) and a long cycle life. An asymmetric SC assembled by the Ni–S/Ti$_3$C$_2$ positive electrode and Ti$_3$C$_2$ negative electrode delivers an energy density of 20.0 Wh kg^{-1} at a voltage range of 0–1.8 V and a good cycling stability (a capacity retention of 71.4% after 10,000 cycles) [68]. Doping heteroatoms (e.g., nitrogen) into the MXene structures could modify their electronic structure, composition, and pseudocapacitance properties. In addition, heteroatom-doping can also lead to a remarkable increase of the interlayer spacing between MXene flakes. As Dai and co-workers found, the c-lattice parameter of the MXene layers could increase from 1.92 to 2.46 nm after the nitrogen-doping process, indicating the doped N atoms expand the interlayer spacing of the MXene sheets. Thereby, the resultant doped materials delivered much higher capacitances of 192 F g^{-1} in 1 M H$_2$SO$_4$ within a potential range of −0.2–0.35 V than those un-doped Ti$_3$C$_2$T$_x$ materials [69].

Previous theoretical and experimental studies reveal that the chemical and physical characters of MXenes are heavily influenced by their surface functional groups. Yury Gogotsi et al. found that the MXene (e.g., Ti$_3$C$_2$T$_x$) with rich surface O-termination exhibited high surface activity and better pseudocapacitance performance [63,70]. Fan et al. enhance the electrochemical performance of MXene by using the ammonium persulfate as the weak oxidant and intercalation agent to partially remove the F terminations coupled with controllably oxidizing the Ti$_3$C$_2$T$_x$ surface to produce rich O-terminations. Recently, electrochemical test results have shown that the modified Ti$_3$C$_2$T$_x$ possess an enhanced capacitance of 303 F g^{-1} and the capacitance retention could be 96.6% after 9000 cycles [71]. MXene can also form a composite with the N, O co-doped carbon materials to expand the interlayer spacing and avoid the re-stacking problems of the MXene sheets. Due to these reasons, a N, O co-doped carbon@MXene composite exhibited a higher specific capacitance of 250.6 F g^{-1} at 1 A g^{-1} compared with pristine Ti$_3$C$_2$ and retained 94% of its initial capacitance after 5000 cycles. Furthermore, a symmetric SC based on this composite electrode displayed an excellent electrochemical performance with an energy density of 10.8 Wh kg^{-1} at 600 W kg^{-1} and desirable cycling stability [28]. Zhenxing Li's group synthesized a dodecaborate/MXene highly conductive composite by surface modified with ammonium ion and inserted the dodecaborate ion into the inner surface of MXene via the electrostatic adsorption. The delocalized negative charge of dodecaborate ion facilitates the cations transfer, which efficiently boosted the ion transfer rate during the charge storage process due to its "lubricant" effect for electrolyte ions. Due to the above reasons, a high specific capacitance of 366 F·g^{-1} can be obtained at 2 mV·s^{-1} [72].

Among all other pseudocapacitive materials, MXenes offer ultrahigh rate capability, excellent conductivity, and volumetric capacitance in SC applications. Actually, MXenes and most MXenes-derived composites commonly are only served as negative electrode materials as their easy oxidation feature at positive potential (anodic oxidation process). To further enhance its antioxidant performance at positive potential, some works tried to improve the work function (WF) of the final product by means of compounding. For instance, Gogotsi and co-workers reported a PANI@MXene cathode material by evenly covering an oxidation resistive PANI layer on the surface of MXene network. The first principle calculation results manifested that the PANI with a large WF (WF = 3.46 eV, the WF for metallic MXene is 1.61 eV) effectively expanded the WF of the composite to 1.97 eV by forming a compact PANI@MXene heterostructure, which helps to enhance the electrochemical stability at a wider positive operating potential (0–0.6 V). Therefore, the as-formed PANI@MXene composite can be used as a positive electrode for steady energy storage with a high volumetric capacitance of 1632 F cm^{-3} and a desirable rate capability. They also claimed that the asymmetric SC assembled by MXene anode and PANI@MXene cathode delivered a volumetric energy density of 50.6 Wh L^{-1} with an ultrahigh volumetric power density of 127 kW L^{-1}. This work also reasonably suggested that other materials (e.g., inorganic compounds or macromolecule materials) with higher WF may enhance the resistance of losing electrons at positive potentials and enlarge the operation working windows for MXenes via forming MXene-based composites [73].

Flexible MXene thin films or free-standing electrodes. It is well known that MXenes are simultaneously capable of conducting electrons and storing charges. In addition, they also possess the compatible structural properties of flexibility, self-assembly, and miniaturization. Meanwhile, the functionality of the MXene macroscopic films or assemblies could be further improved through further optimizing the MXene synthesis conditions, tuning the MXene interlayer spacing, altering the types or amount of surface terminations, controlling the size and thickness of MXene flakes, and compositing with other functional materials (e.g., metal organic framework), and therefore, it is reasonable to suppose that the Ti$_3$C$_2$T$_x$ can be used as free-standing electrodes for SC devices with more functionality. Nicolosi and co-workers adopted a spin-casting method to produce highly aligned and transparent continuous Ti$_3$C$_2$T$_x$ film, which displays remarkable optoelectronic performances, e.g., bulk-like conductivity, high transmittance, and desirable pseudocapacitance performances. Owing to the structural and electronic properties, the as-designed Ti$_3$C$_2$T$_x$ electrodes could function as both active materials and transparent current collector for SCs. Meanwhile, they also found that the as-designed energy devices exhibit high areal/volumetric capacitances with superior energy/power densities and long cycle life [74], demonstrating the application potential for optoelectronics and flexible electronics of MXene materials.

The existence of surface terminations on the MXene sheets are beneficial to form strong bonds between MXenes and specific materials, which help to boost the electrical conductivity and structural/electrochemical stabilities of the active components and the final films. Huang and co-workers employed a vacuum-assisted filtration method to construct the MXene/metal-porphyrin frameworks (MPFs) hybrid flexible free-standing film via forming the interlayer hydrogen bond between the electronegative surface (results from the functional groups of -O, -F, or -OH) of MXene and hydrogen atom of carboxy terminations (-COOH) in MPFs. The composite electrode with interconnected conductive networks and interlayer hydrogen bonds (e.g., F···H-O and O···H-O) eliminated the inferior conductivity and low structural stability problems of MPFs, showing favorable flexibility, high ionic/electronic transport rates, and durability. Meanwhile, the as-designed MXene/MPFs film can normally operate in a potential window of −0.3–0.3 V (vs. Ag/AgCl) and exhibits a specific capacitance of 326.1 F g^{-1} and excellent durability. In addition, a flexible symmetric SC fabricated by the MXene/MPFs film delivers an areal capacitance of 408 mF cm^{-2} and an areal energy density of 20.4 µWh cm^{-2} with a long term stability of 7000 cycles (with a capacitance retention of 95.9%). This work also indicated that the formation of interlayer hydrogen bond between MXene and other components contributes to enhance the chemical stability by effectively avoiding the phase separation or structural collapse problems of the electrode materials during the energy storage processes (Figure 5) [75]. Wu et al. used the Buchwald-hartwig coupling reaction to prepare a decentralized conjugated polymer (PDT)/Ti$_3$C$_2$T$_x$ hybrid flexible freestanding film via an electrophoretic deposition and the following spin coating processes. Similarly, they found that a stable chemical bond (e.g., hydrogen bonds) can be formed between the terminal groups of the Ti$_3$C$_2$T$_x$ sheets and the decentralized chains of PDTs, which enables the final polymer matrix composite film to effectively relieve the volumetric swelling and shrinking problems of polymer chains during the charge/discharge processes [76].

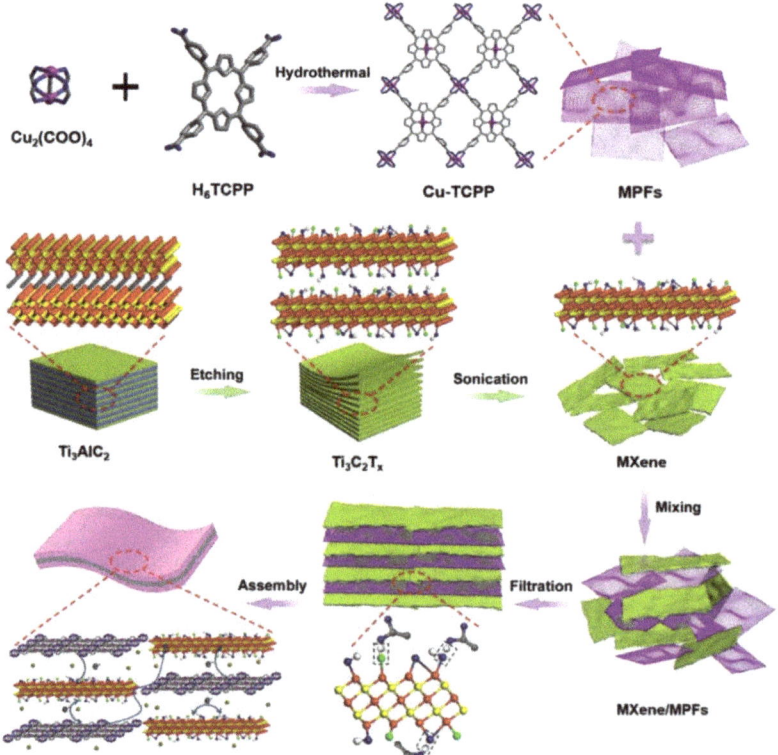

Figure 5. Schematic illustration of the synthesis process for the MXene/MPFs film via interlayer hydrogen bond [75].

The delaminated MXene flakes with monatomic or few-atom-layer thicknesses have shown their capabilities to assemble free-standing and flexible films with outstanding electronic conductivity and excellent electrochemical performance due to their 2D lamellar structures, abundant surface functional groups, and remarkable charge transfer capabilities. Nevertheless, the propensity for "face to face" horizontal stacking with less open tunnels reduces the interlayer space between isolating MXene flakes and suppresses electrolyte ions transport or intercalation, which limits their potential in electrochemical storage applications. Therefore, opportunities to control the MXene nanosheets alignment, especially through designing unique architectures to tune the porosity, expanding the interlayer spacing, and manipulating the surface functional groups, need to be further explored. In order to alleviate the stacking problems of MXenes, Zheng et al. employed the vacuum filtration method with the assistance of entwined metal mesh to control the alignment of MXene sheets and fabricate an "anti-T-shape" MXene film electrode. They claimed that the unique "magazine-bending" structure facilitated the vertical electrolyte ion transport and enhanced the kinetics of the electrochemical process which enable the "anti-T-shape" electrodes in a symmetric SC to display a low interfacial impedance, a high pseudocapacity of 194 F g^{-1}, superior energy/power densities of 11.27 Wh kg^{-1}/699.9 W kg^{-1}, and a favorable capacitance retention of 70.3% after 10,000 cycles [31].

Zhang's group utilized the vacancy ordered $Mo_{1.33}C$ MXene and poly(3,4-ethylenedioxythiophene):poly(styrenesulfonic acid) (PEDOT:PSS) to fabricate the flexible aligned $Mo_{1.33}C$ MXene/PEDOT:PSS composite films by a vacuum filtration process. The electrochemical test results show that the MXene-based films with the optimal ingredients ratio (the mass ratio between $Mo_{1.33}C$ MXene and PEDOT:PSS is 10:1) display a high conductivity of 29,674 S m^{-1} and a maximum volumetric capacitance of 1,310 F cm^{-3} in the potential range from −0.35 to 0.3 V in 1 M H_2SO_4 aqueous electrolyte, which may result from the synergistic effect of expanded interlayer spacing of MXene sheets caused by the PEDOT component and the extra redox activity of the PEDOT. Due to these reasons, a maximum volumetric capacitance of 568 F cm^{-3} and energy density of 33.2 mWh cm^{-3} can be obtained for the film-based flexible all-solid-state SC [77].

The performances and surface chemistries of the MXene film electrode can be further tuned by compositing with various functional nanomaterials or in different types of electrolytes. Xingbin Yan's group adopted the MXene (Ti_3C_2) as charge-transfer pathways and graphite carbon nitride (g-C_3N_4) as the ion-accessible channels to construct a 2D heterogeneous nanospace for dual confining the FeOOH quantum dots via a vacuum filtration to fabricate a freestanding FQDs/CNTC film electrode, which shows superior pseudocapacitive performances with favorable kinetics in a high-voltage ionic liquid (IL) electrolyte (1-ethyl-3-methylimidazolium tetrafluoroborate, abbr., $EMIMBF_4$). The test results indicated that the surface functional groups of Ti_2C_3 and active sites (e.g., N defects) of the g-C_3N_4 in the FQDs/CNTC electrode are crucial for offering sufficient redox reaction by efficiently forming the strong adsorption between the electrolyte ions and the electrode interface (Figure 6a). Furthermore, a 3 V flexible SC was constructed in the IL electrolyte, which exhibited volume energy/power densities of 77.12 mWh cm^{-3}/6000 mW cm^{-3}, and desirable flexible cycling performance (Figure 6b,c) [78]. The as-obtained FQDs/CNTC film based flexible SC can easily power portable electronics under complex motion states (Figure 6d) or store solar energy (Figure 6e), showing a new insight into construction of MXene-based composite film electrode and flexible SC.

Mahiar M. Hamedi's group prepared MXene-based freestanding films via a vacuum filtration by using the carboxymethylated cellulose nanofibrils (CNFs) as functional additive to strengthen the mechanical properties of the films. A hybrid film with a 10% CNF loading displays a high Young's modulus (41.9 GPa), a high mechanical strength (154 MPa), a high electric conductivity (690 S cm^{-1}), and a high specific capacitance (325 F g^{-1}), which shows great potential for flexible functional electronics. They claimed that the superior properties root in the strong interfacial interactions between $Ti_3C_2T_x$ flakes and CNF (Figure 7) [79].

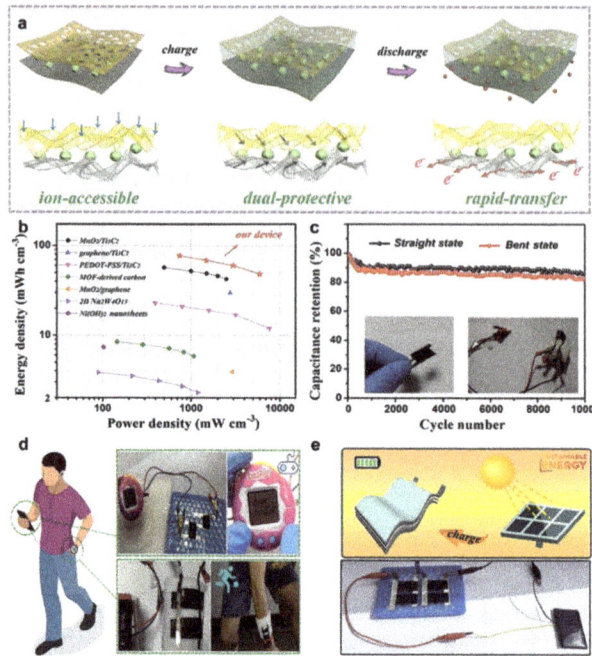

Figure 6. (**a**) Schematic illustration of the charge-discharge processes for FQDs/CNTC film electrode. (**b**) Ragone plot of the as-designed flexible SC. (**c**) Cycling performance of the flexible SC under the bent state. (**d**) Digital photos of the flexible SCs power diverse portable electronics. (**e**) Schematic of the flexible SC charged by harvesting solar energy [78].

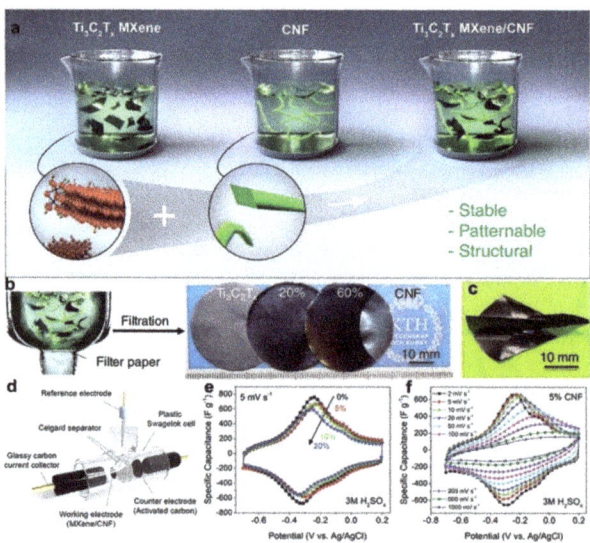

Figure 7. (**a**) Schematics of $Ti_3C_2T_x$/CNF hybrid dispersion. (**b**) The synthesis method and digital photographs of $Ti_3C_2T_x$/CNF nanopapers. (**c**) Photos of a $Ti_3C_2T_x$/CNF nanopaper. (**d**) Schematic of the three-electrode system in this work. (**e**) CV curves of MXene with different ratio of CNF. (**f**) CV curve of $Ti_3C_2T_x$–5% CNF electrode at various scan rates [79].

Given the interconnected networks and high tensile strength, bacterial cellulose (BC) was confirmed as excellent spacers for loading MXene to construct robust film electrodes with strong mechanical strength. In this regard, Guohui Yuan's group prepared a 3D self-supporting film electrode comprising interconnected MXene and BC networks. Due to excellent electron conductivity, large ion-accessible active sites, effective ion diffusion, and robust mechanical properties of the MXene/BC network, the as-obtained film electrode exhibits a high capacitance (416 F g^{-1}, 2084 mF cm^{-2}) coupled with excellent mechanical performance. Moreover, an ultrahigh energy density of 252 µWh cm^{-2} can be obtained in an asymmetric SC, which provided a simple route for fabricating high performance film electrodes in energy storage fields [80].

MXene inks and printable microsupercapacitor (MSC). To date, various construction methodologies to manufacture versatile micro-power systems have been developed. Among them, printing (e.g., laser printing, direct writing, screen printing, and extrusion printing) has been regarded as one of the most revolutionary techniques to construct functional energy storage devices with designed or desirable patterning. MXenes can also be used as attractive printing materials (e.g., conductive inks), which are essential for realizing the fabrication of geometric flexible, printable, and free-standing electrochemical devices. Thus, many works focus on the design of certain formulations of MXene conductive inks for achieving higher compatibility with various patterning methods. In addition, many simple and affordable strategies to obtain the MXene-based coplanar interdigital electrodes for further assembling into printable or direct-write SC devices are progressively developed, which shows extra advances in their versatility, high resolution patterning, simplicity, and desirable performances. Many previous works also indicated that some of the existing micro fabrication techniques often require sophisticated instrumentation and tedious multistep manufacturing processing, and therefore, MXenes inks or MXene-based coplanar electrodes and the corresponding printing techniques exhibit exciting potential for manufacturing of low-cost and easily processing printable electronics.

Yury Gogotsi and co-workers adopted simple and additive-free formulations to prepare solution processable MXene-in-water (e.g., Ti$_3$C$_2$) inks, which can be loaded in pens to direct write conductive MXene electrical circuits and desired shape either by manual drawing or automatic drawing tools on a variety of substrates (e.g., printer paper and polypropylene membranes) for functional energy storage devices. The developed MXene versatile inks are also compatible with other patterns, such as stamping, printing, and painting. Additionally, the as-written MXene collector-free MSC as power sources display an areal capacitance of 5 mF cm^{-2}, and a tandem device can drive a LED to operate normally, demonstrating the practical application of this technique (Figure 8) [81].

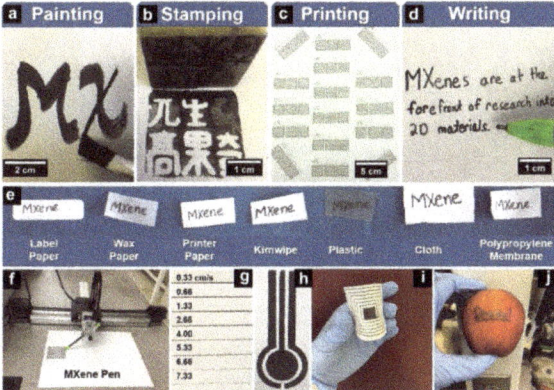

Figure 8. (**a**–**e**) Written MXene on various substrates via various ways, (**f**,**g**) using the automatic drawing tool of AxiDraw for patterning by using the Ti$_3$C$_2$ inks, (**h**,**i**) versatile patterning using AxiDraw, and (**j**) a demonstration of MXene inks drawn on an apple [81].

Screen printing and extrusion printing are two kinds of versatile direct printing techniques to fabricate the SC electrodes or MSC devices with geometric flexibility and desirable architectures. Screen printing is widely used to build up 2D flexible or wearable electronics due to its scalable and facile property, while extrusion printing could implement customized 3D printing through digital processing. In a manner, the rheological properties of the employed inks are pivotal for realizing the different printing modes. Commonly, extrusion printing requires more viscous inks compared to screen printing for keeping the 3D architectures from collapsing [56]. MXenes or their derivates could develop certain functional conductive inks for printed energy storage devices due to their intrinsic favorable hydrophilicity, mechanical flexibility, and modifiability [82]. Sun and co-workers used the melamine formaldehyde as the template to develop the crumpled heteroatom nitrogen-doped MXene flakes (MXene-N). They also found that the nitrogen doping effectively boosts the conductivity and electrochemical activity of the MXene flakes. Accordingly, two kinds of MXene-N inks (e.g., binder-additive aqueous ink and binder-free hybrid ink) are developed via optimizing the viscosity for screen printing and extrusion printing, respectively. The obtained screen printed flexible MXene-N based MSCs display an areal capacitance of 70.1 mF cm^{-2}. The as-fabricated 3D extrusion printed SC using the hybrid ink that includes the N-doping MXene, AC, CNT, and GO, delivers an areal capacitance of 8.2 F cm^{-2} with an areal energy density of 0.42 mWh cm^{-2}. This work also proposed a versatile solution to relieve the restacking issues of MXene flakes within the printed electrodes for further facilitating the ion/electrolyte transport (Figure 9) [56].

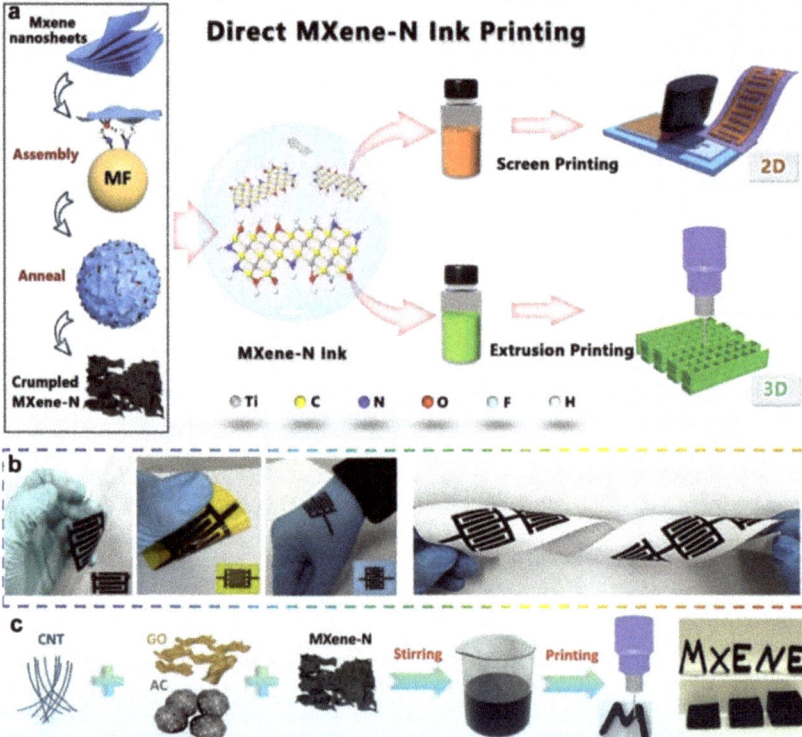

Figure 9. (**a**) Schematic illustration of the synthetic procedure for the crumpled MXene-N and the direct MXene-N ink printing routes of screen printing and extrusion printing. (**b**) Digital photographs of the MSC printed by screen printing. (**c**) Schematic showing of the preparation process of MXene-N based hybrid ink and the extrusion printed 3D architecture [56].

Zhang and co-workers employ the MXene sediments (usually thrown away) of unetched or multilayered MXene to prepare the additive free inks for scalable high-resolution screen printing. By designing different modes, various printed patterns (e.g., conductive tracks, letters, MSCs, and integrated circuits) can be quickly printed by using the as-formulated ink (Figure 10). They found that a low proportion of delaminated MXene flakes in the ink is crucial for realizing a desirable conductivity and excellent mechanical flexibility of the final printed circuitries or the electrodes. They also demonstrated an excellent electrochemical performance of the as-obtained printed MSCs with a high areal capacitance of 158 mF cm^{-2} and a high energy density of 1.64 µWh cm^{-2}. This ink formulation strategy of using MXene etching trash would effectively reduce the cost and waste of MXene-based printed techniques into consideration [83].

Liang et al. developed a thixotropic electrode ink that consisted of RuO$_2$-decorated Ti$_3$C$_2$T$_x$ MXene nanosheets (RuO$_2$·xH$_2$O@MXene) and conductive silver nanowires to fabricate the flexible printable MSC via the screen-printing method (Figure 11). They also found that the in situ growth of RuO$_2$ nanoparticles effectively relieves the agglomeration and restacking problems of the MXene flakes while maintains a suitable rheological nature for printing. Due to the synergistically interplayed effect of ingredients, the as-designed MSCs possess micrometer-scale resolution and desirable electrochemical properties, including a high volumetric capacitance of 864.2 F cm^{-3}, satisfactory energy/power densities of 13.5 mWh cm^{-3}/48.5 W cm^{-3}, a long-term cycling life with a retention of 90% after 10,000 cycles, and high mechanical stability [64].

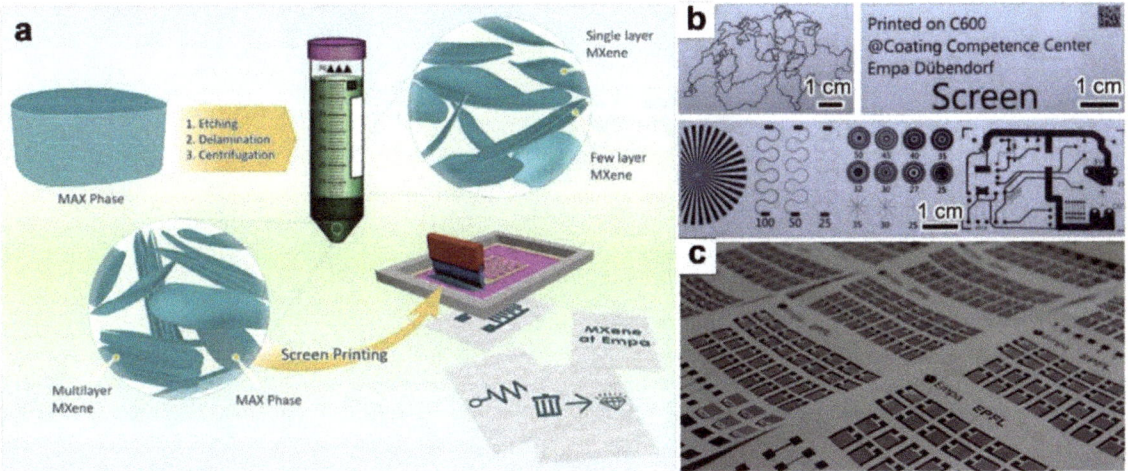

Figure 10. (a) Schematic illustrating of the screen printing using the MXene sediment ink. (b) Various screen-printed patterns. (c) Optical photo of scalable produced screen-printed MXene-based MSC [83].

The traditional micro-fabrication technologies that are based on the conventional silicon-based strategies, such as laser scribing, have also demonstrated feasibility for the fabrication of MXene-based MSC devices. For example, Hua's group employed the laser printing techniques with the assistance of vacuum deposition and physical sputtering to fabricate the MXene-based coplanar interdigital electrodes and planar symmetric on-chip MSC. They also reported that the as-developed fabrication method allows the deposition of fully delaminated MXene sheets with tunable thickness. Due to the good alignment of the MXene flakes, high electrical conductivity, and the unique layered porous structure of the as-designed MXene-based coplanar electrodes, the as-designed all-solid-state MSC displays excellent high-rate capacity, a high areal capacitance of 27.29 mF cm^{-2} and competitive volumetric energy densities of (5.48–6.1 mW h cm^{-3}) [84]. This work also demonstrated that the MXene active materials have great potential for the laser printing process technology for patterning thick coplanar interdigital electrodes or planar SC.

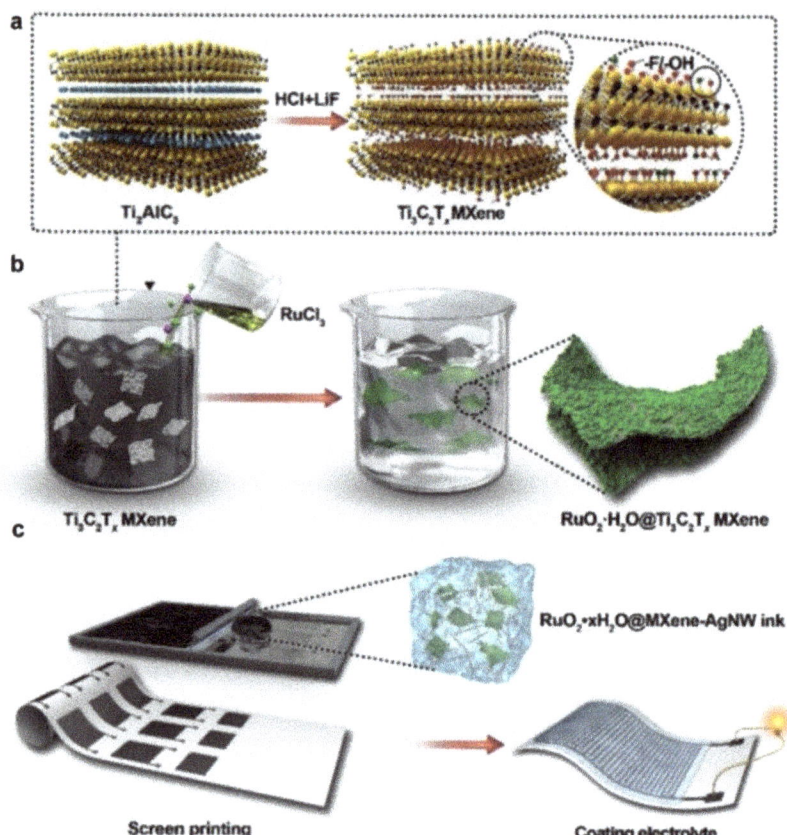

Figure 11. Schematic illustrating of (**a**) the preparation of MXene, (**b**) the synthesis process of the RuO$_2$·xH$_2$O@MXene nanohybrid, and (**c**) the screen-printing process for the fabrication of the flexible MXene-based MSCs [64].

Flexible MXene-based yarn or fiber electrodes. Fiber-shaped SCs are attractive for powering miniaturized or portable wearable electronics due to their compatibility of integrating with textiles. In general, effective and scalable strategies to produce fiber electrodes with desirable electrical storage performances fall within two categories. The first category of deposition technique is based on covering of electrochemical active or conductive materials onto fiber matrix [85]. For instance, Yang and co-workers employed the MXene-polymer nanofibers as the coating layer on the polyester (PET) yarns to construct the PET@MXene nanofiber-based yarn via a modified electrospinning technique, from which the complex yarns display favorable flexibility and certain mechanical strength during the weaving, braiding, or knitting processes, indicating that the obtained yarns can be used as flexible yarn electrodes for wearable SCs. They also reported that two parallel PET@MXene yarn based SC devices delivered an areal capacitance of 18.39 mF cm^{-2} with desirable energy/power densities of 0.38 µWh cm^{-2}/0.39 mW cm^{-2} and long term stability (a retention percentage of 98.2% after 6000 cycles) [86].

Another fabrication approach for fiber production is wet-spinning method, which is more viable for industrial manufacture and applications [85]. Holistically, when considering the "spinnablility" for this approach to construct long continuous fibers, the high conductivity and available electrochemical activity provided by the composite formulations are the critical barrier hindering the electrochemical performances of the fiber electrodes. In addition, the homogeneous dispersion and the suitable spinning behaviors of the composite formulations also have an impact on the ultimate fiber properties (e.g., durability, flexibility, resilient stretching, bending, and twisting). In view of the hydrophilic properties and the excellent conductivity of MXene, which are suitable for wet-spinning processing, Seyedin and co-workers adopted the MXene ($Ti_3C_2T_x$) and poly(3,4-ethylenedioxythiophene):polystyrene sulfonate (PEDOT:PSS) to successfully fabricate an MXene/PEDOT:PSS hybrid fiber electrode. By combining with the flexible PEDOT:PSS, the as-designed optimized $Ti_3C_2T_x$-based fiber electrode (containing 70% MXene and 30% PEDOT:PSS, referred to as M_7P_3) exhibited a record conductivity of 1.489 S cm^{-1}, a high volumetric capacitance of 614.5 F cm^{-3} at 5 mV s^{-1}, and an excellent rate performance of 375.2 F cm^{-3} at 1000 mV s^{-1} in a potential range of -0.8–0.2 V with a three-electrode setup in 1 M H_2SO_4 electrolyte. The as-designed fiber SC device in an operated voltage of 0.6 V could deliver maximum volumetric energy/power densities of 7.13 Wh cm^{-3}/8.249 mW cm^{-3}. Moreover, a stretchable and elastic fiber SC prototype assembled by prestretched silicon rubber and M_7P_3 wrapping with a PVA/H_2SO_4 electrolyte maintains capacitance retention of ≈98% and ≈96% after cyclically stretched under different stretched strains [85].

Flexible stretchability MXene-based electrodes. The unique properties of MXene (e.g., metallic conductivity, strong interactions among large flakes, hydrophilic surface groups, and high volumetric capacitance (up to 1500 F cm^{-3})) enable them particularly attractive for stretchable energy storage devices or electronics. Nevertheless, the fragility and high mechanical modulus (3–75 GPa) at dry state for MXene would increase the device resistance when subjected to larger tensile strains and further limit their stable electrochemical outputs [30]. To address these challenges and constraints, several methods have been developed for constructing robust SC electrodes with high stretchability, superior electrochemical performance, and high chemical stability. Chen and co-workers used a sequential patterning approach to prepare high dimensional MXene nanocoatings with mechanically stable architectures, which can be program crumpling and unfolding. After transferring on the elastomer, MXene/elastomer SC electrodes were obtained, which show high stretchability for reversibly folded/unfolded. A stretchable asymmetric SC fabricated by using the as-obtained electrodes is capable of delivering stable electrochemical output under various mechanical deformations (e.g., bend or stretch), such as favorable capacitances of 395, 390, and 362 F cm^{-3} under 0%, 50%, and 80% strains, efficient energy density of 5.5 Wh kg^{-1}, and long term cycling stability under stretching/bending conditions [87].

Some works also revealed that the combination of $Ti_3C_2T_x$ with other high mechanical robustness nanomaterials (e.g., reduced graphene oxide (RGO)) would generate flexible and stretchable free-standing electrodes with superior electrochemical performance. Cao and co-workers adopted the MXene ($Ti_3C_2T_x$ in this case) and RGO to fabricate a robust and stretchable electrode ($Ti_3C_2T_x$/RGO composite) for flexible SC devices. The test results show that the final composite with 50 wt% RGO would alleviate cracks that result from large strains. Due to strong nanosheet interactions between larger nanoflakes, mechanical flexibility, and excellent electrochemical properties of each component, the as-obtained electrode delivered a capacitance of 49 mF cm^{-2} (~490 F cm^{-3} and ~140 F g^{-1}) with high mechanical stability after various cyclic strains (e.g., uniaxial and biaxial strains). An all-solid-state stretchable symmetric SC fabricated by these electrodes displayed a high capacitance of 18.6 mF cm^{-2} (~90 F cm^{-3} and ~29 F g^{-1}) with high mechanical stretchability [30,87–90]. The comparison of the electrochemical performance of various MXene-based materials for SCs are listed in Table 1.

Table 1. Comparison of the electrochemical properties of different MXene-based materials for SCs.

Materials	Capacity/Current Density	Capacity Retention	Energy Density and Power Density	Reference
1T-MoS$_2$/Ti$_3$C$_2$MXene based flexible ASSS	386.7 F g^{-1} at 1 A g^{-1}	91.1% after 20,000 cycles at 30 mA cm^{-2}	17.4 µW h cm^{-2} (600 µW cm^{-2})	[66]
TC-9//Ti$_3$C$_2$ ASC	49.3 F g^{-1} at 1 A g^{-1}	82.4% after 5000 cycles at 8 A g^{-1}	15.4 W h kg^{-1} (750.2 W kg^{-1})	[65]
MXene/LDH composite	1061 F g^{-1} at 1 A g^{-1}	70% after 4000 cycles at 4 A g^{-1}		[67]
M-Ti$_3$C$_2$T$_x$//PANI	510 F g^{-1} at 10 mV s^{-1}		50.6 W h L^{-1} (1.7 kW L^{-1})	[73]
N-Ti$_3$C$_2$T$_x$-200 °C	192 F g^{-1} at 1 mV s^{-1}	92% after 10,000 cycles at of 50 mV s^{-1}	8.07 W h kg^{-1} (52.8 W kg^{-1})	[69]
all-solid-state symmetric SCs based on MXene/MPFs electrode	408 mF cm^{-2} at 0.5 mA cm^{-2}	95.9% after 7000 cycles at 5 mA cm^{-2}	20.4 µW h cm^{-2} (152.2 µW cm^{-2})	[75]
all-solid-state SC based on PDT/Ti$_3$C$_2$T$_x$	52.4 mF cm^{-2} (3.52 F cm^{-3}) at 0.1 mA cm^{-2}		24 mW h cm^{-3} (502 mW cm^{-3})	[76]
FSC based on FQDs/CNTC film	71.26 F cm^{-3} at 10 mV s^{-1}	80% after 10,000 cycles at 4 A cm^{-3}	77.12 mW h cm^{-3} (750 mW cm^{-3})	[78]
Ti$_3$C$_2$T$_x$/CNF MSC	25.3 mF cm^{-2} at 2 mV s^{-1}	86.8% after 10,000 cycles at 0.57 mA cm^{-2}	0.08 µWh cm^{-2} (145 µW cm^{-2})	[79]
PANI//Ti$_3$C$_2$T$_x$ device	925 mF cm^{-2} (87 F g^{-1}) at 3 mA cm^{-2}	93% after 10,000 cycles at 50 mA cm^{-2}	252 µW h cm^{-2} (2.12 mW cm^{-2})	[80]
M/MoO$_3$ symmetric SC	396 F cm^{-3} (118.8 F g^{-1}) at mV s^{-1}	90% after 5000 cycles at 30 mA cm^{-2}	13.4 W h kg^{-1} (534.6 W kg^{-1})	[82]
MXene NCY SC	18.39 mF cm^{-2} at 5 mV s^{-1}	98.2% after 6000 cycles at 50 mV s^{-1}	0.38 µW h cm^{-2} (0.39 mW cm^{-2})	[86]
MXene-N MSC	70.1 mF cm^{-2} at 10 mV s^{-1}	92% after 7000 cycles at 5 mA cm^{-2}	0.42 mW h cm^{-2} 0.83 mW h cm^{-3}	[56]
MSCs based on this RuO$_2$·xH$_2$O@MXene–AgNW nanocomposite ink	864.2 F cm^{-3} at 1 mV s^{-1}	90% after 10,000 cycles at 100 mV s^{-1}	13.5 mW h cm^{-3} (1.1 W cm^{-3})	[64]

3.2. Li Ions Battery (LIB)

MXenes also display great potential in LIB due to their large available surface areas for electrolyte ion adsorption, satisfactory electrical conductivity, desirable ion transfer within the inter-layers (up to 1.3 nm), and rapid surface redox reaction. Additionally, the MXenes (e.g., Ti$_3$C$_2$) can also be used as functional host matrix for integrating other materials for further enhancing the overall energy storage performance of LIB [91]. For instance, Junjie Niu's group synthesized a hybrid material comprising Ti$_3$C$_2$ MXene wrapped with germanium oxide layer (GeO$_x$ (x = 1.57) @MXene) through the wet-chemical strategy. The high electronic conductivity of MXene and germanium enable a high rate capability, superb capacity retention of ~929.6 mAh g^{-1} at 1.0 C with high Coulombic efficiency of 99.6% after 1000 cycle. Especially, the as-designed LIB also possesses temperature dependence properties [91]. Likun Pan's group prepared a SnS$_2$/Sn$_3$S$_4$ modified multi-layered Ti$_3$C$_2$ MXene hybrid (denoted as S-TC) as anode material for LIB via the solvothermal and calcination strategy. The MXene substrate provides high electronic conductivity and suppresses the aggregation and volume change problems of active components while the nano-sized SnS$_2$/Sn$_3$S$_4$ acts as a "spacer" to effectively inhibit restacking of the Ti$_3$C$_2$ layer. Due to these reasons, the S-TC anode delivers superb rate capability (216.5 mAh g^{-1} at 5 A g^{-1}) coupled with long cycling stability [92]. Wang et al. rationally designed an interconnected MXene hybrid aerogel composited with Fe$_2$O$_3$ nanospheres, which exhibits superb energy density of 216 Wh kg^{-1} at 400 W kg^{-1} for lithium-ion capacitors due to the synergistic effect between the two components [93]. Husam N. Alshareef et al. synthesized a HfO$_2$ coated SnO$_2$/MXene composite anode material via atomic layer deposition. The deposited SnO$_2$ on the MXene can effectively suppress the degradation of MXene while the thin inactive layer of HfO$_2$ would serve as an artificial solid-electrolyte interphase (SEI) layer for further enhancing the cycling stability [94]. Wang et al. found that the integrating of nitrogen and vanadium in the forms of C–V–OH, C–V–O, V–O, and Ti–O–N species into the Ti-deficient $Ti_3C_2T_x$ can further enhance the charge storage capability for approximately 40% [95]. Wang and co-workers modulated the interfacial properties by fabricating a crumpled S-functionalized $Ti_3C_2T_x$ heterostructure embedded with Fe_3O_4/FeS. Due to the tuned electronic properties, the heterostructure displays improved kinetics and structural stability for LIB. The authors claimed that the S terminations boosted the extra (pseudo)capacitive ability of MXene for lithium storage. Due to these reasons, the optimized anode of the heterostructure material delivers a superb long-term stability (913.9 mAh g^{-1} after 1000 cycles) with desirable rate performance. They also found that the heterostructure material exhibits an asymmetric conversion mechanism by experiencing stepwise phase transformations under discharge process coupled with a relatively uniform reconversion under the charge process. This work gives an in-depth understanding about MXene-based heterostructure for Li ion storage [96].

Yan and co-workers improved the properties of MXene by employing a liquid nitrogen quenching method to roll up the MXene sheets into $Ti_3C_2T_x$ scrolls. They found that the prepared scrolls display unclosed topological structure and can effectively relieve the restacking effect. Meanwhile, they also reported that the produced scrolls possess superior electrical conductivity and can be used as buffer matrixes of Si nanoparticles for electrochemical energy storage. Due to these reasons, the as-designed $Ti_3C_2T_x$ scroll anode material exhibits a reversible capacity of 226 mAh g^{-1}, desirable rate performance, and excellent long-term cycling performance with 81.6% capacity retention after 500 cycles in LIB.

Additionally, MXene materials can be directly used as conductive and highly stable hosts by forming artificial solid electrolyte interface (SEI) film for greatly reducing the topical current density on the surface of the electrode, adjusting the electric field, and efficiently inducing the uniform growth of lithium dendrites. As exemplified, Shubin Yang's group demonstrated that the parallelly aligned MXene layers (denoted as PA-MXene-Li) can effectively regulate the uniform nucleation and growth of lithium, forming horizontal-growth of lithium on the surface of MXene. Moreover, the fluorine terminations of MXene make forming a durable and artificial solid electrolyte interface with lithium fluoride and homogenizing the electro-migration for lithium ions during the energy storage process possible [97].

Flexible MXene thin films or free-standing electrodes. The diversification development of LIBs (portable, flexible, wearable, stretchable, and foldable, etc.) is urgently required for ever-increasing demands for future energy-storage devices. In this respect, exploring conductive and flexible substrate with strong coupling of active materials is the key factor for the application of flexible LIBs technology. Given the self-assembly ability, 2D MXene are considered as ideal matrix to load active materials for forming composite paper electrodes in flexible LIBs. For example, Yitai Qian's group developed a MXene/liquid metal paper by confining the low-melting point GaInSnZn liquid metal in the MXene substrate. Due to the excellent conductivity and satisfactory wettability between GaInSnZn and MXene matrix, the obtained flexible anode for LIB exhibits a higher capacity of 638.79 mAh g^{-1} at 20 mA g^{-1} coupled with desirable rate performance (which is better than that of liquid metal coated Cu foil), indicating that the great potential of MXene matrix for compositing with various active materials in flexible LIB [98]. Feng et al. demonstrated that the construction of the Silicon/MXene composite papers via vacuum filtration to covalently anchor silicon nanospheres on MXene can accommodate the volumetric expansion of silicon and restrain the restacking of MXene sheets, which offers superior capacity of 2,118 mAh·g^{-1} at a current density of 200 mA·g^{-1} [18]. The electrochemical performances of different MXene-based materials for LIBs are compared and listed in Table 2.

Table 2. Comparison of the electrochemical properties of different MXene-based materials for LIBs.

Materials	Capacity/Current Density	Cycling Stability	Rate Capacity	Reference
GeO$_2$@MXene	1127.1 mA h g^{-1} at 0.5 C	1048.1 mA h g^{-1} at 0.5 C after 500 cycles	221.2 mAh g^{-1} at 20.0 C	[91]
S-TC	601.3 mA h g^{-1} at 100 mA g^{-1}	426.3 mA h g^{-1} at 100 mA g^{-1} after 100 cycles	216.5 mAh g^{-1} at 5000 mA g^{-1}	[92]
N(V)-modified Ti$_3$C$_2$T$_x$MXene	251.3 mA h g^{-1} at 0.1 C	92 mA h g^{-1} at 3C after 1000 cycles	118.3 mA h g^{-1} at 5 C	[95]
Fe$_3$O$_4$/FeS 30@S-MX	746.6 mAh g^{-1} at 0.1 A g^{-1}	913.9 mA h g^{-1} after 1000 cycles at 1 A g^{-1}	490.4 mA h g^{-1} at 10 A g^{-1}	[96]

3.3. Zn Ions Storage Devices

Recent works suggest that the MXene can also be used as both anode and cathode materials for zinc ion hybrid supercapacitors (ZHSCs). As Qi Yang and co-workers demonstrated, the Zn−MXene capacitor fabricated by Ti$_3$C$_2$ cathode and the Zn@Ti$_3$C$_2$ anode displays 82.5% capacitance retention after 1000 cycles with excellent anti-self-discharge ability of 6.4 mV h^{-1} [99]. The MXene (Ti$_3$C$_2$T$_x$) is directly used as anode coupled with manganese dioxide/carbon nanotubes (MnO$_2$-CNTs) cathode material to fabricate a zinc-ion capacitor (ZIC). The obtained ZIC in aqueous electrolyte possesses a high specific capacitance, energy density of 98.6 Wh kg^{-1}, and high capacitance retention of ~83.6% after 15,000 cycles [100]. Wang et al. reported that when the δ-MnO$_2$ cathode and MXene are grown on the cotton cloth, the assembled device would exhibit a high energy density of 90 Wh kg^{-1} with capacitance retention of ~80.7% after 16,000 cycles [101]. Others would like to employ the MXene as a protective layer for suppressing the growth of Zn dendrites. Qian's group investigated the growth mechanism of metal zinc on the surface of Zn foil and binder-free Ti$_3$C$_2$T$_x$ MXene@Zn hybrid firm and found that the Ti$_3$C$_2$T$_x$ MXene@Zn film effectively restrains the growth of Zn dendrites with reversible plating/stripping of Zn. The developed MXene film would be a valid strategy for dendrite-free electrochemical storage devices [3]. Zeng et al. adopted the MXene with the reduced graphene oxide (rGO) to fabricate a 3D porous MXene-rGO aerogel for preventing the restacking of MXene flakes and enhancing the electrical conductivity and hydrophilicity of the final materials. The electrochemical results show that the ZHSC with the MXene-rGO aerogel cathode delivers a high specific capacitance of 128.6 F g^{-1} coupled with high energy density of 34.9 Wh kg^{-1} and long term stability for 75,000 cycles [2]. The comparison of the electrochemical performance of various MXene-based materials for Zn ions storage devices are listed in Table 3.

Table 3. Comparison of the electrochemical properties of different MXene-based materials for Zn ions storage devices.

Materials	Capacity/Current Density	Cycling Stability	Energy Density and Power Density	Reference
Zn@Ti$_3$C$_2$ anode and Ti$_3$C$_2$ film cathode	132 F g^{-1} at 0.5 A g^{-1}	82.5% after 1000 cycles at 3 A g^{-1}		[99]
MnO$_2$–CNTs cathode and MXene anode	115.1 F g^{-1} at 1 mV s^{-1}	83.6% at 15,000 cycles at 5.224 A g^{-1}	98.6 W h kg^{-1} (77.5 W kg^{-1})	[100]
MXene-rGO2// ZnSO$_4$//Zn	128.6 F g^{-1} at 0.4 A g^{-1}	95% after 75,000 cycles at 5 A g^{-1}	34.9 W h kg^{-1} (279.9 W kg^{-1})	[2]

4. Conclusions

In summary, this review was mainly focused on the synthetic method, the construction of MXene hybrids, and composite films or fibers in terms of their performance for various energy storage devices. The most recent advances of technologies (e.g., vacuum filtration, extrusion printing technique, and directly writing) for fabricating patterned MXene-based composite films or fibers with geometric flexibility in various energy storage devices were described. The multifunctional properties of MXenes enable them to be used as electrodes for SCs and anode materials for both lithium- and zinc-ion batteries. As we introduced above, the electrochemical performance of MXene-based materials can be enhanced by tuning the chemical components, constructing the micro/nano structures, and enlarging the interlayer of MXene flakes. Advanced MXene-based macroscopically assembled films and fibers with dedicated design were also demonstrated in this work, which are expected to significantly enhance the geometric flexibility for portable energy storage devices. Thus, it is expected that the MXene-based nanostructures and advanced architectures for films and fibers will provide intriguing opportunities for next generation energy storage devices.

1. MXene offers attractive properties in SC applications, such as excellent conductivity, ultrahigh rate capability, adjustable composition, hydrophilic, and volumetric capacitance. Still, the stacking and aggregating problems reduce the interlayer spaces and lead to the sluggish kinetics for energy storage. Therefore, many MXene-based composites (e.g., 2D/2D 1T-MoS_2/Ti_3C_2 heterostructures, Ti_3C_2/CuS, MXene/LDH, Ni–S/Ti_3C_2 nanohybrid, N-doped MXene, N, O co-doped carbon@MXene composite, dodecaborate/MXene) were designed for solving the problems via the surface modification, heteroatom doping, or crumpling process. Among the various materials, the Ni–S/Ti_3C_2 composite exhibits a superior capacity of 840.4 C g^{-1} with a retention of 64.3% at 30 A g^{-1} and a long cycle life. In addition, in order to settle the easy oxidation issue at positive potential (anodic oxidation process) of MXenes and most MXene-derived composites, some works focus on the enhancing of the WF to boost the antioxidant ability of MXene-based materials. PANI@MXene cathode material exhibits a larger WF due to the existence of the oxidation-resistive PANI layer on the surface of MXene compared with the pure metallic MXene, which enhance the electrochemical stability at a wider positive operating potential (0–0.6 V).

2. The special flexibility and self-assembly capability of MXenes enable them to be a versatile unit for constructing the macroscopic film and fiber electrodes. Several assembly strategies are developed, including spin-casting method, vacuum-assisted filtration method, and electrophoretic deposition/spin coating coupled method. Additionally, the energy storage performance can be further enhanced by optimizing the MXene synthesis conditions, tuning the MXene interlayer spacing, altering the types or amount of surface terminations, and compositing with other functional materials. Many recent works also revealed that the addition of reinforcement (e.g., carboxymethylated cellulose nanofibrils, bacterial cellulose) also acts as spacers for MXene to fabricate robust film electrodes with strong mechanical strength.

3. MXenes can be employed as the printing material for constructing the geometric flexible, printable, and free-standing electrochemical devices (e.g., MXene-based coplanar interdigital electrodes, printable or direct-write SC devices) due to their hydrophilicity, mechanical flexibility, and modifiability. In this regard, the design of certain formulations of MXene-based conductive inks with specific rheological properties for the compatibility with various patterning methods (e.g., screen printing, direct writing, and extrusion printing) was crucial for the fabrication of the printable energy storage devices.

4. MXenes can also be used for LIBs and Zn ions storage devices due to their layer structure for electrolyte ion adsorption, satisfactory electrical conductivity, and desirable ion transfer ability within the layers. Many recent works reveal that the compositing with other functional materials can further enhance the overall energy storage performance (e.g., GeO_x (x = 1.57)@MXene, SnS_2/Sn_3S_4 modified multi-layered Ti_3C_2 MXene hybrid, SnO_2/MXene composite, MXene/liquid metal, silicon/MXene, and MXene-rGO aerogel). Among the various materials, the GeO_x (x = 1.57)@MXene exhibits a high rate capability, superb capacity retention of ~929.6 mAh g^{-1} at 1.0 C with high Coulombic efficiency of 99.6% after 1000 cycles.

With further substantial research effort on optimized methodologies, MXenes and their derivatives represent a promising platform in scalable and customizable manufacturing of multipurpose electrodes with wearable, flexible, and lightweight properties in the future.

Author Contributions: Conceptualization, H.M. and S.Y.; investigation, C.J. and H.C.; data curation, H.C.; writing—original draft preparation, C.J. and H.C.; writing—review and editing, C.J. and H.C.; visualization, H.C.; supervision, S.Y.; project administration, S.Y.; funding acquisition, S.Y. and C.J. All authors have read and agreed to the published version of the manuscript.

Funding: This research was funded by the National Natural Science Foundation of China, grant number 21805237, U2003132. Fundamental Research Funds for the Central Universities and the Natural Science Foundation of Shaanxi Province, grant number 2021GXLH-Z-082 and 2020JZ-02 (S. Yang). Natural Science Foundation of Xinjiang Uygur Autonomous Region, grant number 2018D01C053. Tianshan Youth Planning Program of Science & Technology Department of Xinjiang Uygur Autonomous Region, grant number 2018Q013. Opening Foundation of the State Key Laboratory of Fine Chemicals, grant number KF2003.

Institutional Review Board Statement: Not applicable.

Informed Consent Statement: Not applicable.

Data Availability Statement: No data support.

Conflicts of Interest: The authors declare no conflict of interest.

References

1. Jakob, M.; Hilaire, J. Unburnable fossil-fuel reserves. *Nature* **2014**, *517*, 150–151. [CrossRef]
2. Wang, Q.; Wang, S.; Guo, X.; Ruan, L.; Wei, N.; Ma, Y.; Li, J.; Wang, M.; Li, W.; Zeng, W. MXene-reduced graphene oxide aerogel for aqueous zinc-ion hybrid supercapacitor with ultralong cycle life. *Adv. Electron. Mater.* **2019**, *5*, 1900537. [CrossRef]
3. Tian, Y.; An, Y.; Wei, C.; Xi, B.; Xiong, S.; Feng, J.; Qian, Y. Flexible and free-standing $Ti_3C_2T_x$ MXene@Zn paper for dendrite-free aqueous zinc metal batteries and nonaqueous lithium metal batteries. *ACS Nano* **2019**, *13*, 11676–11685. [CrossRef] [PubMed]
4. Herou, S.; Bailey, J.J.; Kok, M.; Schlee, P.; Jervis, R.; Brett, D.J.L.; Shearing, P.R.; Ribadeneyra, M.C.; Titirici, M. High-density lignin-derived carbon nanofiber supercapacitors with enhanced volumetric energy density. *Adv. Sci.* **2021**, *8*, 2100016. [CrossRef]
5. Ma, X.; Cheng, J.; Dong, L.; Liu, W.; Mou, J.; Zhao, L.; Wang, J.; Ren, D.; Wu, J.; Xu, C.; et al. Multivalent ion storage towards high-performance aqueous zinc-ion hybrid supercapacitors. *Energy Storage Mater.* **2019**, *20*, 335–342. [CrossRef]
6. Rojaee, R.; Shahbazian-Yassar, R. Two-dimensional materials to address the lithium battery challenges. *ACS Nano* **2020**, *14*, 2628–2658. [CrossRef] [PubMed]
7. Mao, J.; Zhou, T.; Zheng, Y.; Gao, H.; Liua, H.K.; Guo, Z. Two-dimensional nanostructures for sodium-ion battery anodes. *J. Mater. Chem. A* **2018**, *6*, 3284–3303. [CrossRef]
8. Lin, L.; Lei, W.; Zhang, S.Y.; Liu, G.G.; Wallace, J.C. Two-dimensional transition metal dichalcogenides in supercapacitors and secondary batteries. *Energy Storage Mater.* **2019**, *19*, 408–423. [CrossRef]
9. Zhu, H.; Zhang, F.; Li, J.; Tang, Y. Penne-like MoS_2/carbon nanocomposite as anode for sodium-ion-based dual-ion battery. *Small* **2018**, *14*, 1703951. [CrossRef] [PubMed]
10. Haseeb, H.H.; Li, Y.; Ayub, S.; Fang, Q.; Yu, L.; Xu, K.; Ma, F. Defective phosphorene as a promising anchoring material for lithium–sulfur batteries. *J. Phys. Chem. C* **2020**, *124*, 2739–2746. [CrossRef]
11. Leng, K.; Li, G.; Guo, J.; Zhang, X.; Wang, A.; Liu, X.; Luo, J. A safe polyzwitterionic hydrogel electrolyte for long-life quasi-solid state zinc metal batteries. *Adv. Funct. Mater.* **2020**, *30*, 2001317. [CrossRef]
12. Zhang, H.; Liu, Q.; Fang, Y.; Teng, C.; Liu, X.; Fang, P.; Tong, Y.; Lu, X. Boosting Zn-ion energy storage capability of hierarchically porous carbon by promoting chemical adsorption. *Adv. Mater.* **2019**, *31*, 1904948. [CrossRef]
13. Xiao, Z.; Li, Z.; Meng, X.; Wang, R. MXene-engineered lithium-sulfur batteries. *J. Mater. Chem. A* **2019**, *7*, 22730–22743. [CrossRef]
14. Zhang, X.; Zhang, Z.; Zhou, Z. MXene-based materials for electrochemical energy storage. *J. Energy Chem.* **2018**, *27*, 73–85. [CrossRef]
15. Lei, J.C.; Zhang, X.; Zhou, Z. Recent advances in MXene: Preparation, properties, and applications. *Front. Phys.* **2015**, *10*, 276–286. [CrossRef]
16. Jiang, Q.; Lei, Y.; Liang, H.; Xi, K.; Xia, C.; Alshareef, H.N. Review of MXene electrochemical microsupercapacitors. *Energy Storage Mater.* **2020**, *27*, 78–95. [CrossRef]
17. Zhao, Q.; Zhu, Q.; Liu, Y.; Xu, B. Status and prospects of MXene-based lithium–sulfur batteries. *Adv. Funct. Mater.* **2021**, *31*, 2100457. [CrossRef]
18. Tian, Y.; An, Y.; Feng, J. Flexible and freestanding silicon/MXene composite papers for high-performance lithium-ion batteries. *ACS Appl. Mater. Interfaces* **2019**, *11*, 10004–10011. [CrossRef] [PubMed]
19. Wang, H.; Wu, Y.; Yuan, X.; Zeng, G.; Zhou, J.; Wang, X.; Chew, J.W. Clay-inspired MXene-based electrochemical devices and photo-electrocatalyst: State-of-the-art progresses and challenges. *Adv. Mater.* **2018**, *30*, 1704561. [CrossRef] [PubMed]
20. Yu, H.; Wang, Y.; Jing, Y.; Ma, J.; Du, C.F.; Yan, Q. Surface modified MXene-based nanocomposites for electrochemical energy conversion and storage. *Small* **2019**, *15*, 1901503. [CrossRef]
21. Gogotsi, Y.; Anasori, B. The rise of MXenes. *ACS Nano* **2019**, *13*, 8491–8494. [CrossRef]

22. Yang, Q.; Xu, Z.; Fang, B.; Huang, T.; Cai, S.; Chen, H.; Liu, Y.; Gopalsamy, K.; Gao, W.; Gao, C. MXene/graphene hybrid fibers for high performance flexible supercapacitors. *J. Mater. Chem. A* **2017**, *5*, 22113–22119. [CrossRef]
23. Li, X.; Ma, Y.; Shen, P.; Zhang, C.; Yan, J.; Xia, Y.; Luo, S.; Gao, Y. Self-healing microsupercapacitors with size-dependent 2D MXene. *ChemElectroChem* **2020**, *7*, 821–829. [CrossRef]
24. Liu, J.; Zhang, H.B.; Sun, R.; Liu, Y.; Liu, Z.; Zhou, A.; Yu, Z.Z. Hydrophobic, flexible, and lightweight MXene foams for high-performance electromagnetic-interference shielding. *Adv. Mater.* **2017**, *29*, 1702367. [CrossRef] [PubMed]
25. Zha, X.-H.; Zhou, J.; Zhou, Y.; Huang, Q.; He, J.; Francisco, J.S.; Luoa, K.; Du, S. Promising electron mobility and high thermal conductivity in Sc_2CT_2 (T = F, OH) MXenes. *Nanoscale* **2016**, *8*, 6110–6117. [CrossRef]
26. Xu, N.; Li, H.; Gan, Y.; Chen, H.; Li, W.; Zhang, F.; Jiang, X.; Shi, Y.; Liu, J.; Wen, Q.; et al. Zero-dimensional MXene-based optical devices for ultrafast and ultranarrow photonics applications. *Adv. Sci.* **2020**, *7*, 2002209. [CrossRef]
27. Wu, Y.; Hu, H.; Yuan, C.; Song, J.; Wu, M. Electrons/ions dual transport channels design: Concurrently tuning interlayer conductivity and space within re-stacked few-layered MXenes film electrodes for high-areal-capacitance stretchable micro-supercapacitor-arrays. *Nano Energy* **2020**, *74*, 104812. [CrossRef]
28. Pan, Z.; Ji, X. Facile synthesis of nitrogen and oxygen co-doped $C@Ti_3C_2$ MXene for high performance symmetric supercapacitors. *J. Power Source* **2019**, *439*, 227068. [CrossRef]
29. Li, M.; Li, X.; Qin, G.; Luo, K.; Lu, J.; Li, Y.; Liang, G.; Huang, Z.; Zhou, J.; Hultman, L.; et al. Halogenated Ti_3C_2 MXenes with electrochemically active terminals for high-performance zinc ion batteries. *ACS Nano* **2021**, *15*, 1077–1085. [CrossRef]
30. Zhou, Y.; Maleski, K.; Anasori, B.; Thostenson, J.O.; Pang, Y.; Feng, Y.; Zeng, K.; Parker, C.B.; Zauscher, S.; Gogotsi, Y.; et al. $Ti_3C_2T_x$ MXene-reduced graphene oxide composite electrodes for stretchable supercapacitors. *ACS Nano* **2020**, *14*, 3576–3586. [CrossRef]
31. Lu, M.; Han, W.; Li, H.; Li, H.; Zhang, B.; Zhang, W.; Zheng, W. Magazine-bending-inspired architecting anti-T of MXene flakes with vertical ion transport for high-performance supercapacitors. *Adv. Mater. Interfaces* **2019**, *6*, 1900160. [CrossRef]
32. Meng, J.; Zhang, F.; Zhang, L.; Liu, L.; Chen, J.; Yang, B.; Yan, X. Rolling up MXene sheets into scrolls to promote their anode performance in lithium-ion batteries. *J. Energy Chem.* **2020**, *46*, 256–263. [CrossRef]
33. Shi, H.; Yue, M.; Zhang, C.J.; Dong, Y.; Lu, P.; Zheng, S.; Huang, H.; Chen, J.; Wen, P.; Xu, Z.; et al. 3D flexible, conductive, and recyclable $Ti_3C_2T_x$ MXene-melamine foam for high-areal-capacity and long-lifetime alkali-metal anode. *ACS Nano* **2020**, *14*, 8678–8688. [CrossRef]
34. Fan, Z.; Wei, C.; Yu, L.; Xia, Z.; Cai, J.; Tian, Z.; Zou, G.; Dou, S.X.; Sun, J. 3D printing of porous nitrogen-doped Ti_3C_2 MXene scaffolds for high-performance sodium-ion hybrid capacitors. *ACS Nano* **2020**, *14*, 867–876. [CrossRef]
35. Dong, Y.; Shi, H.; Wu, Z.S. Recent advances and promise of MXene-based nanostructures for high-performance metal ion batteries. *Adv. Funct. Mater.* **2020**, *30*, 2000706. [CrossRef]
36. Huang, H.; Jiang, R.; Feng, Y.; Ouyang, H.; Zhou, N.; Zhang, X.; Wei, Y. Recent development and prospects of surface modification and biomedical applications of MXenes. *Nanoscale* **2020**, *12*, 1325–1338. [CrossRef] [PubMed]
37. Pang, J.; Mendes, R.G.; Bachmatiuk, A.; Zhao, L.; Ta, H.Q.; Gemming, T.; Liu, H.; Liu, Z.; Rummeli, M.H. Applications of 2D MXenes in energy conversion and storage systems. *Chem. Soc. Rev.* **2019**, *48*, 72–133. [CrossRef]
38. Nan, J.; Guo, X.; Xiao, J.; Li, X.; Chen, W.; Wu, W.; Liu, H.; Wang, Y.; Wu, M.; Wang, G. Nanoengineering of 2D MXene-based materials for energy storage applications. *Small* **2021**, *17*, 1902085. [CrossRef] [PubMed]
39. Xiong, D.; Li, X.; Bai, Z.; Lu, S. Recent advances in layered $Ti_3C_2T_x$ MXene for electrochemical energy storage. *Small* **2018**, *14*, 1703419. [CrossRef] [PubMed]
40. Alhabeb, M.; Maleski, K.; Anasori, B.; Lelyukh, P.; Clark, L.; Sin, S.; Gogotsi, Y. Guidelines for synthesis and processing of two-dimensional titanium carbide ($Ti_3C_2T_x$ MXene). *Chem. Mater.* **2017**, *29*, 7633–7644. [CrossRef]
41. Barsoum, B.M.W. *MAX Phases: Properties of Machinable Ternary Carbides and Nitrides*; Wiley-VCH Verlag GmbH & Co. KGaA: Weinheim, Germany, 2013.
42. Naguib, M.; Kurtoglu, M.; Presser, V.; Lu, J.; Niu, J.; Heon, M.; Hultman, L.; Gogotsi, Y.; Barsoum, M.W. Two-dimensional nanocrystals produced by exfoliation of Ti_3AlC_2. *Adv. Mater.* **2011**, *23*, 4248–4253. [CrossRef] [PubMed]
43. Halim, J.; Lukatskaya, M.R.; Cook, K.M.; Lu, J.; Smith, C.R.; Naslund, L.A.; May, S.J.; Hultman, L.; Gogotsi, Y.; Eklund, P.; et al. Transparent conductive two-dimensional titanium carbide epitaxial thin films. *Chem. Mater.* **2014**, *26*, 2374–2381. [CrossRef] [PubMed]
44. Shahzad, F.; Alhabeb, M.; Hatter, C.B.; Anasori, B.; Hong, S.M.; Koo, C.M.; Gogotsi, Y. Electromagnetic interference shielding with 2D transition metal carbides (MXenes). *Science* **2016**, *353*, 1137–1140. [CrossRef] [PubMed]
45. Lipatov, A.; Alhabeb, M.; Lukatskaya, M.R.; Boson, A.; Gogotsi, Y.; Sinitskii, A. Effect of synthesis on quality, electronic properties and environmental stability of individual monolayer Ti_3C_2 MXene flakes. *Adv. Electron. Mater.* **2016**, *2*, 1600255. [CrossRef]
46. Vahid Mohammadi, A.; Moncada, J.; Chen, H.; Kayali, E.; Orangi, J.; Carrero, C.A.; Beidaghi, M. Thick and freestanding MXene/PANI pseudocapacitive electrodes with ultrahigh specific capacitance. *J. Mater. Chem. A* **2018**, *6*, 22123–22133. [CrossRef]
47. Yang, S.; Zhang, P.; Wang, F.; Ricciardulli, A.G.; Lohe, M.R.; Blom, P.W.M.; Feng, X. Fluoride-free synthesis of two-dimensional titanium carbide (MXene) using a binary aqueous system. *Angew. Chem. Int. Ed.* **2018**, *57*, 15491–15495. [CrossRef]
48. Zhao, Y.; Zhang, M.; Yan, H.; Feng, Y.; Zhang, X.; Guo, R. Few-layer large $Ti_3C_2T_x$ sheets exfoliated by $NaHF_2$ and applied to the sodium-ion battery. *J. Mater. Chem. A* **2021**, *9*, 9593–9601. [CrossRef]

49. Feng, A.; Yu, Y.; Wang, Y.; Jiang, F.; Yu, Y.; Mi, L.; Song, L. Two-dimensional MXene Ti_3C_2 produced by exfoliation of Ti_3AlC_2. *Mater. Des.* **2017**, *114*, 161–166. [CrossRef]
50. Levitt, A.S.; Alhabeb, M.; Hatter, C.B.; Sarycheva, A.; Dionb, G.; Gogotsi, Y. Electrospun MXene/carbon nanofibers as supercapacitor electrodes. *J. Mater. Chem. A* **2019**, *7*, 269–277. [CrossRef]
51. Huang, X.; Wu, P. A facile, high-yield, and freeze-and-thaw-assisted approach to fabricate MXene with plentiful wrinkles and its application in on-chip micro-supercapacitors. *Adv. Funct. Mater.* **2020**, *30*, 1910048. [CrossRef]
52. Mendoza-Sanchez, B.; Gogotsi, Y. Synthesis of two-dimensional materials for capacitive energy storage. *Adv. Mater.* **2016**, *28*, 6104–6135. [CrossRef] [PubMed]
53. Liu, T.; Liu, S.; Tu, K.H.; Schmidt, H.; Chu, L.; Xiang, D.; Martin, J.; Eda, G.; Ross, C.A.; Garaj, S. Crested two-dimensional transistors. *Nat. Nanotechnol.* **2019**, *14*, 223–226. [CrossRef] [PubMed]
54. Naguib, M.; Mashtalir, O.; Carle, J.; Presser, V.; Lu, J.; Hultman, L.; Gogotsi, Y.; Barsoum, M.W. Two-dimensional transition metal carbides. *ACS Nano* **2012**, *6*, 1322–1331. [CrossRef]
55. Lukatskaya, M.R.; Mashtalir, O.; Ren, C.E.; Dall'Agnese, Y.; Rozier, P.; Taberna, P.L.; Naguib, M.; Simon, P.; Barsoum, M.W.; Gogotsi, Y. Cation intercalation and high volumetric capacitance of two-dimensional titanium carbide. *Science* **2013**, *341*, 1502–1505. [CrossRef] [PubMed]
56. Yu, L.; Fan, Z.; Shao, Y.; Tian, Z.; Sun, J.; Liu, Z. Versatile N-doped MXene ink for printed electrochemical energy storage application. *Adv. Energy Mater.* **2019**, *9*, 1901839. [CrossRef]
57. Persson, I.; El Ghazaly, A.; Tao, Q.; Halim, J.; Kota, S.; Darakchieva, V.; Palisaitis, J.; Barsoum, M.W.; Rosen, J.; Persson, P.O.A. Tailoring structure, composition, and energy storage properties of MXenes from selective etching of in-plane, chemically ordered max phases. *Small* **2018**, *14*, 1703676. [CrossRef]
58. Wang, F.; Wu, X.; Yuan, X.; Liu, Z.; Zhang, Y.; Fu, L.; Zhu, Y.; Zhou, Q.; Wu, Y.; Huang, W. Latest advances in supercapacitors: From new electrode materials to novel device designs. *Chem. Soc. Rev.* **2017**, *46*, 6816–6854. [CrossRef]
59. Shao, Y.; El-Kady, M.F.; Sun, J.; Li, Y.; Zhang, Q.; Zhu, M.; Wang, H.; Dunn, B.; Kaner, R.B. Design and mechanisms of asymmetric supercapacitors. *Chem. Rev.* **2018**, *118*, 9233–9280. [CrossRef]
60. Zhong, C.; Deng, Y.; Hu, W.; Qiao, J.; Zhang, L.; Zhang, J. A review of electrolyte materials and compositions for electrochemical supercapacitors. *Chem. Soc. Rev.* **2018**, *44*, 7484–7539. [CrossRef]
61. Muzaffar, A.; Ahamed, M.B.; Deshmukh, K.; Thirumalai, J. A review on recent advances in hybrid supercapacitors: Design, fabrication and applications. *Renew. Sustain. Energy Rev.* **2019**, *101*, 123–145. [CrossRef]
62. Liu, H.; Liu, X.; Wang, S.; Liu, H.-K.; Li, L. Transition metal based battery-type electrodes in hybrid supercapacitors: A review. *Energy Storage Mater.* **2020**, *28*, 122–145. [CrossRef]
63. Lukatskaya, M.R.; Bak, S.-M.; Yu, X.; Yang, X.-Q.; Barsoum, M.W.; Gogotsi, Y. Probing the mechanism of high capacitance in 2D titanium carbide using in situ x-ray absorption spectroscopy. *Adv. Energy Mater.* **2015**, *5*, 1500589. [CrossRef]
64. Li, H.; Li, X.; Liang, J.; Chen, Y. Hydrous RuO_2-decorated MXene coordinating with silver nanowire inks enabling fully printed micro-supercapacitors with extraordinary volumetric performance. *Adv. Energy Mater.* **2019**, *9*, 1803987. [CrossRef]
65. Pan, Z.; Cao, F.; Hua, X.; Ji, X. A facile method for synthesizing CuS decorated Ti_3C_2 MXene with enhanced performance for asymmetric supercapacitors. *J. Mater. Chem. A* **2019**, *7*, 8984–8992. [CrossRef]
66. Wang, X.; Li, H.; Li, H.; Lin, S.; Ding, W.; Zhu, X.; Sheng, Z.; Wang, H.; Zhu, X.; Sun, Y. 2D/2D 1T-MoS_2/Ti_3C_2 MXene heterostructure with excellent supercapacitor performance. *Adv. Funct. Mater.* **2020**, *30*, 0190302. [CrossRef]
67. Wang, Y.; Dou, H.; Wang, J.; Ding, B.; Xu, Y.; Chang, Z.; Hao, X. Three-dimensional porous MXene/layered double hydroxide composite for high performance supercapacitors. *J. Power Source* **2016**, *327*, 221–228. [CrossRef]
68. Luo, Y.; Yang, C.; Tian, Y.; Tang, Y.; Yin, X.; Que, W. A long cycle life asymmetric supercapacitor based on advanced nickel-sulfide/titanium carbide (MXene) nanohybrid and MXene electrodes. *J. Power Source* **2020**, *450*, 227694. [CrossRef]
69. Wen, Y.; Rufford, T.E.; Chen, X.; Li, N.; Lyu, M.; Dai, L.; Wang, L. Nitrogen-doped $Ti_3C_2T_x$ MXene electrodes for high-performance supercapacitors. *Nano Energy* **2017**, *38*, 368–376. [CrossRef]
70. Mu, X.; Wang, D.; Du, F.; Chen, G.; Wang, C.; Wei, Y.; Gogotsi, Y.; Gao, Y.; Dall'Agnese, Y. Revealing the pseudo-intercalation charge storage mechanism of MXenes in acidic electrolyte. *Adv. Funct. Mater.* **2019**, *29*, 1902953. [CrossRef]
71. Zhang, M.; Chen, X.; Sui, J.; Abraha, B.S.; Li, Y.; Peng, W.; Zhang, G.; Zhang, F.; Fan, X. Improving the performance of titanium carbide MXene in supercapacitor by partial oxidation treatment. *Inorg. Chem. Front.* **2020**, *7*, 1205–1211. [CrossRef]
72. Li, Z.; Ma, C.; Wen, Y.; Wei, Y.; Xing, X.; Chu, J.; Yu, C.; Wang, K.; Wang, Z.-K. Highly conductive dodecaborate/MXene composites for high performance supercapacitors. *Nano Res.* **2019**, *13*, 196–202. [CrossRef]
73. Li, K.; Wang, X.; Li, S.; Urbankowski, P.; Li, J.; Xu, Y.; Gogotsi, Y. An ultrafast conducting polymer@MXene positive electrode with high volumetric capacitance for advanced asymmetric supercapacitors. *Small* **2020**, *16*, 1906851. [CrossRef]
74. Zhang, C.J.; Anasori, B.; Seral-Ascaso, A.; Park, S.H.; McEvoy, N.; Shmeliov, A.; Duesberg, G.S.; Coleman, J.N.; Gogotsi, Y.; Nicolosi, V. Transparent, flexible, and conductive 2D titanium carbide (MXene) films with high volumetric capacitance. *Adv. Mater.* **2017**, *29*, 1702678. [CrossRef] [PubMed]
75. Zhao, W.; Peng, J.; Wang, W.; Jin, B.; Chen, T.; Liu, S.; Zhao, Q.; Huang, W. Interlayer hydrogen-bonded metal porphyrin frameworks/MXene hybrid film with high capacitance for flexible all-solid-state supercapacitors. *Small* **2019**, *15*, 1901351. [CrossRef]

76. Wu, X.; Huang, B.; Lv, R.; Wang, Q.; Wang, Y. Highly flexible and low capacitance loss supercapacitor electrode based on hybridizing decentralized conjugated polymer chains with MXene. *Chem. Eng. J.* **2019**, *378*, 122246. [CrossRef]
77. Qin, L.; Tao, Q.; El Ghazaly, A.; Fernandez-Rodriguez, J.; Persson, P.O.Å.; Rosen, J.; Zhang, F. High-performance ultrathin flexible solid-state supercapacitors based on solution processable $Mo_{1.33}C$ MXene and PEDOT:PSS. *Adv. Funct. Mater.* **2018**, *28*, 1703808. [CrossRef]
78. Shi, M.; Xiao, P.; Lang, J.; Yan, C.; Yan, X. Porous $g-C_3N_4$ and MXene dual-confined FeOOH quantum dots for superior energy storage in an ionic liquid. *Adv. Sci.* **2020**, *7*, 1901975. [CrossRef]
79. Tian, W.; Vahid Mohammadi, A.; Reid, M.S.; Wang, Z.; Ouyang, L.; Erlandsson, J.; Pettersson, T.; Wagberg, L.; Beidaghi, M.; Hamedi, M.M. Multifunctional nanocomposites with high strength and capacitance using 2D MXene and 1D nanocellulose. *Adv. Mater.* **2019**, *31*, 1902977. [CrossRef] [PubMed]
80. Wang, Y.; Wang, X.; Li, X.; Bai, Y.; Xiao, H.; Liu, Y.; Liu, R.; Yuan, G. Engineering 3D ion transport channels for flexible MXene films with superior capacitive performance. *Adv. Funct. Mater.* **2019**, *29*, 1900326. [CrossRef]
81. Quain, E.; Mathis, T.S.; Kurra, N.; Maleski, K.; van Aken, K.L.; Alhabeb, M.; Alshareef, H.N.; Gogotsi, Y. Direct writing of additive-free MXene-in-water ink for electronics and energy storage. *Adv. Mater. Technol.* **2019**, *4*, 1800256. [CrossRef]
82. Wang, Y.; Wang, X.; Li, X.; Liu, R.; Bai, Y.; Xiao, H.; Liu, Y.; Yuan, G. Intercalating ultrathin MoO_3 nanobelts into MXene film with ultrahigh volumetric capacitance and excellent deformation for high-energy-density devices. *Nano-Micro Lett.* **2020**, *12*, 115. [CrossRef] [PubMed]
83. Abdolhosseinzadeh, S.; Schneider, R.; Verma, A.; Heier, J.; Nuesch, F.; Zhang, C.J. Turning trash into treasure: Additive free MXene sediment inks for screen-printed micro-supercapacitors. *Adv. Mater.* **2020**, *32*, 2000716. [CrossRef] [PubMed]
84. Hu, H.; Hua, T. An easily manipulated protocol for patterning of MXenes on paper for planar micro-supercapacitors. *J. Mater. Chem. A* **2017**, *5*, 19639–19648. [CrossRef]
85. Zhang, J.; Seyedin, S.; Qin, S.; Wang, Z.; Moradi, S.; Yang, F.; Lynch, P.A.; Yang, W.; Liu, J.; Wang, X.; et al. Highly conductive $Ti_3C_2T_x$ MXene hybrid fibers for flexible and elastic fiber-shaped supercapacitors. *Small* **2019**, *15*, 1804732. [CrossRef]
86. Shao, W.; Tebyetekerwa, M.; Marriam, I.; Li, W.; Wu, Y.; Peng, S.; Ramakrishna, S.; Yang, S.; Zhu, M. Polyester@MXene nanofibers-based yarn electrodes. *J. Power Source* **2018**, *396*, 683–690. [CrossRef]
87. Chang, T.H.; Zhang, T.; Yang, H.; Li, K.; Tian, Y.; Lee, J.Y.; Chen, P.Y. Controlled crumpling of two-dimensional titanium carbide (MXene) for highly stretchable, bendable, efficient supercapacitors. *ACS Nano* **2018**, *12*, 8048–8059. [CrossRef] [PubMed]
88. Lipatov, A.; Lu, H.; Alhabeb, M.; Anasori, B.; Gruverman, A.; Gogotsi, Y.; Sinitskii, A. Elastic properties of 2D $Ti_3C_2T_x$ MXene monolayers and bilayers. *Sci. Adv.* **2018**, *4*, 491. [CrossRef]
89. An, H.; Habib, T.; Shah, S.; Gao, H.; Radovic, M.; Green, M.J.; Lutkenhaus, J.L. Surface-agnostic highly stretchable and bendable conductive MXene multilayers. *Sci. Adv.* **2018**, *4*, 118. [CrossRef] [PubMed]
90. Come, J.; Xie, Y.; Naguib, M.; Jesse, S.; Kalinin, S.V.; Gogotsi, Y.; Kent, P.R.C.; Balke, N. Nanoscale elastic changes in 2D $Ti_3C_2T_x$ (MXene) pseudocapacitive electrodes. *Adv. Energy Mater.* **2016**, *6*, 1502290. [CrossRef]
91. Shang, M.; Chen, X.; Li, B.; Niu, J. A fast charge/discharge and wide-temperature battery with a germanium oxide layer on a $Ti_3C_2T_x$ MXene matrix as anode. *ACS Nano* **2020**, *14*, 3678–3686. [CrossRef] [PubMed]
92. Li, J.; Han, L.; Li, Y.; Li, J.; Zhu, G.; Zhang, X.; Lu, T.; Pan, L. MXene-decorated SnS_2/Sn_3S_4 hybrid as anode material for high-rate lithium-ion batteries. *Chem. Eng. J.* **2020**, *380*, 122590. [CrossRef]
93. Tang, X.; Liu, H.; Guo, X.; Wang, S.; Wu, W.; Mondal, A.K.; Wang, C.; Wang, G. A novel lithium-ion hybrid capacitor based on the aerogellikeMXene wrapped Fe_2O_3nanosphere anode and the 3D nitrogen sulphur dual-doped porous carbon cathode. *Mater. Chem. Front.* **2018**, *2*, 1811–1821. [CrossRef]
94. Ahmed, B.; Anjum, D.H.; Gogotsi, Y.; Alshareef, H.N. Atomic layer deposition of SnO_2 on MXene for Li-ion battery anodes. *Nano Energy* **2017**, *34*, 249–256. [CrossRef]
95. Cheng, R.; Wang, Z.; Cui, C.; Hu, T.; Fan, B.; Wang, H.; Liang, Y.; Zhang, C.; Zhang, H.; Wang, X. One-step incorporation of nitrogen and vanadium between $Ti_3C_2T_x$ MXene interlayers enhances lithium ion storage capability. *J. Phys. Chem. C* **2020**, *124*, 6012–6021. [CrossRef]
96. Ruan, T.; Wang, B.; Yang, Y.; Zhang, X.; Song, R.; Ning, Y.; Wang, Z.; Yu, H.; Zhou, Y.; Wang, D.; et al. Interfacial and electronic modulation via localized sulfurization for boosting lithium storage kinetics. *Adv. Mater.* **2020**, *32*, 2000151. [CrossRef]
97. Zhang, D.; Wang, S.; Li, B.; Yang, Y.G.S. Horizontal growth of lithium on parallelly aligned MXene layers towards dendrite-free metallic lithium anodes. *Adv. Mater.* **2019**, *31*, 1901820. [CrossRef]
98. Wei, C.; Fei, H.; Tian, Y.; An, Y.; Zeng, G.; Feng, J.; Qian, Y. Room-temperature liquid metal confined in MXene paper as a flexible, freestanding, and binder-free anode for next-generation lithium-ion batteries. *Small* **2019**, *15*, 1903214. [CrossRef]
99. Yang, Q.; Huang, Z.; Li, X.; Liu, Z.; Li, H.; Liang, G.; Wang, D.; Huang, Q.; Zhang, S.; Chen, S.; et al. A wholly degradable, rechargeable $Zn-Ti_3C_2$ MXene capacitor with superior anti-self-discharge function. *ACS Nano* **2019**, *13*, 8275–8283. [CrossRef] [PubMed]
100. Wang, S.; Wang, Q.; Zeng, W.; Wang, M.; Ruan, L.; Ma, Y. A new free-standing aqueous zinc-ion capacitor based on MnO_2–CNTs cathode and MXene anode. *Nano-Micro Lett.* **2019**, *11*, 70. [CrossRef]
101. Shi, J.; Wang, S.; Wang, Q.; Chen, X.; Du, X.; Wang, M.; Zhao, Y.; Dong, C.; Ruan, L.; Zeng, W. A new flexible zinc-ion capacitor based on δ-MnO_2@carbon cloth battery-type cathode and MXene@cotton cloth capacitor-type anode. *J. Power Source* **2020**, *446*, 227345. [CrossRef]

Article

A Comparison of Electrical Breakdown Models for Polyethylene Nanocomposites

Zhaoliang Xing [1,2], Chong Zhang [1,2], Mengyao Han [2], Ziwei Gao [2], Qingzhou Wu [3] and Daomin Min [2,*]

1. State Key Laboratory of Advanced Power Transmission Technology, Global Energy Interconnection Research Institute Co., Ltd., Beijing 102209, China; 15811444029@163.com (Z.X.); zhangc@sgcc.com.cn (C.Z.)
2. State Key Laboratory of Electrical Insulation and Power Equipment, Xi'an Jiaotong University, Xi'an 710049, China; hmy1042286601@stu.xjtu.edu.cn (M.H.); gaoziwei@stu.xjtu.edu.cn (Z.G.)
3. Institute of Fluid Physics, China Academy of Engineering Physics, Mianyang 621900, China; wuqingzhou@163.com
* Correspondence: forrestmin@xjtu.edu.cn; Tel.: +86-29-8266-3781

Abstract: The development of direct current high-voltage power cables requires insulating materials having excellent electrically insulation properties. Experiments show that appropriate nanodoping can improve the breakdown strength of polyethylene (PE) nanocomposites. Research indicates that traps, free volumes, and molecular displacement are key factors affecting the breakdown strength. This study comprehensively considered the space charge transport, electron energy gain, and molecular chain long-distance movement during the electrical breakdown process. In addition, we established three simulation models focusing on the electric field distortion due to space charges captured by traps, the energy gain of mobile electrons in free volumes, the free volume expansion caused by long-distance movement of molecular chains under the Coulomb force, and the energy gained by the electrons moving in the enlarged free volumes. The three simulation models considered the electrical breakdown modulated by space charges, with a maximum electric field criterion and a maximum electron energy criterion, and the electrical breakdown modulated by the molecular displacement (EBMD), with a maximum electron energy criterion. These three models were utilized to simulate the breakdown strength dependent on the nanofiller content of PE nanocomposites. The simulation results of the EBMD model coincided best with the experimental results. It was revealed that the breakdown electric field of PE nanodielectrics is improved synergistically by both the strong trapping effect of traps and the strong binding effect of molecular chains in the interfacial regions.

Keywords: polymer nanocomposites; traps; DC breakdown; energy gain; molecular motion; free volume

1. Introduction

Power cables are the main equipment in urban transmission grids and offshore wind power transmission [1–6]. Direct current (DC) power cables have the advantages of long transmission distance, large transmission capacity, and low power loss, and they are the key electrical equipment for large-scale reception of new energy power generation. Under the action of DC voltage, the power cable has no capacitive current and can realize long-distance power transmission. The aging of the insulating material of power cables under the DC electric field is slow, and its lifespan is greatly prolonged [1,3–6]. Moreover, the breakdown electric field of the insulating material under DC voltage is 2–3 times higher than that under AC voltage [7], which improves the safety margin of the DC power cable. Low-density polyethylene (LDPE) is the main insulating material of power cables, and its electrical insulation performance is important for the safe and reliable operation of power cables [1–6]. Polymer nanocomposites (PNCs) have excellent properties, such as lower electrical conductivity, higher breakdown electric field, less space charges, higher thermal stability, and higher mechanical strength [5,6,8–12]. PNCs, known as third-generation insulating materials, have broad application prospects. The experimental results demonstrate that doping a relatively low content of nanoscale fillers in polyethylene can form deep traps, leading to the reduction in the electrical conductivity of the nanocomposites and the improvement in the breakdown strength [5,6,9,11–14]. Advanced LDPE nanocomposites can be used as insulating materials for DC power cables [5,6,12] and energy storage capacitors [15,16], improving the capacity of power cables to transmit electrical energy and the energy storage density of capacitors. Power cables and energy storage capacitors are key equipment for the centralized transmission of large-scale offshore wind power, providing support for the supply of clean energy to cities.

It is generally believed that the excellent electrical properties of polymer nanocomposites originate from the interfacial region between the nanoparticles and the polymer matrix [6,12]. The multi-core interfacial region model proposed by Tanaka et al. [17,18] and the multi-region structure model proposed by Li et al. [19] show that deep traps are formed in nanocomposites when a small amount of doping is used. Deep traps near the electrodes trap more charges and reduce the number of charges injected. This can suppress the space charge accumulation, reduce the electric field concentration, and improve the breakdown electric field. By comparison, traps with higher energies can reduce the effective charge carrier mobility, reduce electrical conductivity and Joule heating, and improve breakdown performance. The interfacial region models proposed by Nelson et al. [20] and Min et al. [21] show that the interfacial regions not only form deep traps, but also constrain the motion of molecular chains. Under the action of the Coulomb force, the molecular chains undergo directional displacement, which affects the size of the surrounding free volumes. This, in turn, changes the breakdown performance of nanodielectrics. The interfacial region models show that increasing the trap level and/or the interaction between molecular chains in the interfacial regions can improve the breakdown strength of polymer nanocomposites.

Experiments and simulations indicate that the breakdown of polymer materials under a strong electric field is related to physical processes such as electric field distortion and electron energy gain. Tanaka et al. [22] used pulsed electroacoustic equipment to test the space charge distributions and the electric field distributions of LDPE under the action of a strong electric field. It was found that positive space charge packets are formed in LDPE when the electric field is higher than a threshold value. As the positive space charge packets move toward the cathode, the electric field in front of the charge packets gradually increases, and the material is broken down when the maximum distorted electric field in LDPE reaches the breakdown electric field. Chen et al. [23] considered the formation and migration of space charges in LDPE, and established a polymer breakdown model based on the accumulation of space charges and the corresponding electric field distortion. The relationship between the DC breakdown electric field of LDPE and the thickness of samples was calculated. It was found that the breakdown electric field has an inverse power function relationship with the thickness. Choi et al. [24] also used a breakdown model based on the electric field distorted by the space charges to calculate the breakdown characteristics of multilayer polymers with partial barrier contact, and found that partial barrier contact between multilayer structures enhanced the breakdown strength of multilayer dielectrics. From the viewpoint that the electric field force acts on the trapped charges and affects the molecular chain motion, and to comprehensively consider the charge transport, the long-range motion of molecular chains, and the electron energy accumulation, we established a charge trapping and molecular displacement breakdown (CTMD) model for polymer nanocomposites [13,25]. The energy accumulation process of electrons in the free volume expanded by the long-range motion of the molecular chain was simulated, and the relation between the breakdown electric field of the polyethylene nanocomposites with the nanofiller content, the applied pressure, the thickness of the sample, and the ramping rate was obtained. The results are consistent with the results of the electric breakdown experiments.

The above analysis shows that the strong trapping effect of traps and the strong interaction between molecular chains in the interface regions are the two key factors to improve the breakdown strength. However, which of the trap trapping effect and the molecular chain interaction is more influential is still unclear. To clarify the factors of the breakdown characteristics of polyethylene nanocomposites having the greatest influence, this study compared three breakdown models, namely, the electric field distortion, the electron energy gain in a fixed-scale free volume, and the energy gain of electrons in an expanded free volume caused by the motion of molecular chains. By comparing the simulation results with the experiments, the electric breakdown mechanism of polyethylene nanocomposites was clarified. In the present work, we determined the quantitative roles of trapping effects and molecular chain interactions on breakdown strength. This paper provides simulation methods and data support for the improvement in the breakdown strength of polymer nanocomposites.

2. Electrical Breakdown Models

Generally, a ramp voltage with a constant rising rate is applied to the electrodes on both sides of the polymer nanocomposite to investigate the electrical breakdown properties. Firstly, when the voltage is gradually increased, the electrons and holes in the cathode and anode, respectively, are injected into the nanocomposite. After these electrons and holes enter the nanocomposite, they drift under the driving of the electric field. Since there are deep traps formed by many polar groups inside the nanocomposite, the deep traps capture electrons and holes, thereby accumulating space charges of the same polarity inside the nanocomposite [26]. The space charges cause the electric field inside the nanocomposite to be distorted. When the distorted electric field reaches a certain level, it may lead to the breakdown of the nanocomposite. Nanodoping changes the trap properties inside the nanocomposite, which in turn changes the charge injection, the space charge accumulation, and the electric field concentration properties; these changes ultimately lead to the variation in the breakdown strength with the nanofiller content. This breakdown model is called the electrical breakdown modulated by space charges with a maximum electric field criterion (EBEF).

Secondly, there are spaces inside the polymer nanocomposite that are not occupied by atoms, known as free volumes. When charges are transported inside the nanocomposite, they may enter the free volumes. Charges in the free volumes can be rapidly accelerated by electric fields to obtain high energies. If the energy gained by the charges in the free volumes exceeds the trapping ability of the deep traps, the high-energy charges will result in the breaking of the molecular chain and cause the electric breakdown of the nanocomposite [27]. Nanodoping changes the trap energy, which in turn affects the ability of traps to trap the high-energy charges, and ultimately changes the breakdown strength of the nanocomposite. In this process, the effects of space charge accumulation and electric field concentration on the breakdown strength also need to be considered. This breakdown model is called the electrical breakdown modulated by space charges with a maximum electron energy criterion (EBEG).

Thirdly, since the substance entities trapped inside the polymer nanocomposites are polar groups, when the traps capture charges, the Coulomb force on the trapped charges must be transferred to the molecular chains. The molecular chains will undergo directional displacement under the effect of the Coulomb force, resulting in the expansion of the free volumes. An increase in the size of free volumes allows the charges entering them to accumulate more energies. When the energies accumulated by the charges exceed the trapping ability of the deep traps, it causes the breakdown of the polymer nanocomposite. Nanodoping affects both the molecular chain motion and trap properties in nanocomposites, which in turn affects the energy gain properties of charges and the ability of traps to capture high-energy charges. Finally, the electric breakdown strength of the nanodielectric is changed. This breakdown model is called the EBMD model [13,25].

The three breakdown models of EBEF, EBEG, and EBMD all include charge carrier injection and charge carrier transport processes. The mathematical equations of charge carrier injection and charge carrier transport processes are introduced first. Then, the breakdown criteria of EBEF and EBEG models are given. Finally, the equation of molecular chain displacement under the action of electric force in EBMD model is introduced, and the breakdown criterion is given.

2.1. Charge Injection and Charge Transport in Nanocomposites

Figure 1 presents a schematic diagram of the charge transport inside a polymer nanocomposite [25]. The one-dimensional coordinate x is set in the thickness direction of the nanocomposite, and the thickness of the material is L. The left side of the nanocomposite is the cathode and the right side is the anode. Under the action of an applied voltage, the cathode injects electrons into the nanocomposite, and the anode injects holes. Potential barriers exist between the cathode and anode and the nanocomposite interface, which are $u_{in(e)}$ and $u_{in(h)}$, respectively. The interfacial barrier hinders the transfer of charges from the electrode into the nanocomposite. Under an externally applied strong electric field, a potential barrier lowering u_{Sch} appears in the interface barrier, which promotes the transfer of charges to the bulk of the nanocomposite. When the applied electric field is E, the Schottky barrier reduction u_{Sch} is proportional to the square root of the electric field, $u_{Sch} = (eE/4\pi\varepsilon_0\varepsilon_r)^{1/2}$. Here, e is the electron charge, ε_0 is the permittivity of the vacuum in F/m, and ε_r is the dielectric constant of the nanocomposite. We adopt the Schottky thermal emission model to describe the charge injections, $j_{in(e)}$ and $j_{in(h)}$, per unit time and unit area of the cathode and anode, into the nanodielectrics [28].

$$j_{in(e)}(0,t) = AT^2 \exp\left(-u_{in(e)}/k_BT\right) \exp(u_{Sch}(0,t)/k_BT) \quad (1)$$

$$j_{in(h)}(L,t) = AT^2 \exp\left(-u_{in(h)}/k_BT\right) \exp(u_{Sch}(L,t)/k_BT) \quad (2)$$

where A is the Richardson constant, T is the temperature of materials, and t is the elapsed time of applied voltage.

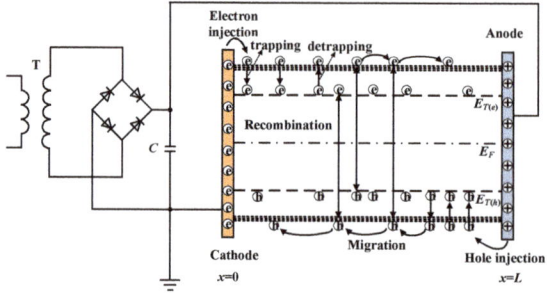

Figure 1. Schematic of the bipolar charge carrier injection and charge carrier transport in polymer nanocomposites under a ramp voltage.

Under an electric field E, there will be a certain concentration of mobile electrons and mobile holes in extended states after the charges in the electrodes are injected into the nanocomposite. Their concentrations are $n_{M(e)}$ and $n_{M(h)}$, respectively. These mobile electrons and mobile holes migrate in the extended states driven by the applied electric field. The mobilities of electrons and holes in the extended states are $\mu_{0(e)}$ and $\mu_{0(h)}$, respectively. Due to the existence of polar groups in polymer nanodielectrics, a certain concentration of deep traps is formed. It is assumed that the energy levels of deep traps of electrons and holes are $u_{T(e)}$ and $u_{T(h)}$, respectively, and their concentrations are $N_{T(e)}$ and $N_{T(h)}$, respectively. Due to the strong trapping ability of deep traps, mobile electrons and mobile holes may be captured by them during the migration process. The trapping probabilities of electrons and holes in extended states in the electron deep traps and hole deep traps are $P_{T(e)}$ and $P_{T(h)}$, respectively. After a period of time, the deep trapped electron and deep trapped hole densities are $n_{T(e)}$ and $n_{T(h)}$, respectively. In addition, the charges in the deep traps gradually gain energy from the heat bath due to thermal excitation. When the trapped charges gain sufficient energy, trapped electrons and trapped holes can transition to extended states with probabilities of $P_{D(e)}$ and $P_{D(h)}$, respectively. The detrapping probabilities $P_{D(e)}$ and $P_{D(h)}$ decrease exponentially with the increase in the deep trap levels $u_{T(e)}$ and $u_{T(h)}$. When positive and negative charges meet inside the sample, the recombination between these charges occurs. Mobile electrons may recombine with mobile and trapped holes, and mobile holes may recombine with mobile and trapped electrons. The trapped electrons mainly recombine with the mobile holes, and the trapped holes mainly recombine with the mobile electrons.

The time-dependent change in the charge densities of mobile electrons and mobile holes per unit volume is related to the current density flowing into and out of the control volume. Additionally, trap capturing and recombination lead to a decrease in mobile electrons and mobile holes, while the detrapping of trapped charges results in an increase in mobile electrons and mobile holes. The time-dependent changes in trapped electrons and trapped holes per unit volume are related to charge trapping, charge detrapping, and recombination. Charge detrapping and recombination lead to a decrease in deep trapped electrons and deep trapped holes, while charge trapping leads to an increase in deep trapped electrons and deep trapped holes. Four partial differential equations are needed to describe the dynamic processes of mobile electrons, deep trap electrons, mobile holes, and deep trap holes in nanocomposites, respectively [25,29,30]:

$$\frac{\partial n_{M(e)}}{\partial t} + \frac{\partial \left(n_{M(e)} \mu_{0(e)} E\right)}{\partial x} = -P_{T(e)} n_{M(e)} \left(1 - \frac{n_{T(e)}}{N_{T(e)}}\right) + P_{D(e)} n_{T(e)} - R_{Me,Mh} n_{M(e)} n_{M(h)} - R_{Me,Th} n_{M(e)} n_{T(h)} \quad (3)$$

$$\frac{\partial n_{T(e)}}{\partial t} = P_{T(e)} n_{M(e)} \left(1 - \frac{n_{T(e)}}{N_{T(e)}}\right) - P_{D(e)} n_{T(e)} - R_{Te,Mh} n_{T(e)} n_{M(h)} \quad (4)$$

$$\frac{\partial n_{M(h)}}{\partial t} + \frac{\partial \left(n_{M(h)} \mu_{0(h)} E\right)}{\partial x} = -P_{T(h)} n_{M(h)} \left(1 - \frac{n_{T(h)}}{N_{T(h)}}\right) + P_{D(h)} n_{T(h)} - R_{Me,Mh} n_{M(e)} n_{M(h)} - R_{Te,Mh} n_{T(e)} n_{M(h)} \quad (5)$$

$$\frac{\partial n_{T(h)}}{\partial t} = P_{T(h)} n_{M(h)} \left(1 - \frac{n_{T(h)}}{N_{T(h)}}\right) - P_{D(h)} n_{T(h)} - R_{Me,Th} n_{M(e)} n_{T(h)} \quad (6)$$

The detrapping probabilities $P_{D(e)}$ and $P_{D(h)}$ of the deep trapped charges in the nanocomposite are related to the trap energy levels $u_{T(e)}$ and $u_{T(h)}$ as $P_{D(e,h)} = \nu_0 \exp(-u_{T(e,h)}/k_B T)$. Here, ν_0 is the attempt-to-escape frequency. $R_{Me,Mh}$ is the recombination coefficient between mobile electrons and mobile holes. According to the Langevin recombination model, the recombination coefficient $R_{Me,Mh}$ is determined by $e(\mu_{0(e)} + \mu_{0(h)})/\varepsilon_0\varepsilon_r$ [31]. $R_{Me,Th}$ and $R_{Te,Mh}$ are the recombination coefficient between mobile electrons and trapped holes, and that between trapped electrons and mobile holes, respectively. According to the Shockley–Read–Hall model, the recombination rates $R_{Me,Th}$ and $R_{Te,Mh}$ are determined by $e\mu_{0(e)}/\varepsilon_0\varepsilon_r$ and $e\mu_{0(h)}/\varepsilon_0\varepsilon_r$, respectively [32].

When space charges accumulate inside the nanocomposite, the space charges can distort the electric potential and the electric field. The potential φ inside the nanocomposite can be calculated by Poisson's equation:

$$\partial^2 \varphi / \partial x^2 = -e \left(n_{M(h)} + n_{T(h)} - n_{M(e)} - n_{T(e)}\right) / \varepsilon_0 \varepsilon_r \quad (7)$$

Then, the electric field distribution inside the nanocomposite can be calculated through the negative gradient of the electric potential, namely $E = -\nabla \phi$.

2.2. Electrical Breakdown Criteria

In the EBEF breakdown model, the electric field concentration effect caused by the accumulation of space charges is mainly considered. It is assumed that when the highest electric field E_{max} inside the nanocomposite reaches a certain threshold value E_C, the material is broken down. The breakdown criterion is $E_{max} \geq E_C$ [23].

In the EBEG breakdown model, the process of electron energy gain under the applied electric field after the mobile charge enters the free volume is mainly considered. When the energy gained by the mobile charges is greater than the trapping ability of the deep traps, the nanocomposite is broken down. Assuming that the free volume length is λ_0, the energy gained by the electron in the free volume is $e\lambda_0 E(x)$. The maximum energy of electrons inside the nanocomposite is $e\lambda_0 E_{max}$. As shown in Figure 2a, the breakdown criterion of the EBEG model can be obtained as $e\lambda_0 E_{max} \geq u_T$ [27].

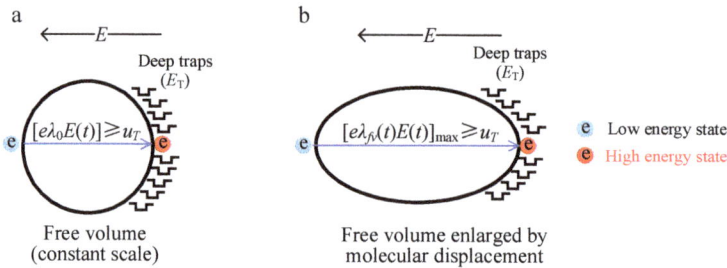

Figure 2. Electrical breakdown criteria of EBEG (**a**) and EBMD (**b**) models under a ramp voltage.

In the EBMD breakdown model, the charge carrier injection and transport processes inside the nanocomposite under a strong electric field are considered. Furthermore, after the charges are captured by the deep traps on the molecular chain, the electric force acts on the molecular chain to cause its directional displacement. The time-dependent relationship of the displacement λ_{mol} of the molecular chain is calculated by the following equation [33]:

$$\frac{d\lambda_{mol}}{dt} = \mu_{mol} E - \frac{\lambda_{mol}}{\tau_{mol}} \qquad (8)$$

where μ_{mol} is the molecular chain mobility and τ_{mol} is the relaxation time constant of the molecular chain. In addition, μ_{mol} decreases with the increase in the deep trap energy, $\mu_{mol} = \mu_0 \exp(-u_T/k_B T)$, and τ_{mol} increases with the increase in the deep trap level, $\tau_{mol} = \nu_0^{-1} \exp(u_T/k_B T)$.

The directional movement of the molecular chain will cause the expansion of the free volume. It is assumed that the free volume length is equal to the molecular chain displacement, that is, $\lambda_{fv}(t) = \lambda_{mol}(t)$. The mobile charges gain energy in the expanded free volume. If the energy gained by the electrons exceeds the trapping ability of the deep trap, the nanocomposite will be broken down. As shown in Figure 2b, the breakdown criterion of the EBMD model is expressed as $[e\lambda_{fv}(t)E(t)]_{max} \geq u_T$ [25].

3. Results

Both the trap energies and the breakdown strengths of LDPE nanodielectrics doped with various fillers first increased and then decreased with the increase in doping content [5,6,11,12]. It is generally believed that the interfacial region is responsible for the excellent electrical properties of polymer nanocomposites. The trap levels in the interfacial region and the interaction between molecular chains are two key factors to improving the breakdown strength of polymer nanocomposites. However, which factor is more influential remains unclear. The main influencing factors of the breakdown strength are determined by comparing the above-mentioned breakdown simulation models with the experimental results. The quantitative relationship between the influencing factors and the breakdown strength was obtained. We took the LDPE/Al$_2$O$_3$ nanocomposite as an example to compare the three breakdown models. In order to easily compare the simulation results with the experimental results, the parameters in the simulation model were derived from the experimental results [13]. The thickness of the LDPE/Al$_2$O$_3$ nanocomposite was 200 μm and the temperature of the sample was 300 K. We investigated the breakdown properties of pure LDPE and LDPE/Al$_2$O$_3$ nanodielectrics with nanofiller contents of 0.1, 0.5, 2, and 5 wt%. Since the numerical simulation of the charge drift equation should obey the Courant–Friedrich–Levy (CFL) law, the films were discretized into 300 elements and the computational time interval Δt was set as 1 ms [34]. Figure 3 shows the trap density, trap energy level, and attempt-to-escape frequency of pure LDPE and LDPE/Al$_2$O$_3$ nanodielectrics as a function of nanofiller content [13]. N_T, u_T, and v_0 were obtained from the experimental results of thermally stimulated depolarization currents, and they all first increased and then decreased with the increase in nanofiller content. Charges in the electrodes may first be transferred to the surface traps in the nanocomposite and then injected into the bulk of the material. In this case, the charge injection barriers can be set to the deep trap levels. The mobilities of mobile electrons and mobile holes in the extended states of the nanocomposite were both set to be 1×10^{-13} m^{2}V^{-1}s^{-1}. The trapping probabilities of the deep traps were calculated from the trap densities and the carrier mobilities of the mobile charges. The detrapping probabilities of trapped charges were calculated from the trap levels and the attempt-to-escape frequencies. The recombination coefficients of positive and negative charges were calculated from the mobilities of mobile electrons and mobile holes.

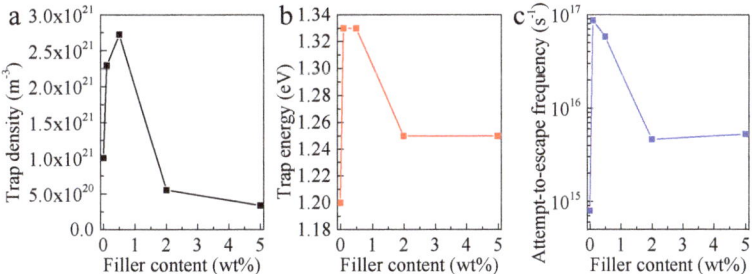

Figure 3. Trap density (**a**), trap energy (**b**), and attempt-to-escape frequency (**c**) versus filler content of LDPE/Al$_2$O$_3$ nanocomposites.

It can be seen that the trap energy and density are positively correlated with the breakdown strength of the polymer nanocomposites. However, quantitative studies are still lacking about how much the trap energy and density can change the breakdown strength by affecting the space charge accumulation and electric field concentration. In addition, quantitative studies are also lacking about the extent to which the trap energy and density change the breakdown strength by affecting the electric field distortion and the energy accumulation of electrons in a fixed free volume. Moreover, whether the increase in breakdown strength is caused by the binding effect of nanofillers on molecular chains should also be considered. In order to obtain these quantitative relationships, it is necessary to carry out simulation studies of different breakdown models and compare them with the experimental results to determine the primary and secondary influencing factors. Finally, the quantitative relationship between each influencing factor and the variation range in the breakdown strength can be obtained.

3.1. Simulation Resutls of EBEF Model

A voltage having a ramping rate of k_{ramp} was applied to the electrodes on both sides of the nanocomposite. As the time t increases, the voltage V_{appl} applied to the electrodes on both sides of the nanocomposite increases gradually, that is, $V_{appl} = k_{ramp}t$. The electric field inside the nanocomposite increases gradually, and the charges injected into the material from the electrodes gradually increase. Due to the slow charge transport inside the nanocomposite, the space charges of the same polarity gradually accumulate inside the material. Homogeneous space charges can distort the electric potential and electric field in the nanocomposite. Figure 4 demonstrates the time-dependent changes in space charge distribution and electric field distribution in the pure LDPE and LDPE/Al$_2$O$_3$ nanocomposites. As the voltage increases gradually, the accumulated space charges in all samples increase greatly, and the distortions of the electric fields become stronger and stronger. When the maximum electric field in the material reaches a certain threshold, that is, $E_{max} \geq E_C$, the sample will be broken down. By comparing the experimental results, E_C was set to 290 kV/mm.

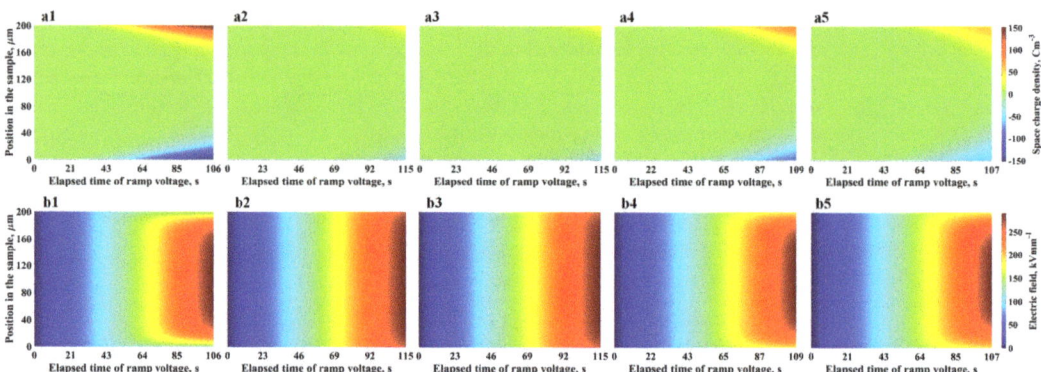

Figure 4. Distributions of space charges (**a1–a5**) and electric fields (**b1–b5**) in pure LDPE and LDPE/Al$_2$O$_3$ nanodielectrics with nanofiller contents of 0.1 wt%, 0.5 wt%, 2.0 wt%, and 5.0 wt%.

With a small amount of doping, the trap density and energy levels of LDPE/Al$_2$O$_3$ nanocomposites increase. The ability of the traps to capture charges is enhanced, and more mobile charges are captured by the traps in the insulating material near the electrodes. The trapped charges of the same polarity near the electrodes increase, so that the electric field near the electrodes is greatly reduced. It can be seen from the Schottky emission equation that the charge carriers injected by the electrodes into the nanocomposite are greatly reduced at low electric fields. This can decrease the degree of electric field distortion inside the nanocomposite. Under the same voltage, the maximum electric field E_{max} in the nanocomposite becomes smaller. This increases the breakdown electric field of the nanocomposite. When a large amount of doping is used, the trap density and energy level of the LDPE/Al$_2$O$_3$ nanocomposite gradually decrease, the space charge distribution becomes longer, and the electric field concentration becomes larger, leading to a larger maximum electric field E_{max} in the nanocomposite. This can lead to a reduction in the breakdown strength of nanocomposites having larger filler contents.

3.2. Simulation Results of EBEG Model

It is assumed that a certain scale of free volume exists in the LDPE nanocomposites. Charges in the extended states enter the free volumes and are accelerated by the electric field to gain energy. Figure 5 depicts the time-dependent changes in space charge distributions, electric field distributions, and electron energy distributions in pure LDPE and LDPE/Al$_2$O$_3$ nanocomposites. As time passes, the voltage applied to the electrodes on both sides of the materials increases gradually. The space charge accumulation inside the materials increases, the electric field distortion becomes more serious, and the energy gained by the electrons in the free volume also increases. When the charges gain energy beyond the trapping ability of the deep traps, breakdown of the material will occur.

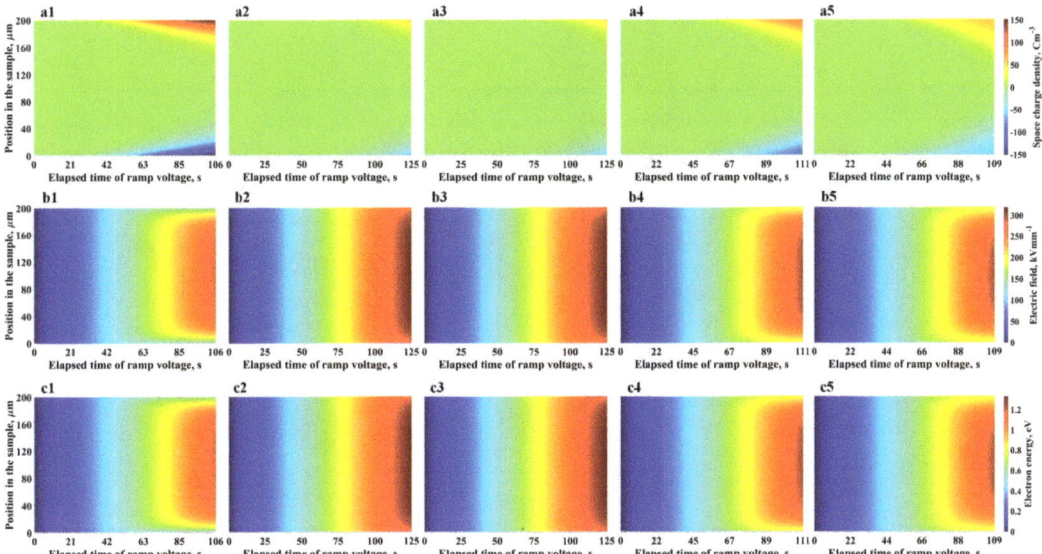

Figure 5. Distributions of space charges (**a1–a5**), electric fields (**b1–b5**), and electron energy gains (**c1–c5**) in pure LDPE and LDPE/Al$_2$O$_3$ nanocomposites with nanofiller contents of 0.1 wt%, 0.5 wt%, 2.0 wt%, and 5.0 wt%.

When doped with a small amount of nanofillers, the LDPE/Al$_2$O$_3$ nanocomposites produce more traps having deeper energy levels. More homogeneous space charges accumulate near the electrodes, weakening the electric field distortion inside the nanocomposites. In addition, as the traps become deeper, the trapping ability of the deep traps is increased. Two factors work together to increase the breakdown strength of the nanodielectrics having lower nanofiller weight fractions. After doping a large amount of nanofillers, the trap density and energy levels of LDPE/Al$_2$O$_3$ nanocomposites decrease due to the overlapping of the interfacial regions. At this time, the electric field in the material is seriously distorted, and the trapping ability of the deep traps is weakened. Their combined effects lead to a decrease in the breakdown electric field of the nanocomposites having larger nanofiller contents.

3.3. Simulation Results of EBMD Model

Figure 6 depicts the variation in the distributions of space charges, electric fields, molecular chain displacements, and electron energies with time in pure LDPE and LDPE/Al$_2$O$_3$ nanodielectrics. With the increase in nanofiller content, under the same voltage, the accumulation of space charges in the nanocomposites first decreases and then increases, while the electric field concentration first decreases and then increases. It can be obtained from the dynamic equation of molecular chain motion that, with the increase in nanofiller content, the displacement of molecular chains within nanocomposites first decreases and then increases under the same voltage. This causes the energy accumulated by the electrons to decrease first and then increase. At the same time, it is considered that the trapping ability of deep traps in nanocomposites first increases and then decreases. The energy accumulation of charge carriers in the dynamic free volume and the trapping ability of deep traps together determine the breakdown electric field of nanodielectrics as a function of nanofiller content. With the increase in the mass percentage of nanofillers, the breakdown electric field of the nanodielectrics increased first and then decreased.

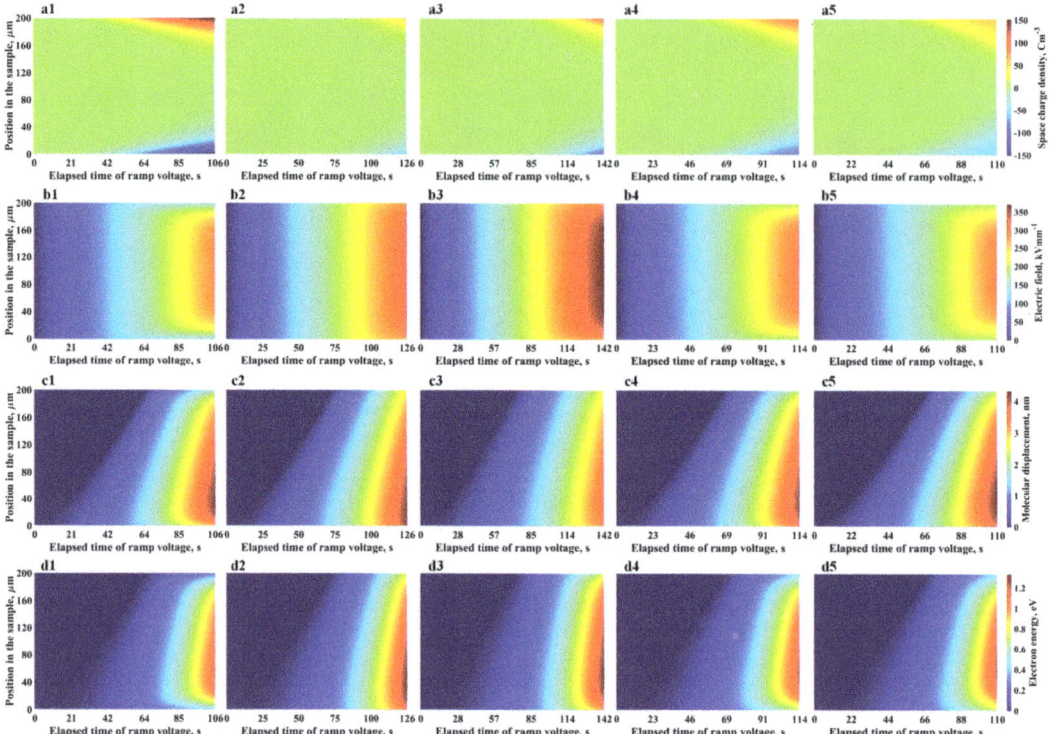

Figure 6. Distributions of space charges (**a1**–**a5**), electric fields (**b1**–**b5**), molecular displacements (**c1**–**c5**), and electron energy gains (**d1**–**d5**) in pure LDPE and LDPE/Al$_2$O$_3$ nanocomposites with nanofiller contents of 0.1 wt%, 0.5 wt%, 2.0 wt%, and 5.0 wt%.

4. Discussion

In polymer nanocomposites, traps are potential wells formed by polar groups on molecular chains, which can capture mobile charges and then hinder the motion of mobile charges [17,18]. Some of the electrons or holes may be caught in shallow traps and the extended states, and others may be trapped by the deep traps on molecular chains. Free volumes account for some of the space that is not occupied by atoms in polymers. In free volumes, mobile electrons are accelerated. Electrons captured by traps will lead to local space charge accumulation and then distort the local electric field. However, those that are not captured by traps will keep moving, leading to local current multiplication and Joule heating. Nanodoping can change the trap properties and expansion dynamics of free volumes in nanocomposites [20,21]. Figure 7 summarizes the logical block diagrams of EBEF, EBEG, and EBMD models, illustrating the concept of traps and free volumes, and comparing differences among these three criteria.

After charges are injected into the polymer nanocomposites, some mobile charge carriers are captured by deep traps on molecular chains, resulting in the space charge accumulation and electric field distortion. When the highest electric field exceeds a threshold value, namely $E(x,t)_{max} \geq E_C$ [23], DC electric breakdown occurs, which is the criterion of the EBEF model. The accelerated mobile charges in free volumes that are not captured by deep traps will gain energy from the local electric field. When the highest gained electron energy from the constant scale free volume exceeds the deep trap energy, namely $[e\lambda_0 E(x,t)]_{max} \geq u_T$ [27], DC electric breakdown occurs, which is the criterion of the EBEG model. The property of interfacial region is extremely vital for the distribution of traps in the bulk of nanocomposites. Accordingly, the electrical breakdown fields calculated by the EBEF and EBEG models change with the increase in nanofiller contents.

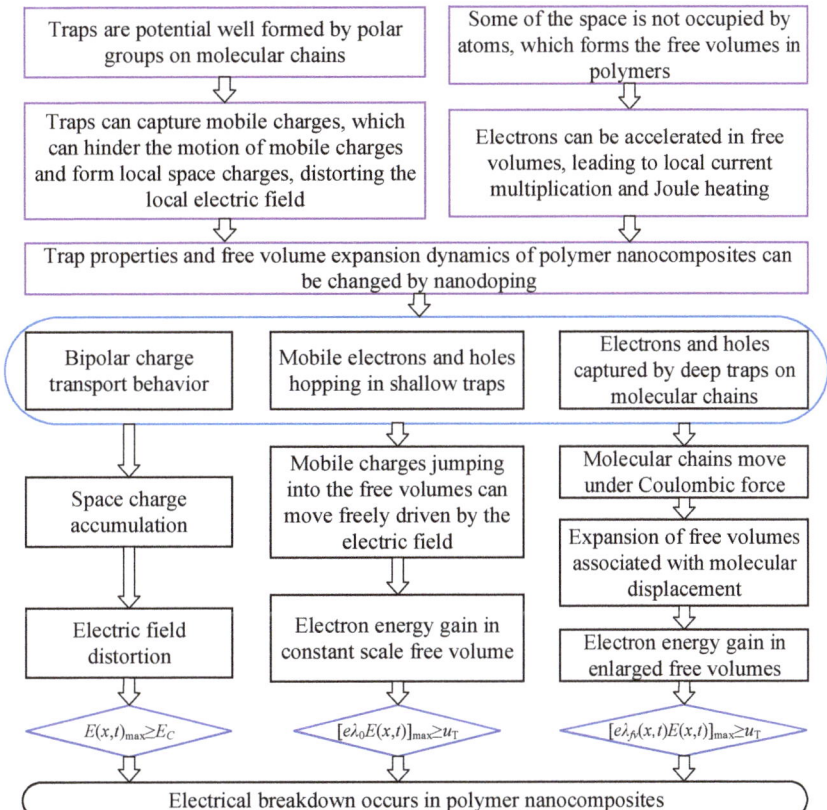

Figure 7. Schematic diagram of EBEF, EBEG, and EBMD models.

With incorporation of different types of nanofiller, the motion behavior of molecular chains also changes. At a relatively low content of nanoparticles, molecular chains are arranged in an orderly manner in interfacial zones. The mean distance between nanoparticles is smaller than the entanglement tube diameter of the polymer with an increase in nanofiller content, leading to continuous overlapping of Gouy–Chapman layers; then, the nanocomposite system changes from polymer-like to network-like [35,36]. In DC electric breakdown experiments, the electric field is sufficiently strong to force molecular chains to move and rotate if they have a dipole moment. Otherwise, the Coulomb force will act on the molecular chains with occupied deep traps and enlarge the local free volume, leading to larger energy accumulation of accelerated electrons. If the electron energy gain in this expanded free volume is higher than the deep trap energy, namely $[e\lambda_{fv}(x,t)E(x,t)]_{max} > u_T$ [13,25], the electric breakdown may be triggered, which is the criterion of the EBMD model. This model focuses on the molecular chain movement with the deep traps occupied by charges to investigate the influence of charge carrier transport and molecular chain displacement on the DC breakdown strength.

Figure 8 depicts the comparison between the simulated electric breakdown strengths obtained by the three models of EBEF, EBEG, and EBMD and the experimental results of the LDPE/Al_2O_3 nanodielectrics. It demonstrates that, with the increase in nanofiller content, the breakdown strengths obtained by EBEF, EBEG, and EBMD models all show a trend of increasing first and then decreasing. The general trends of the simulation and experimental results are similar. However, the simulation results of the EBMD model are in best agreement with the experiments. When the nanofiller content is around 0.5 wt%, the simulation results of EBEG and EBEF deviate greatly from the experimental results. According to the experimental results, the maximum electric breakdown strength is 355.8 kVmm^{-1}, which appears at the nanofiller content of 0.5wt%, while the simulation results of EBEF, EBEG, and EBMD models are 286.3, 312.2, and 356.1 kVmm^{-1}, respectively. It is apparent that the simulation results of the EBMD model are more consistent with the experiments. This indicates that the synergistic effect of deep trap centers in interfacial zones and the tight binding of molecular chains enhance the breakdown performance of LDPE nanocomposites.

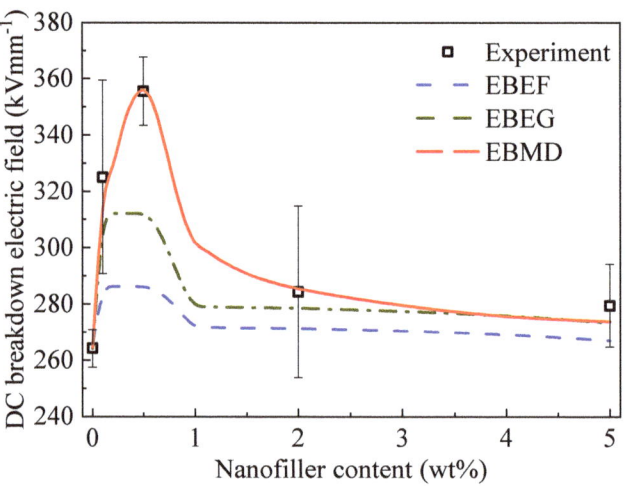

Figure 8. Comparison of EBEF, EBEG, and EBMD simulation electrical breakdown fields with experimental results.

The larger trap energy and density in nanodielectrics doped with small amounts of nanofillers reduce space charge accumulation and electric field concentration, resulting in the increase in the breakdown strength of nanocomposites. However, the changes in trap energy and density have a limited effect on the breakdown electric field. It is necessary to simultaneously consider the space charge accumulation formed by the trapping of deep traps, the free volume expansion caused by the long-range displacement of the molecular chain driven by the electric field force, and the effect of the energy accumulation of electrons in the expanded free volume on the breakdown strength. Comparative studies show that the energy accumulation of electrons in the expanded free volume due to the long-range displacement of the molecular chain dominates the breakdown strength. When the interaction between the molecular chains in the interface region between the nanofiller and the polymer matrix is enhanced, it is difficult for the molecular chains to undergo long-range displacement under the driving of the electric field force. In this case, the free volume expansion is small and it is difficult for electrons to obtain sufficient energy, so the breakdown strength is increased. Therefore, when designing the structure of the interface region, the interaction between the molecular chains in the interface region can be enhanced by the surface modification method, so that the breakdown strength can be greatly improved. LDPE is a key insulating material for power cables and energy storage dielectric capacitors. Revealing the breakdown mechanism of LDPE nanodielectrics can better develop insulating materials with high breakdown strength. This will provide theoretical and simulation model support for the development of high-performance power cables and energy storage dielectric capacitors.

In addition, the aggregated structure of polymer nanocomposites can be changed to some extent, compared to that of pure polymers. The interface between the crystalline region and the amorphous region, and the interface region between the nanofiller and the polymer matrix, may form different trap distributions. The change in crystallinity and the change in lamellar length may change the interaction strength between molecular chains. Because the molecular chains in the interface region are bound by nanoparticles, the interaction strength between the molecular chains also changes. In future studies, we will correlate aggregated structures with trapping effects and molecular chain interactions through density functional theory [37] and molecular dynamics simulations [38]. Then, the EBMD model will be used to determine the effect of the trapping effect and molecular chain interactions on the breakdown strength. Ultimately, a multiscale model will be established to study the relationship among the aggregated structure, the trapping effect and molecular chain interactions, and the electric breakdown performance.

5. Conclusions

The electrical breakdown electric fields simulated by three breakdown models at various nanofiller contents were compared, and the breakdown mechanism of LDPE nanocomposites was illustrated. Doping of Al_2O_3 nanoparticles into LDPE can change the trap, conductivity, and electrical breakdown properties. The results indicated that the charge trapping, molecular motion, and electron energy gain in free volumes are important factors for the electrical breakdown of polyethylene nanocomposites. Then, space charge transport, electron energy gain, and molecular chain long-distance movement were comprehensively considered to investigate the electrical breakdown mechanism of nanocomposites. EBEF, EBEG, and EBMD models were established via focusing on space charge accumulation due to deep trappings and associated electric field distortion, mobile electrons gaining energy in free volumes, long-distance movement of molecular chains under the Coulomb force expanding free volumes, and accelerating electrons in the enlarged free volumes. Simulation results showed that the EBMD model fits the experimental results much better than the EBEF and EBEG models. The correlation between the long-distance molecular chain movement under the Coulomb force and the electric breakdown characteristics of LDPE was established. The comparisons between simulation results of different models and experiments showed that the electric breakdown electric field of LDPE nanodielectrics is synergistically enhanced by both the strong trapping effect of traps in interfacial regions and the strong binding effect of molecular chains.

Author Contributions: Conceptualization, D.M.; Formal analysis, Z.X. and C.Z.; Funding acquisition, D.M.; Resources, Z.X. and Q.W.; Supervision, Q.W. and D.M.; Writing–original draft, Z.X., C.Z., M.H. and Z.G.; Writing–review and editing, M.H., Z.G. and D.M. All authors have read and agreed to the published version of the manuscript.

Funding: This work was supported by State Key Laboratory of Advanced Power Transmission Technology (Grant No. GEIRI-SKL-2018-010).

Institutional Review Board Statement: Not applicable.

Informed Consent Statement: Not applicable.

Conflicts of Interest: The authors declare no conflict of interest.

References

1. Zhou, Y.; Peng, S.M.; Hu, J.; He, J.L. Polymeric insulation materials for HVDC cables: Development, challenges and future perspective. *IEEE Trans. Dielectr. Electr. Insul.* **2017**, *24*, 1308–1318. [CrossRef]
2. Orton, H. Power cable technology review. *High Volt. Eng.* **2015**, *41*, 1057–1067.
3. Chen, G.; Hao, M.; Xu, Z.Q.; Vaughan, A.S.; Cao, J.Z.; Wang, H.T. Review of high voltage direct current cables. *CSEE J. Power Energy Syst.* **2015**, *1*, 9–20. [CrossRef]
4. Meng, F.B.; Chen, X.R.; Shi, Y.W.; Zhu, H.S.; Hong, Z.L.; Muhammad, A.; Paramane, A.; Chen, L.; Zhang, Y.M.; Huang, R.B.; et al. Temperature-dependent charge dynamics of double layer interface in 500 kV HVDC XLPE cable factory joint with different interfacial roughness. *IEEE Trans. Dielectr. Electr. Insul.* **2022**, *29*, 655–662. [CrossRef]

5. Gupta, R.; Smith, L.; Njuguna, J.; Deighton, A.; Pancholi, K. Insulating MgO-Al$_2$O$_3$-LDPE nanocomposites for offshore medium-voltage DC cables. *ACS Appl. Electron. Mater.* **2020**, *2*, 1880–1891. [CrossRef]
6. Mansour, D.E.A.; Abdel-Gawad, N.M.K.; El Dein, A.Z.; Ahmed, H.M.; Darwish, M.M.F.; Lehtonen, M. Recent advances in polymer nanocomposites based on polyethylene and polyvinylchloride for power cables. *Materials* **2020**, *14*, 66. [CrossRef]
7. Li, S.T.; Wang, W.W.; Yu, S.H.; Sun, H.G. Influence of Hydrostatic Pressure on Dielectric Properties of Polyethylene/aluminum Oxide Nanocomposites. *IEEE Trans. Dielectr. Electr. Insul.* **2014**, *21*, 519–528. [CrossRef]
8. Li, Q.; Chen, L.; Gadinski, M.R.; Zhang, S.H.; Zhang, G.Z.; Li, H.Y.U.; Iagodkine, E.; Haque, A.; Chen, L.Q.; Jackson, T.N.; et al. Flexible high-temperature dielectric materials from polymer nanocomposites. *Nature* **2015**, *523*, 576–580. [CrossRef]
9. Pourrahimi, A.M.; Olsson, R.T.; Hedenqvist, M.S. The role of interfaces in polyethylene/metal-oxide nanocomposites for ultrahigh-voltage insulating materials. *Adv. Mater.* **2017**, *30*, 1703624–1703625. [CrossRef]
10. Li, Z.L.; Du, B.X.; Han, C.L.; Xu, H. Trap modulated charge carrier transport in polyethylene/graphene nanocomposites. *Sci. Rep.* **2017**, *7*, 4011–4015. [CrossRef]
11. Cheng, Y.J.; Yu, G.; Yu, B.Y.; Zhang, X.H. The research of conductivity and dielectric properties of ZnO/LDPE composites with different particles size. *Materials* **2020**, *13*, 4136. [CrossRef] [PubMed]
12. Thomas, J.; Joseph, B.; Jose, J.P.; Maria, H.J.; Main, P.; Rahman, A.A.; Francis, B.; Ahmad, Z.; Thomas, S. Recent advances in cross-linked polyethylene-based nanocomposites for high voltage engineering applications: A critical review. *Ind. Eng. Chem. Res.* **2019**, *58*, 20863–20879. [CrossRef]
13. Min, D.M.; Cui, H.Z.; Wang, W.W.; Wu, Q.Z.; Xing, Z.L.; Li, S.T. The coupling effect of interfacial traps and molecular motion on the electrical breakdown in polyethylene nanocomposites. *Compos. Sci. Technol.* **2019**, *184*, 107873. [CrossRef]
14. Nelson, J.K. *Dielectric Polymer Nanocomposites*; Springer: Spring, NY, USA, 2010; pp. 1–285.
15. Dou, L.Y.; Lin, Y.H.; Nan, C.W. An overview of linear dielectric polymers and their nanocomposites for energy storage. *Molecules* **2021**, *26*, 6148. [CrossRef]
16. Zhao, X.C.; Bi, Y.J.; Xie, J.H.; Hu, J.; Sun, S.L.; Song, S.X. Enhanced dielectric, energy storage and tensile properties of BaTiO$_3$–NH$_2$/low-density polyethylene nanocomposites with POE-GMA as interfacial modifier. *Polym. Test.* **2021**, *95*, 107094. [CrossRef]
17. Tanaka, T.; Kozako, M.; Fuse, N.; Ohki, Y. Proposal of a multi-core model for polymer nanocomposite dielectrics. *IEEE Trans. Dielectr. Electr. Insul.* **2005**, *12*, 669–681. [CrossRef]
18. Tanaka, T. Dielectric nanocomposites with insulating properties. *IEEE Trans. Dielectr. Electr. Insul.* **2005**, *12*, 914–928. [CrossRef]
19. Li, S.; Yin, G.; Chen, G.; Li, J.; Bai, S.; Zhong, L.; Zhang, Y.; Lei, Q.Q. Short-term Breakdown and Long-term Failure in Nanodielectrics: A Review. *IEEE Trans. Dielectr. Electr. Insul.* **2010**, *17*, 1523–1535. [CrossRef]
20. Smith, R.C.; Liang, C.; Landry, M.; Nelson, J.K.; Schadler, L.S. The mechanisms leading to the useful electrical properties of polymer nanodielectrics. *IEEE Trans. Dielectr. Electr. Insul.* **2008**, *1*, 187–196. [CrossRef]
21. Min, D.M.; Ji, M.Z.; Li, P.X.; Gao, Z.W.; Liu, W.F.; Li, S.T.; Liu, J. Entropy reduced charge transport and energy loss in interfacial zones of polymer nanocomposites. *IEEE Trans. Dielectr. Electr. Insul.* **2021**, *28*, 2011–2017. [CrossRef]
22. Matsui, K.; Tanaka, Y.; Takada, T.; Fukao, T. Space Charge Behavior in Low-density Polyethylene at Pre-breakdown. *IEEE Trans. Dielectr. Electr. Insul.* **2005**, *12*, 406–415. [CrossRef]
23. Chen, G.; Zhao, J.; Li, S.; Zhong, L. Origin of thickness dependent dc electrical breakdown in dielectrics. *Appl. Phys. Lett.* **2012**, *100*, 222904. [CrossRef]
24. Choi, D.H.; Randall, C.; Lanagan, M. Combined electronic and thermal breakdown models for polyethylene and polymer laminates. *Mater. Lett.* **2015**, *141*, 14–19. [CrossRef]
25. Min, D.M.; Li, S.T.; Ohki, Y. Numerical Simulation on Molecular Displacement and DC Breakdown of LDPE. *IEEE Trans. Dielectr. Electr. Insul.* **2016**, *23*, 507–516. [CrossRef]
26. Laurent, C.; Teyssedre, G.; Le Roy, S.; Baudoin, F. Charge Dynamics and its Energetic Features in Polymeric Materials. *IEEE Trans. Dielectr. Electr. Insul.* **2013**, *20*, 357–381. [CrossRef]
27. Artbauer, J. Electric strength of polymers. *J. Phys. D Appl. Phys.* **1996**, *29*, 446–456. [CrossRef]
28. Kao, K.C. *Dielectric Phenomena in Solids*; Elsevier Academic Press: San Diego, CA, USA, 2004; pp. 327–572.
29. Boufayed, F.; Teyssedre, G.; Laurent, C.; Le Roy, S.; Dissado, L.A.; Segur, P.; Montanari, G.C. Models of bipolar charge transport in polyethylene. *J. Appl. Phys.* **2006**, *100*, 104105–104110. [CrossRef]
30. Le Roy, S.; Segur, P.; Teyssedre, G.; Laurent, C. Description of Bipolar Charge Transport in Polyethylene Using a Fluid Model with a Constant Mobility: Model Prediction. *J. Phys. D: Appl. Phys.* **2004**, *37*, 298–305. [CrossRef]
31. Sessler, G.M.; Figueiredo, M.T.; Ferreira, G.F.L. Models of Charge Transport in Electron-beam Irradiated Insulators. *IEEE Trans. Dielectr. Electr. Insul.* **2004**, *11*, 192–202. [CrossRef]
32. Shockley, W.; Read, W.T. Statistics of the Recombinations of Holes and Electrons. *Phys. Rev.* **1952**, *87*, 835–842.
33. Lowell, J. Absorption and conduction currents in polymers: A unified model. *J. Phys. D: Appl. Phys.* **1990**, *23*, 205–210. [CrossRef]
34. Courant, R.; Friedrichs, K.; Lewy, H. On the partial difference equations of mathematical physics. *IBM J. Res. Dev.* **1967**, *11*, 215–234. [CrossRef]
35. Baeza, G.P.; Dessi, C.; Costanzo, S.; Zhao, D.; Gong, S.; Alegría, A.; Colby, R.H.; Rubinstein, M.; Vlassopoulos, D.; Kumar, S.K. Network dynamics in nanofilled polymers. *Nat. Commun.* **2016**, *7*, 1–6. [CrossRef]

36. Sen, S.; Xie, Y.P.; Kumar, S.K.; Yang, H.; Bansal, A.; Ho, D.L.; Hall, L.; Hooper, J.B.; Schweizer, K.S. Chain conformations and bound-layer correlations in polymer nanocomposites. *Phys. Rev. Lett.* **2007**, *98*, 128302. [CrossRef]
37. Song, S.W.; Zhao, H.; Yao, Z.H.; Yan, Z.Y.; Yang, J.M.; Wang, X.; Zhao, X.D.; Zhao, X.D. Enhanced electrical properties of polyethylene-graft-polystyrene/LDPE composites. *Polymers* **2020**, *12*, 124. [CrossRef]
38. Yu, C.H.; Hu, K.; Yang, Q.L.; Wang, D.D.; Zhang, W.G.; Chen, G.X.; Kapyelata, C. Analysis of the storage stability property of carbon nanotube/recycled polyethylene-modified asphalt using molecular dynamics simulations. *Polymers* **2021**, *13*, 1658. [CrossRef]

Article

Theory of Electrical Breakdown in a Nanocomposite Capacitor

Vladimir Bordo * and Thomas Ebel

Centre for Industrial Electronics, Department of Mechanical and Electrical Engineering, University of Southern Denmark, Alsion 2, DK-6400 Sønderborg, Denmark; ebel@sdu.dk
* Correspondence: bordo@sdu.dk

Abstract: The electrostatic field in a nanocomposite represented by spherical nanoparticles (NPs) embedded into a dielectric between two parallel metallic electrodes is derived from first principles. The NPs are modeled by point dipoles which possess the polarizability of a sphere, and their image potential in the electrodes is found using a dyadic Green's function. The derived field is used to obtain the parameters which characterize the electrical breakdown in a nanocomposite capacitor. It is found, in particular, that for relatively low volume fractions of NPs, the breakdown voltage linearly decreases with the volume fraction, and the slope of this dependence is explicitly found in terms of the dielectric permittivities of the NPs and the dielectric host. The corresponding decrease in the maximum energy density accumulated in the capacitor is also determined. A comparison with the experimental data on the breakdown strength in polymer films doped with BaTiO$_3$ NPs available in the literature reveals a dominant role of the interface polarization at the NP-polymer interface and an existence of a nonferroelectric surface layer in NPs. This research provides a rigorous approach to the electrical breakdown phenomenon and can be used for a proper design of nanocomposite capacitors.

Keywords: electrical breakdown; breakdown voltage; nanocomposite; nanocomposite capacitor

Citation: Bordo, V.; Ebel, T. Theory of Electrical Breakdown in a Nanocomposite Capacitor. *Appl. Sci.* 2022, 12, 5669. https://doi.org/10.3390/app12115669

Academic Editor: Alberto Vomiero

Received: 19 May 2022
Accepted: 31 May 2022
Published: 2 June 2022

Publisher's Note: MDPI stays neutral with regard to jurisdictional claims in published maps and institutional affiliations.

Copyright: © 2022 by the authors. Licensee MDPI, Basel, Switzerland. This article is an open access article distributed under the terms and conditions of the Creative Commons Attribution (CC BY) license (https://creativecommons.org/licenses/by/4.0/).

1. Introduction

Recently, the polymer film capacitors have received growing attention due to their advantages of a low cost, facile fabrication, excellent flexibility and high operating voltage [1]. To also attain high energy storage density, one employs nanosized inclusions with a high dielectric constant, thus increasing the effective dielectric permittivity of such polymer-based nanocomposites [2–5]. This can be performed, however, at the expense of a decrease in the capacitor breakdown strength [6–8] that diminishes to some extent the advantages of this approach.

To properly design nanocomposite capacitors, one needs a deep understanding of the factors which control the electrical breakdown in them. For relatively low volume fractions of inclusions, which do not create deep traps for electrons [9–13], the primary effect of their embedding is a modification of the electric field in the capacitor. Among different approaches which aim to account for this influence, only the Maxwell Garnett approximation [14,15] follows from first principles, while the other ones are rather phenomenological models (see, for example, a review of different models in Ref. [7]). This approximation treats the spherical inclusions as point dipoles with a dipole moment which is determined by the sphere polarizability. It is applicable, however, to composite media provided they extend throughout a space of dimensions which are much larger than the wavelength of the external field or to unbounded media in the static case. For a nanocomposite capacitor, this criterion is not fulfilled, and one must take into account the polarization of electrodes and, resulting from it, the image potential [16].

In the present paper, we develop a first-principles approach to the static electric field in a nanocomposite capacitor which is based on a rigorous account of the image potential for point dipoles between parallel metallic electrodes. The obtained results are used to find the parameters which characterize the electrical breakdown in a nanocomposite capacitor.

The predicted dependencies are compared with the available experimental data on the breakdown field strength as a function of the volume fraction of inclusions and temperature.

2. Methods

2.1. Model

We consider a parallel-plate capacitor in which the gap of thickness d between the metallic electrodes is filled with a nanocomposite dielectric. The rectangular electrodes are assumed identical to each other and have the lateral dimensions L_1 and L_2 along the x and y axes, respectively. The nanocomposite is represented by identical spherical nanoparticles (NPs) of radius R with the dielectric permittivity ϵ_i randomly dispersed in the host dielectric with the dielectric permittivity ϵ_h (see Figure 1).

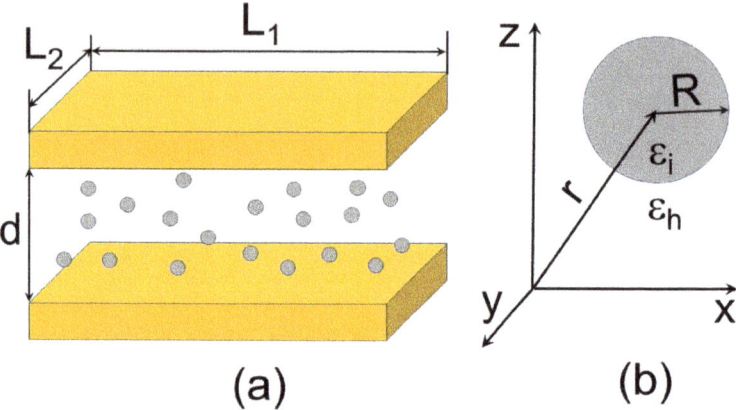

Figure 1. (**a**) The model of a nanocomposite capacitor. (**b**) The model of an NP.

We assume that the NP radius is much less than the capacitor dimensions, i.e., $R \ll d, L_1, L_2$, and treat the NPs as point dipoles which possess the polarizability of a dielectric sphere. Such an approach follows the assumption made by Maxwell Garnett [14] which has been widely used for the modeling of the dielectric properties of composites. The applicability of this approximation is limited to relatively small volume fractions of NPs (see Section 2.3).

2.2. Electrical Breakdown in Conventional Capacitors

We assume that the main dependencies which govern the electrical breakdown in capacitors remain valid when a small volume fraction of NPs is embedded into the host dielectric. Namely, the avalanching effect in the dielectric in a strong electric field leads to an exponential increase in the electron current density with the dielectric film thickness as follows [17]

$$j_e(d) = j_e(0) \exp(\gamma E d), \qquad (1)$$

where $j_e(0)$ is the injected current density at the electrode with a negative bias (at $z = 0$), the constant γ can be determined in terms of the electron energy sufficient for impact ionization and the recombination rate, E is the electric field strength inside the dielectric and d is the dielectric film thickness. The breakdown occurs when the electron current density reaches the critical value j_B which is sufficient for the dielectric destruction. For a given thickness, this happens at a certain value of the field strength, E_B, which determines the breakdown voltage $V_B = E_B d$.

The quantity $j_e(0)$ is dictated, in its turn, by the work function of the electrode and the potential barrier for electrons formed by both the electric field potential and the image potential for an electron originating from the polarization of the electrode. Two processes contribute to this current: the quantum-mechanical tunneling through the barrier and the thermionic current above the barrier.

For relatively weak electric fields, the latter one prevails over the first one and is given by the Richardson–Schottky equation [18]

$$j_e(0) = AT^2 \exp\left(-\frac{W}{kT}\right) \exp\left(\frac{BE^{1/2}}{kT}\right), \qquad (2)$$

where A and B are constants, A being known as the Richardson constant, W is the work function of the electrode, E is the electric field strength inside the dielectric film, k is the Boltzmann constant and T is the temperature.

In strong electric fields, the tunneling mechanism is the dominant one and the injection current is given by the Fowler–Nordheim formula [19]

$$j_e(0) = CE^2 \exp\left(-\frac{D}{E}\right), \qquad (3)$$

where the constants C and D are determined by the work function of the electrode and do not depend on the temperature.

2.3. Local Field in a Nanocomposite

When applying the above equations to a nanocomposite capacitor, one should regard the field strength E as the microscopic (local) field which is determined not only by the applied voltage but also by the field of the polarized NPs. For extended media and relatively small volume fractions of NPs, this field can be found in the Maxwell Garnett approximation which models the inclusions by polarizable spheres [14,15]. However, this approach is not applicable to the operating wavelengths λ which are comparable with or larger than the capacitor dimensions and especially for static fields which correspond to the limit $\lambda \to \infty$ [16].

A proper approach to this problem involves an account of the image potential which originates from the polarization of the electrodes and can be obtained by an infinite summation of all the induced NP dipole images in the two electrodes [20]. However, a more convenient way of calculation is using the Fourier transform of the field derived for a dipole oscillating with frequency ω between two reflecting surfaces and taking the limit $\omega \to 0$ [21].

In the adopted approximation, the induced dipole moment of an NP located at the point $\mathbf{r} = (x, y, z)$ is given by [22]

$$\mathbf{p}(\mathbf{r}) = \alpha \mathbf{E}(\mathbf{r}) = \epsilon_h R^3 \frac{\epsilon_i - \epsilon_h}{\epsilon_i + 2\epsilon_h} \mathbf{E}(\mathbf{r}), \qquad (4)$$

where α is the sphere polarizability and $\mathbf{E}(\mathbf{r})$ is the microscopic (local) field [23] at the position of the dipole. A random distribution of NPs allows one to formally consider their polarization, $\mathbf{P}(\mathbf{r}) = N\mathbf{p}(\mathbf{r})$, as a continuous function of the radius vector $\mathbf{r}$. Then, the local field can be written in the form [23,24]

$$\mathbf{E}(\mathbf{r}) = \mathbf{E}_0 + N \int_{V'} \bar{\mathbf{F}}(\mathbf{r}, \mathbf{r}'; \omega) \mathbf{p}(\mathbf{r}') d\mathbf{r}', \qquad (5)$$

where $\mathbf{E}_0$ is the external electric field applied to the capacitor and directed along the z axis and the integral term represents a collective action of the induced dipoles. Here, N is the volume number density of NPs, the quantity $\bar{\mathbf{F}}(\mathbf{r}, \mathbf{r}'; \omega)$ is the so-called field susceptibility tensor that relates the electric field at the point $\mathbf{r}$ generated by a classical dipole, oscillating at frequency ω, with the dipole moment itself, located at $\mathbf{r}'$ [25], and the symbol V' denotes the gap volume after removal of a small volume around the NP under consideration that excludes its self-action. As far as the dipole moment in the integrand depends on the local field, Equation (5) is an integral equation which provides a self-consistent solution for the electric field in the capacitor.

The tensor $\bar{\mathbf{F}}(\mathbf{r}, \mathbf{r}'; \omega)$ can be expressed in terms of the dyadic Green's function for Maxwell's equations [26] and allows a decomposition into the direct contribution, which describes the field of a dipole in free space [27], and the reflected contribution, which provides the dipole field reflected from the parallel plates [21]. For a dipole near a flat surface, it is convenient to write it in the form of the 2D spatial Fourier integral as follows [28]

$$\bar{\mathbf{F}}(\mathbf{r}, \mathbf{r}'; \omega) = \frac{1}{(2\pi)^2} \times \int_{-\infty}^{\infty} \int_{-\infty}^{\infty} \bar{\mathbf{f}}(z, z'; k_x, k_y; \omega) e^{ik_x(x-x')} e^{ik_y(y-y')} dk_x dk_y, \tag{6}$$

where $\bar{\mathbf{f}}(z, z'; k_x, k_y; \omega)$ is the Fourier transform of the tensor $\bar{\mathbf{F}}(\mathbf{r}, \mathbf{r}'; \omega)$, x', y' and z' are the Cartesian coordinates of the point $\mathbf{r}'$, k_x and k_y are the spatial frequencies along the x and y axes, respectively.

For the purposes of the present derivation, one needs the limiting value of the tensor component $f_{zz}(z, z'; k_x, k_y; \omega)$ when $\omega \to 0$, which we denote as $f_{zz}(z, z'; k_x, k_y; 0)$. This quantity determines the z-component of the local field which dictates the potential barrier for injected electrons. In this limit, the reflected field is reduced to the field originating from the images of the NP-induced dipole in the two electrodes.

Let us introduce the Fourier transform of the local field in the capacitor,

$$E_z(\mathbf{r}) = \frac{1}{(2\pi)^2} \int_{-\infty}^{\infty} \int_{-\infty}^{\infty} e_z(z; k_x, k_y) e^{ik_x x} e^{ik_y y} dk_x dk_y, \tag{7}$$

and substitute it into Equation (5). The obtained equation involves the quantity $f_{zz}(z, z'; k_x, k_y; 0)$ whose variation with z and z' is determined by the factors $\exp(\pm \kappa z)$ and $\exp(\pm \kappa z')$ with $\kappa = (k_x^2 + k_y^2)^{1/2}$. Assuming the inequality $\kappa d \ll 1$, which will be justified in what follows, and the model of a perfect conductor for the electrodes, one obtains (see Ref. [29] for the detail)

$$f_{zz}(0, 0; k_x, k_y; 0) \approx \frac{4\pi}{\epsilon_h d}. \tag{8}$$

We then obtain

$$e_z(k_x, k_y) = \frac{e_0(k_x, k_y)}{1 - (4\pi/\epsilon_h) N \alpha}, \tag{9}$$

where

$$e_0(k_x, k_y) = \int_{-L_2/2}^{L_2/2} \int_{-L_1/2}^{L_1/2} E_0 e^{-ik_x x} e^{-ik_y y} dx dy$$

$$= E_0 L_1 L_2 \operatorname{sinc}\left(\frac{k_x L_1}{2}\right) \operatorname{sinc}\left(\frac{k_y L_2}{2}\right) \tag{10}$$

is the Fourier transform of the applied electric field. Here, $\operatorname{sinc} x \equiv \sin x / x$, L_1 and L_2 are the lateral dimensions of the electrodes along the x and y axes, respectively, and we have assumed that the field is zero outside the capacitor. Taking into account that the function $\operatorname{sinc} x$ is essentially nonzero within the range $|x| \leq 3$, one concludes that the essential range of integration in Equation (7) is limited to the values $|k_x| \leq 6/L_1$ and $|k_y| \leq 6/L_2$. This means that the condition $\kappa d \ll 1$, which we have used above, is justified provided $d \ll L_1, L_2$.

Finally, the local field, Equation (7), takes the form

$$E_z = \frac{E_0}{1 - 3f\beta}, \tag{11}$$

with $f = (4\pi/3) N R^3$ being the volume fraction of NPs and $\beta = (\epsilon_i - \epsilon_h)/(\epsilon_i + 2\epsilon_h)$. Let us note that the applicability of this approach which models NPs by point dipoles is limited to the range $f \leq 0.2$ [16].

3. Results

3.1. Breakdown Parameters in a Nanocomposite Capacitor

Equations (1)–(3), which describe the electrical breakdown, can be written in terms of the applied voltage, V, and the applied field strength, $E_0 = V/d$, as

$$j_e(d) = j_e(0) \exp\left(\frac{\gamma V}{1 - 3f\beta}\right), \tag{12}$$

$$j_e(0) = AT^2 \exp\left(-\frac{W}{kT}\right) \exp\left(\frac{B' E_0^{1/2}}{kT}\right) \tag{13}$$

and

$$j_e(0) = C' E_0^2 \exp\left(-\frac{D'}{E_0}\right), \tag{14}$$

respectively. Here, the new parameters

$$B' = \frac{B}{(1 - 3f\beta)^{1/2}}, \tag{15}$$

$$C' = \frac{C}{(1 - 3f\beta)^2} \tag{16}$$

and

$$D' = D(1 - 3f\beta) \tag{17}$$

determine the field and temperature dependencies of the injection current in a nanocomposite capacitor.

The breakdown voltage for a nanocomposite capacitor, V_B', is related with the one for an NPs free capacitor, V_B, as follows

$$V_B' = (1 - 3f\beta) V_B. \tag{18}$$

For NPs with $\epsilon_i > \epsilon_h$, the quantity β is positive and therefore Equation (18) describes a lowering of the breakdown voltage with an addition of NPs.

3.2. Maximum Energy Density

An important characteristic of a capacitor is the electromagnetic energy density, U, which it can accumulate. This quantity is given by

$$U = \frac{1}{2} \epsilon_0 \epsilon E^2 \tag{19}$$

with ϵ_0 being the permittivity of the vacuum and ϵ being an effective permittivity of the dielectric in the capacitor. The maximum value of the energy density which can be attained in a nanocomposite capacitor is determined by the breakdown field strength $E_B' = V_B'/d$, i.e.,

$$U_{max}' = \frac{1}{2} \epsilon_0 \epsilon E_B'^2. \tag{20}$$

For a nanocomposite capacitor in the static limit [16]

$$\epsilon = \epsilon_h \frac{1 + 3f\beta}{1 - 3f\beta} \tag{21}$$

that, together with Equation (18), gives

$$U_{max}' = [1 - (3f\beta)^2] U_{max}, \tag{22}$$

where U_{max} is the maximum energy density in an NPs free capacitor. Equation (22) predicts a slower decrease in the maximum energy density with f than the decrease in the breakdown voltage, Equation (18).

4. Comparison with Experiments and Discussion

The above theoretical findings can be compared with the experimental results available in the literature. One should note, however, that the quantities γ, W and j_B, which determine the breakdown field strength, are unknown and do not allow to calculate the *absolute* value of E'_B. Instead, one can consider the *relative* value E'_B/E_B with E_B being the breakdown field strength in an NP-free capacitor and compare it with the predicted trend which follows from Equation (18).

First, we consider the experiments on the breakdown field in nanocomposites represented by BaTiO$_3$ NPs with an average size of 100 nm dispersed in a polyvinylidene fluoride (PVDF) matrix [6]. We restrict ourselves by the range of relatively small volume fractions of BaTiO$_3$ NPs ($f \leq 0.2$) where our approach can be applied. One can see that the observed decrease with f in the breakdown strength E'_B measured at different temperatures can be well fitted by straight lines (see Figure 2) which is in agreement with Equation (18). The quantity β found from the slopes of these dependencies nonmonotonically varies between 0.34 and 0.46 when the temperature changes from 20 to 120 °C which can be attributed to the error bars of the measurements.

These results can be compared with the value $\beta \approx 0.86$ calculated under the assumption that $\epsilon_i = 200$, which is typical for thin BaTiO$_3$ films [30], and $\epsilon_h = 10$ [31]. A significant difference between the calculated and fitted values of β can be ascribed to the fact that the dielectric properties of BaTiO$_3$ NPs are distinct from those of BaTiO$_3$ films [32,33].

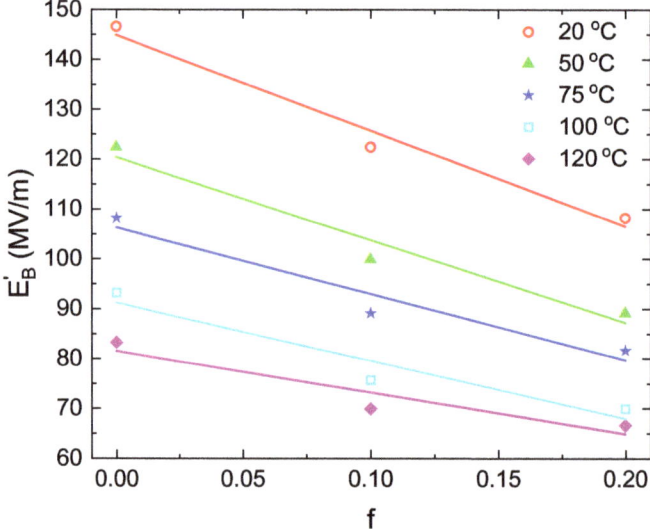

Figure 2. The dependence of the breakdown field strength on the volume fraction of BaTiO$_3$ NPs embedded in PVDF matrix for different temperatures indicated in the inset. The straight lines are the best linear fits to the experimental data shown by symbols and taken from Figure 5a in Ref. [6]. The experimental error bars are not available.

A remarkable temperature dependence of the breakdown field indicates that the thermionic current, Equation (13), is the dominant mechanism of the electron injection. As it follows from Equations (12) and (13), the temperature dependence of the quantity $E'_B/(1-3f\beta)$ should be the same for different f. This can indeed be obtained from the experimental plots of E'_B versus T which, being recalculated to $E'_B/(1-3f\beta)$ with the mean value $\beta = 0.4$, coincide with each other within the possible error bars (see Figure 3). Taking $\beta = 0.4$ and the dielectric permittivity of PVDF $\epsilon_h = 10$ [31], one finds for the effective dielectric permittivity of BaTiO$_3$ NPs $\epsilon_i \approx 30$.

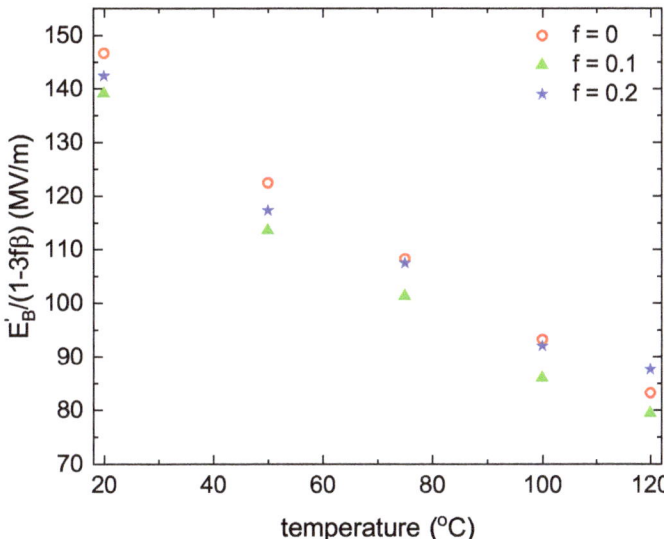

Figure 3. The dependence of the quantity $E'_B/(1-3f\beta)$ on the temperature for different volume fractions of BaTiO$_3$ NPs embedded in PVDF matrix indicated in the inset. The experimental data are taken from Figure 5b in Ref. [6]. The experimental error bars are not available.

In another set of experiments, BaTiO$_3$ NPs of 30–50 nm in diameter were surface modified using a pentafluorobenzyl phosphonic acid (PFBPA) and incorporated into a poly(vinylidene fluoride-cohexafluoropropylene) (P(VDF-HFP)) matrix [7]. Again, for relatively small volume fractions of NPs ($f \leq 0.2$), the breakdown field strength linearly decreases with f (see Figure 4). The value of β deduced from the slope of this dependence is about 0.57, and taking $\epsilon_h = 12.6$ [7], one obtains $\epsilon_i \approx 70$. For comparison, the calculated value of β obtained with $\epsilon_i = 200$ is about 0.83.

Finally, a decrease in the breakdown field strength was also observed for BaTiO$_3$ NPs of 7 nm in diameter dispersed in polystyrene (PS) [8]. A linear fit of this dependence (Figure 5) gives $\beta \approx 0.54$ which, together with the value $\epsilon_h = 2.4$ for the PS [8], provides $\epsilon_i \approx 11$. The calculated value with $\epsilon_i = 200$ is $\beta \approx 0.96$.

As one can see from the above consideration, the values of the effective dielectric permittivity of BaTiO$_3$ NPs derived from different measurements differ from each other significantly which points at a dominant role of the interface polarization at the boundary between an NP and a polymer matrix. The obtained values of ϵ_i are in the range between 10 and 70 which is much less than the values typical for thin BaTiO$_3$ films [30]. This discrepancy can be attributed to the existence of a nonferroelectric surface layer in the NPs which has a low dielectric permittivity, its relative contribution being that it is increasing with a decrease in the NP size [32,33].

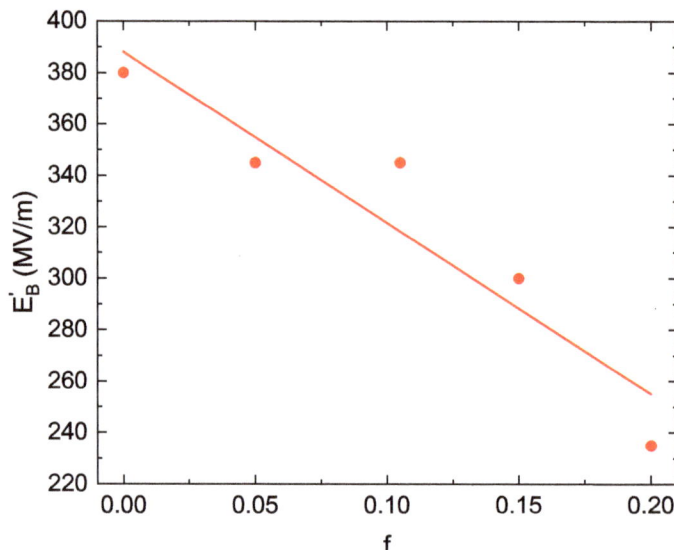

Figure 4. The dependence of the breakdown field strength on the volume fraction of BaTiO$_3$ NPs embedded in P(VDF-HFP) matrix. The straight line is the best linear fit to the experimental data shown by circles and taken from Figure 8b in Ref. [7]. The experimental error bars are not available.

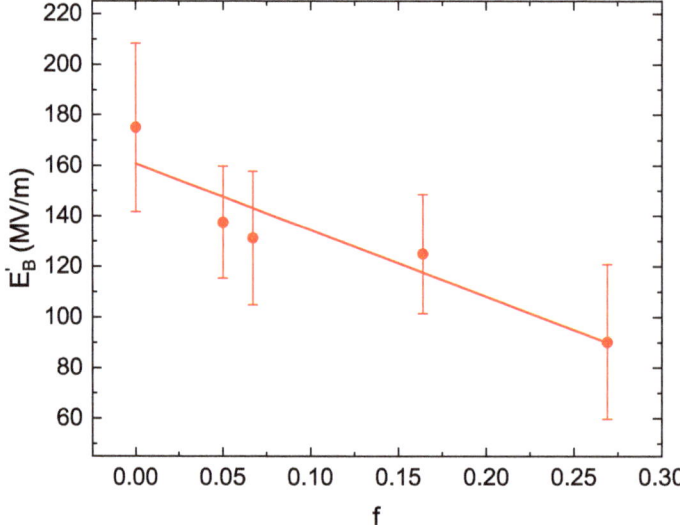

Figure 5. The dependence of the breakdown field strength on the volume fraction of BaTiO$_3$ NPs embedded in PS matrix. The straight line is the best linear fit to the experimental data shown by circles and taken from Figure 5 in Ref. [8].

5. Conclusions

In the present paper, we have developed a first-principles approach which allows one to calculate the local electric field in a nanocomposite capacitor and the parameters which characterize the electrical breakdown in it. It is found that for relatively low volume fractions of NPs ($f \leq 0.2$), the breakdown voltage linearly decreases with f. The slope of this dependence is determined by the parameter β which is related to the NP polarizability.

A comparison with the available experimental data on the breakdown strength in polymer films doped with BaTiO$_3$ NPs has revealed that the effective dielectric permittivity of NPs differs remarkably for different host polymers and is significantly lower than the values typical for ferroelectrics. These findings point at a dominant role of the interface polarization at the NP-polymer interface and an existence of a nonferroelectric surface layer.

Author Contributions: Conceptualization, V.B. and T.E.; funding acquisition, T.E.; investigation, V.B.; methodology, V.B.; project administration, T.E.; writing—original draft, V.B.; writing—review and editing, T.E. All authors have read and agreed to the published version of the manuscript.

Funding: This research is kindly supported by the IE Industrial Elektronik project (SFD-17-0036), which has received EU co-financing from the European Social Fund.

Institutional Review Board Statement: Not applicable.

Informed Consent Statement: Not applicable.

Data Availability Statement: Not applicable.

Conflicts of Interest: The authors declare no conflict of interest.

References

1. Li, H.; Liu, F.; Fan, B.; Ai, D.; Peng, Z.; Wang, Q. Nanostructured ferroelectric-polymer composites for capacitive energy storage. *Small Methods* **2018**, *2*, 1700399. [CrossRef]
2. Cao, Y.; Irwin, P.C.; Younsi, K. The future of nanodielectrics in the electrical power industry. *IEEE Trans. Dielectr. Electr. Insul.* **2004**, *11*, 797–807.
3. Dang, Z.-M.; Yuan, J.-K.; Zha, J.-W.; Zhou, T.; Li, S.-T.; Hu, G.-H. Fundamentals, processes and applications of high-permittivity polymer-matrix composites. *Prog. Mater. Sci.* **2012**, *57*, 660–723. [CrossRef]
4. Dang, Z.-M.; Yuan, J.-K.; Yao, S.-H.; Liao, R.-J. Flexible nanodielectric materials with high permittivity for power energy storage. *Adv. Mater.* **2013**, *25*, 6334–6365. [CrossRef]
5. Prateek Thakur, V.K.; Gupta, R.K. Recent progress on ferroelectric polymer-based nanocomposites for high energy density capacitors: Synthesis, dielectric properties, and future aspects. *Chem. Rev.* **2016**, *116*, 4260–4317. [CrossRef]
6. Rajib, M.; Shuvo, M.A.I.; Karim, H.; Delfin, D.; Afrin, S.; Lin, Y. Temperature influence on dielectric energy storage of nanocomposites. *Ceram. Int.* **2015**, *41*, 1807–1813. [CrossRef]
7. Kim, P.; Doss, N.M.; Tillotson, J.P.; Hotchkiss, P.J.; Pan, M.-J.; Marder, S.R.; Li, J.; Calame, J.P.; Perry, J.W. High energy density nanocomposites based on surface-modified BaTiO$_3$ and a ferroelectric polymer. *ACS Nano* **2009**, *3*, 2581–2592. [CrossRef]
8. Grabowski, C.A.; Fillery, S.P.; Koerner, H.; Tchoul, M.; Drummy, L.; Beier, C.W.; Brutchey, R.L.; Durstock, M.F.; Vaia, R.A. Dielectric performance of high permitivity nanocomposites: Impact of polystyrene grafting on BaTiO$_3$ and TiO$_2$. *Nanocomposites* **2016**, *2*, 117–124. [CrossRef]
9. Smith, R.C.; Liang, C.; Landry, M.; Nelson, J.K.; Schadler, L.S. The mechanisms leading to the useful electrical properties of polymer nanodielectrics. *IEEE Trans. Dielectr. Electr. Insul.* **2008**, *15*, 187–196. [CrossRef]
10. Dou, X.; Liu, X.; Zhang, Y.; Feng, H.; Chen, J.-F.; Du, S. Improved dielectric strength of barium titanate-polyvinylidene fluoride nanocomposite. *Appl. Phys. Lett.* **2009**, *95*, 132904. [CrossRef]
11. Bi, K.; Bi, M.; Hao, Y.; Luo, W.; Cai, Z.; Wang, X.; Huang, Y. Ultrafine core-shell BaTiO$_3$@SiO$_2$ structures for nanocomposite capacitors with high energy density. *Nano Energy* **2018**, *51*, 513–523. [CrossRef]
12. Zhou, Y.; Yuan, C.; Wang, S.; Zhu, Y.; Cheng, S.; Yang, X.; Yang, Y.; Hu, J.; He, J.; Li, Q. Interface-modulated nanocomposites based on polypropylene for high-temperature energy storage. *Energy Storage Mater.* **2020**, *28*, 255–263. [CrossRef]
13. Cheng, L.; Liu, W.; Zhang, Z.; Zhou, Y.; Li, S. Enhanced breakdown strength and restrained dielectric loss of polypropylene/maleic anhydride grafted polypropylene/core-shell ZrO$_2$@SiO$_2$ nanocomposites. *Polym. Compos.* **2022**, *43*, 2175–2183. [CrossRef]
14. Maxwell Garnett, J.C. Colours in metal glasses and in metallic films. *Phil. Trans. R. Soc. A* **1904**, *203*, 385–420.
15. Markel, V.A. Introduction to the Maxwell Garnett approximation: Tutorial. *J. Opt. Soc. Am. A* **2016**, *33*, 1244–1256. [CrossRef] [PubMed]
16. Bordo, V.; Ebel, T. How to determine the capacitance of a nanocomposite capacitor. *AIP Adv.* **2022**, *12*, 045107. [CrossRef]
17. Forlani, F.; Minnaja, N. Thickness influence in breakdown phenomena of thin dielectric films. *Phys. Stat. Sol.* **1964**, *4*, 311–324. [CrossRef]
18. Nordheim, L. Zur Theorie der thermischen Emission und der Reflexion von Elektronen an Metallen. *Z. Phys.* **1928**, *46*, 833–855. [CrossRef]
19. Fowler, R.H.; Nordheim, L. Electron emission in intense electric fields. *Proc. R. Soc. A* **1928**, *119*, 173–181.
20. Schmidlin, F.W. Enhanced tunneling through dielectric films due to ionic defects. *J. Appl. Phys.* **1966**, *37*, 2823–2832. [CrossRef]
21. Nha, H.; Jhe, W. Cavity quantum electrodynamics between parallel dielectric surfaces. *Phys. Rev. A* **1996**, *54*, 3505–3513. [CrossRef] [PubMed]

22. Stratton, J.A. *Electromagnetic Theory*; McGraw-Hill: New York, NY, USA, 1941; p. 206.
23. Born, M.; Wolf, E. *Principles of Optics*, 6th ed.; Pergamon Press: Oxford, UK, 1980; p. 100.
24. Bordo, V.G. Local field in finite-size metamaterials: Application to composites of dielectrics and metal nanoparticles. *Phys. Rev. B* **2018**, *97*, 115410. [CrossRef]
25. Wylie, J.M.; Sipe, J.E. Quantum electrodynamics near an interface. *Phys. Rev. A* **1984**, *30*, 1185–1193. [CrossRef]
26. Bordo, V.G. Self-excitation of surface plasmon polaritons. *Phys. Rev. B* **2016**, *93*, 155421. [CrossRef]
27. Sipe, J.E. The ATR spectra of multipole surface plasmons. *Surf. Sci.* **1979**, *84*, 75–105. [CrossRef]
28. Sipe, J.E. The dipole antenna problem in surface physics: A new approach. *Surf. Sci.* **1981**, *105*, 489–504. [CrossRef]
29. Bordo, V.G. Dicke superradiance from a plasmonic nanocomposite slab. *J. Opt. Soc. Am. B* **2021**, *38*, 2104–2111. [CrossRef]
30. Bajac, B.; Vukmirović, J.; Tripković, D.; Djurdjić, E.; Stanojev, J.; Cvejić, Ž.; Škorić, B.; Srdić, V.V. Structural characterization and dielectric properties of $BaTiO_3$ thin films obtained by spin coating. *Process. Appl. Ceram.* **2014**, *8*, 219–224. [CrossRef]
31. Wang, S.; Liu, L.; Zeng, Y.; Zhou, B.; Teng, K.; Ma, M.; Chen, L.; Xu, Z.J. Improving dielectric properties of poly(vinylidene fluoride) composites: Effects of surface functionalization of exfoliated graphene. *Adhes. Sci. Technol.* **2015**, *29*, 678–690. [CrossRef]
32. Goswami, A.K. Dielectric properties of unsintered barium titanate. *J. Appl. Phys.* **1969**, *40*, 619–624. [CrossRef]
33. Hoshina, T. Size effect of barium titanate: Fine particles and ceramics. *J. Ceram. Soc. Jpn.* **2013**, *121*, 156–161. [CrossRef]

Article

The Influence of Nanoparticles' Conductivity and Charging on Dielectric Properties of Ester Oil Based Nanofluid

Konstantinos N. Koutras [1,*], Ioannis A. Naxakis [1], Eleftheria C. Pyrgioti [1,*], Vasilios P. Charalampakos [2], Ioannis F. Gonos [3], Aspasia E. Antonelou [4] and Spyros N. Yannopoulos [4]

1. High Voltage Laboratory, Department of Electrical & Computer Engineering, University of Patras, 26500 Patras, Greece; naxakis@ece.upatras.gr
2. Department of Electrical & Computer Engineering, University of the Peloponnese, 26334 Patras, Greece; charalambakos@uop.gr
3. School of Electrical & Computer Engineering, National Technical University of Athens, 15780 Athens, Greece; igonos@cs.ntua.gr
4. Foundation for Research and Technology Hellas, Institute of Chemical Engineering Sciences, 26504 Patras, Greece; antonelou@iceht.forth.gr (A.E.A.); sny@iceht.forth.gr (S.N.Y.)
* Correspondence: k.koutras@ece.upatras.gr (K.N.K.); e.pyrgioti@ece.upatras.gr (E.C.P.)

Received: 13 November 2020; Accepted: 9 December 2020; Published: 11 December 2020

Abstract: This study addresses the effect of nanoparticles' conductivity and surface charge on the dielectric performance of insulating nanofluids. Dispersions of alumina and silicon carbide nanoparticles of similar size (~50 nm) and concentration (0.004% *w/w*) were prepared in natural ester oil. The stability of the dispersions was explored by dynamic light scattering. AC, positive and negative lightning impulse breakdown voltage, as well as partial discharge inception voltage of the nanofluid samples were measured and compared with the respective properties of the base oil. The obtained results indicate that the addition of SiC nanoparticles can lead to an increase in AC breakdown voltage and also enhance the resistance of the liquid to the appearance of partial discharge. On the other hand, the induction of positive charge from the Al_2O_3 nanoparticles could be the main factor leading to an improved positive Lightning Impulse Breakdown Voltage and worse performance at negative polarity.

Keywords: nanofluid; conductivity; alumina nanoparticles; silicon carbide; AC breakdown; lightning impulse breakdown; partial discharge

1. Introduction

The lifetime of the power transformer is a major factor for the reliability and uninterrupted operation of the electricity grid. Liquid insulation, as well as the paper immersed in dielectric liquid, should provide protection of its windings under the influence of electrical, thermal, and even environmental effects to prevent short circuits and leakage currents [1,2]. The mineral oil typically used as dielectric and thermal coolant has certain disadvantages, such as toxicity, high flammability, and reduced lifetime [3–7]. Therefore, the first step towards the improvement of transformer insulation is to turn the attention of research to the study of alternatives, such as natural or synthetic ester oils [7–10]. The use of ester oils entails, beyond improved properties like biodegradability and higher moisture tolerance, benefits for the transformer itself, essentially aging characteristics, enhanced lifetime, and loading capability [3,11,12].

Nanotechnology has already been used for various subsurface applications [13–16]. In liquid insulation, the integration of nanoparticles (NPs) in a base conventional oil volume has been part of recent research in an effort to achieve enhanced dielectric and thermal performance with optimum insulation quantity to recede the size of the transformer [17]. The term "nanofluid" (NF) was firstly proposed by Choi et al. [18], indicating a mixture where both the NPs and the base oil contribute to the application providing enhanced thermal conductivity. Since then, a number of reports have appeared in the literature regarding the integration of many different types of NPs in mineral or ester oils achieving improvement not only in thermal but also in dielectric properties based on their type, concentrations, shapes, and sizes [1–9,19–22].

It has been noticed that the addition of nitrides mainly leads to improvement of thermal properties [20,21], while Thomas et al. [3] indicated that the integration of $CaCu_3Ti_4O_{12}$ NPs in 0.050% vol. concentration could lead to thermal conductivity enhancement of the synthetic ester oil by 10% at room temperature. The integration of metal oxides, on the other hand, could have beneficial effect on dielectric ones. Towards this direction, Zhong et al. [8] and Du et al. [2,22] have reported improvement of AC and Lightning Impulse Breakdown characteristics following the addition of semi-conductive NPs in natural ester and mineral oil volumes, respectively. The addition of magnetic [4,23] and dielectric [19,23,24] NPs, even at low weight fractions, has also shown increase in the dielectric strength with respect to the base oil. Khaled et al. [23] studied the effect of conductive Fe_3O_4, and dielectric Al_2O_3 and SiO_2 NPs on AC breakdown voltage (AC BDV) of synthetic ester oil reporting 48% improvement with the addition of conductive Fe_3O_4 NPs (50 nm) at a concentration of 0.4 g/L, while the integration of insulating Al_2O_3 NPs of 13 nm size at a concentration of 0.05 g/L led to a 35% improvement. Last but not least, the viscosity of the oil is also affected based on the type and the loading of the NPs inside the matrix. Fontes et al. [25] reported a 25% rise in viscosity adding diamond NPs in transformer oil at 0.050 vol.%. Ilyas et al. [26] concluded that the dynamic viscosity of the mineral oil-based alumina NFs had decreased with the rise of temperature.

With regard to the mechanisms leading to the beneficial effects of the NPs on the aforementioned properties, a number of theories have been proposed [27–29]. Most of them consider the operation of NPs as "electron traps" under the application of an external electric field hindering the streamer propagation between the electrodes, adopting different explanations [27,28]. Conductive NPs can capture the fast electrons in shallow traps through charge induction [27,28]. Semi-conductive and dielectric NPs could have the same effect as the conductive ones, because they are polarized under the influence of an external electric field.

Despite the number of publications devoted to the study of the metal oxide NPs' effect on dielectric and thermal properties of the transformer oil, there is limited literature concerning the influence of metal carbides in transformer oils, although they are widely used in other industrial sectors. The main contribution of this article is to compare the dielectric performance of insulating Al_2O_3 and semi-conducting SiC-ester based NFs in terms of AC breakdown voltage (AC BDV), positive and negative Lightning Impulse breakdown voltage (LI BDV) and partial discharge (PD) activity. The choice of these specific types of NPs, having similar nominal size, is based on the fact that they have similar permittivities, therefore conductivity should play a major role in the dielectric performance of the corresponding NFs.

In this study, insulating alumina and semi-conducting SiC NPs were dispersed into natural ester oil Envirotemp^{TM} FR3^{TM} in concentration of 0.004% w/w. The choice of this particular concentration is based on previous work [30]. After the preparation of the NF samples, dynamic light scattering was used to assess their agglomeration behavior in the dispersions. AC, positive and negative LI BDV, as well as PDIV, have been measured, analyzed, and compared to the base oil's characteristics in an effort to conclude about their effect on dielectric performance of transformer oil. The results of the conducted measurements have indicated that the NF-containing SiC NPs possesses improved dielectric properties which is attributed to the higher conductivity of these NPs with respect to the Al_2O_3 NPs for the same nominal size and weight percentages.

2. Materials and Methods

This section includes the steps followed for the preparation of NF dispersions and the experiments undertaken for the study the stability and the conduction of measurements of dielectric properties.

Envirotemp^{TM} FR3^{TM}, purchased by Cargill, was used as matrix. It is a renewable, bio-based natural ester dielectric coolant, which is formulated from seed oils and performance enhancing additives with a density of 0.92 g·cm^{-3} at 20 °C, relative permittivity 3.2, conductivity 5×10^{-14} S·m^{-1}, and flash and fire points greater than 250 and 300 °C, respectively. The commercially obtained dielectric alumina and semi-conducting SiC NPs were used as received without any further treatment. These NPs were selected considering their similar nominal size and permittivities. Some of the most important dielectric and thermal properties of the NPs are presented in Table 1.

Table 1. Dielectric and thermal properties of the NPs.

Property	Al_2O_3	SiC
Nominal average diameter (nm)	50	50
Relative permittivity	9.8	9.7
Conductivity (S·m^{-1})	10^{-12}	3.16×10^{-3}
Thermal Conductivity (W·m^{-1}/°C at 25 °C)	25.5	350

2.1. Preparation of Nanofluid Samples

A two-step method was implemented for the production of NF samples. Primarily, a volume of 500 mL of FR3^{TM} was picked as base oil for the formation of NFs. Each matrix sample was filtered and dried in hot air oven at 120 °C overnight in order to achieve moisture reduction within the prescribed limits. The level of moisture within each sample after drying was confirmed with Karl Fischer titration method, following the procedure reported elsewhere [31]. In brief, Mitsubishi Chemicals Co. CA-100 moisture meter and Metrohm KF Coulometer 831 measured and confirmed, respectively, that the remaining moisture was below the recommended values.

Subsequently, 18.4 mg of the two types of NPs under study were immersed in appropriate base oil volumes to achieve the desired *w/w* concentration. As NPs have high surface areas [1,32,33] they tend to agglomerate shortly after their addition in the matrix forming clusters, where primary particles are held together by van der Waals interaction forces [32,33].

In order to avoid the agglomeration of NPs, which diminishes their beneficial effects on dielectric strength [1,7,32,33], each mixture was subjected to ultrasonication and vigorous magnetic stirring to achieve good dispersion of the NPs and homogenization of the final NF samples. In fact, each sample was sonicated for 30 min with the assistance of a Raypa UCI-50 ultrasonic bath, followed by a 15 min stirring with the use of a Biosan MSH-300 magnetic stirrer. The maximum input power of the sonicator is about 300 W with a frequency of 35 kHz. Each NF sample was subjected to ultrasonication under full power and the amount of ultrasonication energy was in the range of 540 kJ. This cycle was repeated three times for both NF samples, ensuring uniform dispersion of the colloidal suspensions, as indicated in [7]. Table 2 depicts the labeling of the NF samples, as well as the base oil, which will be submitted in the same experiments as well, while images of the two NF samples are shown in Figure 1 immediately after their preparation.

Table 2. Details of nanofluids studied in this work.

Sample Name	NPs Concentration (% *w/w*)
Base	0
iNF	NF with 0.004% Al_2O_3 NPs
sNF	NF with 0.004% SiC NPs

Figure 1. The NF samples shortly after their synthesis: sNF left; iNF right.

2.2. Experimental Part

There are two major categories, according to which the dielectric measurements regarding any High Voltage (HV) equipment are classified: the destructive and non-destructive tests. In the context of the destructive tests, AC, positive and negative LI BDV are measured for the two NF samples along with a base oil sample. In regard to the non-destructive ones, PDIV, as well as the variation of the apparent charge for various voltage levels, have been measured.

AC BDV is the fundamental parameter for the characterization of any insulation system, as long the transformer nominally operates under AC voltage. For the determination of this property, each one of the samples under investigation is stressed under AC voltage/60 Hz using a Baur DTA 100 C generator, as illustrated in Figure 2a, according to the conditions of IEC 60,156 standard. The generator can produce up to 100 kV increasing the voltage at a rate of 2 kV/s until breakdown occurs between two Rogowski electrodes, which ensure a uniform electric field distribution, with the gap distance set at 2.5 mm. At the beginning of the experimental procedure, the electrodes are polished and cleaned. 10 sets of 6 breakdown events (60 in total) are performed, with the applied interval being 2 min between two successive breakdowns and 5 min between two sets. The results are fit to Weibull and normal distributions in order the AC BDV in low probability levels to be estimated [7,23,34].

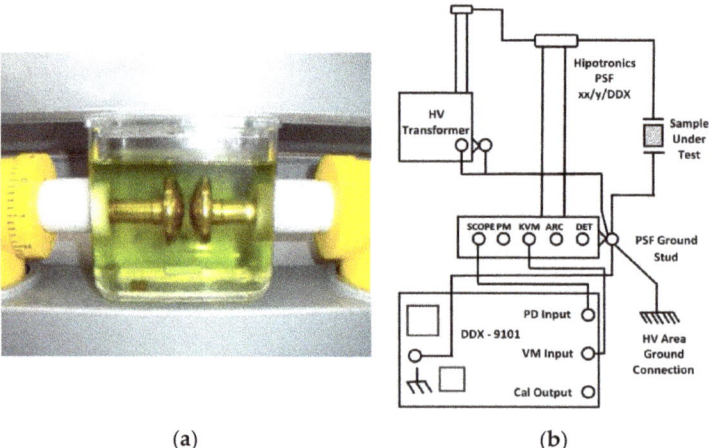

Figure 2. AC BDV and PD apparatuses: (**a**) BAUR DTA test cell with Rogowski electrodes. (**b**) Schematic diagram of the PD setup.

Regarding the measurement of LI BDV, this is another property of major importance as it is an indication for the strength of the insulation in emergency conditions, such as lightning strikes.

A two-stage impulse generator is used, producing up to 400 kV, with positive and negative polarity, according to IEC 60,897 standard. A point–sphere geometry is used, ensuring a highly divergent field geometry, with the gap distance set at 25 mm. The radius of curvature is 50 µm, while the diameter of the sphere is 12.7 mm. Five voltage applications are conducted for each voltage level with each sample being replaced with a new one after every BD event, ensuring a good reliability for the processing of the results. The applied lightning impulse voltage 1.2/50 µs, is monitored on oscilloscope Tektronix DPO4104; 1 GHz/5 GS s-1.

In terms of the conduction of PD measurements, a HV test transformer (HIGH VOLT GmbH Transformer PEOI 40/100,100 kV) is used with a TETTEX Instruments PD Detector DDX-9101 to measure the apparent charge to a nominal capacitor (HIPOTRONICS capacitor PSF 100-1). The schematic diagram of the configuration is displayed in Figure 2b. The test cell consists of two plate electrodes with the gap between them set at 0.75 mm. Nomex® Dupont™ insulating paper with 0.75 mm thickness impregnated to the NFs and base oil samples, is put in the gap. The AC voltage value for which the measured apparent charge exceeds the value of 10 pC is considered as the PDIV for each sample under investigation. The insulating papers impregnated to iNF, sNF, and base oil samples are stressed under AC voltage while the transformer is increasing the voltage at a rate of 0.1 kV/s until the measured apparent charge exceeds the threshold value. Each experiment is repeated 4 times for ensuring the reliability of the obtained results.

3. Results

3.1. Particle Size and Agglomeration Dynamics in Dispersions

Field emission scanning electron microscopy (FE-SEM) was used to study the particle size and particle size distribution of the dry powders. The NPs were drop-casted by ethanol dispersion onto Si substrates. Figure 3 illustrates typical images of the Al_2O_3 Figure 3a and SiC particles Figure 3b at high magnifications. For both samples, the particles seem agglomerated, as it typically occurs when particles are drop casted from a dispersion to the substrate. The images reveal that the particles sizes have the nominal values, being, in particular, ~45–50 nm in the case of SiC and in the range 30–35 for Al_2O_3. For both materials, image analysis over broader areas by FE-SEM, incorporating a larger number of primary particles, show very narrow particle size distributions.

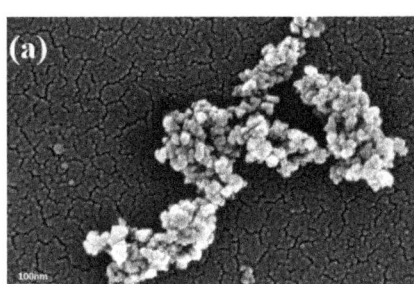 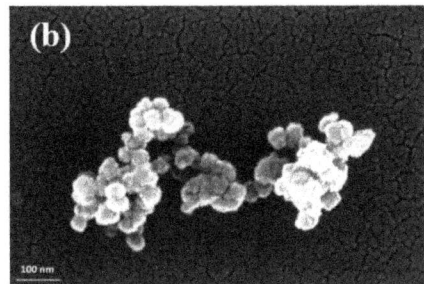

Figure 3. Representative FE-SME images of (a) Al_2O_3 and (b) SiC NPs at high magnification.

The dynamics of NPs in dispersion are of paramount importance since their agglomeration rate essentially determines the high-performance lifetime of the NF. DLS is a versatile technique able to provide the rate of agglomeration process in situ. Apart from the primary particle size, the technique can sensitively furnish information on the formation of small agglomerates ad their evolution upon aging. Details about the method and data analysis applied to a similar system, namely, a TiO_2-based NF, have been presented elsewhere [7]. Figure 4 shows the hydrodynamic radii of the two NFs measured at two different times, immediately after their dispersion preparation and after 5 days. The R_h data reveal that for each NF there are two different population of particles. One population with R_h around

40 nm is related to the dynamics of the primary NPs, while agglomerates with R_h of few hundreds of nm also exist in the dispersions.

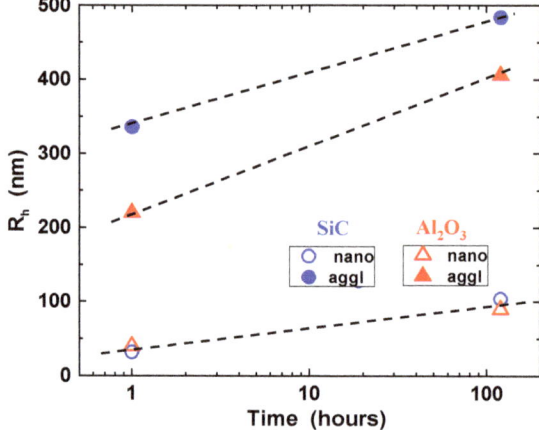

Figure 4. Hydrodynamic radii estimated by DLS data in dispersion, measured at short and long times after the dispersion preparation.

The evolution of R_h against time show that a weak agglomeration, in relation to TiO_2-based oil dispersion [7] takes place. Both the primary particles (nano) and their agglomerates (aggl) exhibit an increase in size following similar rates among the two types of NPs. These findings demonstrate that the proclivity toward agglomeration of SiC and Al_2O_3 is rather weak, e.g., compared to TiO_2 NFs. This finding lends support to the accuracy of the electrical characterization, considering that possible fast agglomeration process may affect the accurate monitoring of electrical parameters.

3.2. AC Breakdown Voltage

According to the mean AC BDVs displayed in Figure 5 and Table 3, iNF and sNF samples present enhanced dielectric strength by 4.1% and 16.3%, respectively. The standard deviation of the collected AC BDV events is approximately the same for the iNF with respect to that of the base oil. On the contrary, the standard deviation of the results regarding sNF is increased.

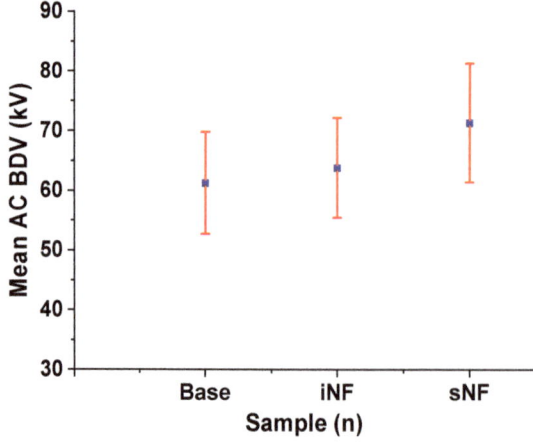

Figure 5. Mean AC BDVs and standard deviation of the NF and base oil samples.

Table 3. Descriptive statistics of the compared samples.

Sample Name (n)	Mean BDV (kV)	Standard Deviation (kV)	Scale	Shape
Base	61.3	8.5	8.7	64.8
iNF	63.8	8.3	8.9	67.4
sNF	71.3	9.9	8.5	75.5

The estimation of the BDV in low probability levels is of major importance as well. The BDV can be considered as a random variable that follows normal or/and Weibull distribution [7,33–36]. The cumulated distribution functions (CDF) of the normal and Weibull random variable x, expressing the breakdown voltage, are given by (1) and (2), respectively:

$$F_{(x)} = \frac{1}{2}\left[1 + erf\left(\frac{x-\mu}{\sigma\sqrt{2}}\right)\right] \quad (1)$$

$$F_{(x)} = 1 - e^{-\frac{x}{\beta}\alpha} \quad (2)$$

where, μ is the mean value; σ is the standard deviation; α is the shape parameter and β is the scale parameter.

Firstly, Anderson–Darling goodness of fit test is performed to determine whether the sample of the BD events are normally distributed. The same procedure is followed for Weibull distribution too, because it is expected to give more precise analysis as it does not make assumptions of the skewness and kurtosis. At 5% significance level, the p-value is higher for all the samples under investigation both for normal and Weibull distribution, as shown in Table 4.

Table 4. Goodness of fit test (Anderson-Darling).

Sample Name (n)	p-Value Normal	p-Value Weibull
Base	0.07	>0.25
iNF	0.12	>0.25
sNF	0.32	>0.25

Figure 4 depicts the frequency density plot of the BD events along with adjustment to normal distribution per each sample under investigation. Such a display is necessary for the detection of possible anomalies in the distribution of the AC BDV [7,23,34]. From the plots of Figure 6, along with the goodness of fit test results; it is concluded that none of them could be rejected for the following statistical analysis.

The breakdown voltages at 1, 10, and 50% probability levels of base, iNF and sNF (Table 5) were calculated from the probability density plots for both distributions, which are demonstrated in Figure 7. The choice of these probability levels corresponds to their importance for the operation of the transformer. On the one hand, $U_{50\%}$ gives an assessment for the expected breakdown voltage, while on the other hand, $U_{10\%}$ represents an indication on the initiation threshold of ionization, and therefore assists in concluding about the reliability of the transformer oil sample [7,23,32–35]. Last but not least, the BDV at 1% cumulative probability is necessary information for the designing of electric equipment. $U_{1\%}$ corresponds to the limit of voltage for safety and continuous operation of the appliance [37].

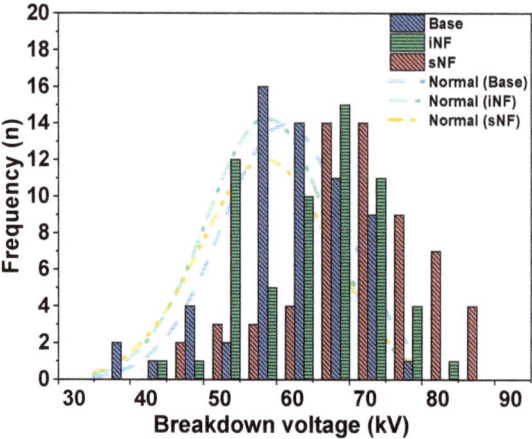

Figure 6. Frequency density plots of the breakdown events for each sample in question with conformity to Normal distribution.

Table 5. Breakdown Voltages at 1, 10 and 50% probability levels for normal and Weibull distributions.

Sample Name (n)	Distribution	$U_{50\%}$ (kV)	$U_{10\%}$ (kV)	$U_{1\%}$ (kV)
Base	Normal	61.3	50.3	41.3
	Weibull	62.2	50.1	38.6
iNF	Normal	63.8	53.1	44.4
	Weibull	64.7	52.3	40.3
sNF	Normal	71.3	58.6	48.2
	Weibull	72.3	57.9	43.9

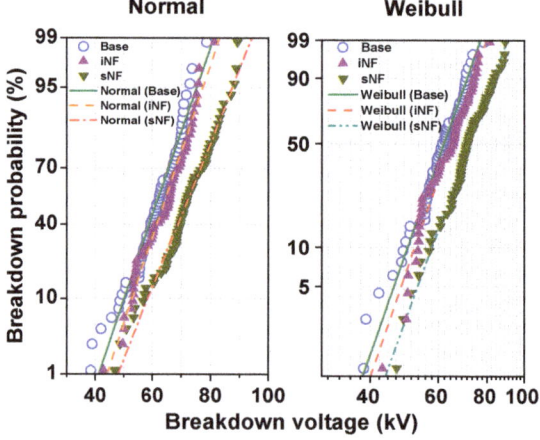

Figure 7. Probability density plots of the breakdown events for each sample in question according to Normal and Weibull distributions.

Based on the results of Table 5, both iNF and sNF demonstrate enhanced BDVs in low cumulative probability levels, with sNF showing the highest improvement with respect to the natural ester oil sample by around 16.2, 15.6, and 13.7% for $U_{50\%}$, $U_{10\%}$, and $U_{1\%}$, respectively.

From the $U_{10\%}$ and $U_{1\%}$ BD probability results, it is evident that the addition of SiC NPs could assure more reliable nominal operation of the transformer and delay the initiation of streamer propagation [23,35,37].

3.3. Positive and Negative Lightning Impulse Breakdown Voltage

The up-and-down method was implemented in order to estimate the mean LI BDV under positive and negative polarity of base oil and the two NF samples after conducting measurements in various voltage levels according to the demands of IEC 60,897 standard. The mean LI BDVs as well as the average times to breakdown are exhibited in Table 6. Figures 8 and 9 depict the oscillograms of the LI BDVs at 50% breakdown probability under positive and negative polarity, respectively.

Table 6. Mean positive and negative LI BDVs and time to breakdown.

Sample Name (n)	LI BDV (kV)		Time to Breakdown (µs)	
	Positive	Negative	Positive	Negative
Base	73.4	−118.7	16.6	22.3
iNF	94.3	−120.5	14.8	15.7
sNF	77.5	−123.4	15.7	24.2

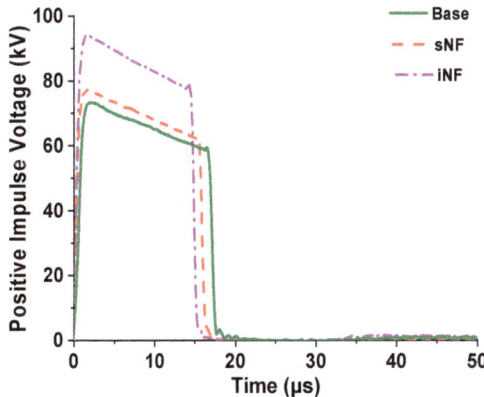

Figure 8. Oscillogram of the positive LI BDVs of the under-study samples at 50% breakdown probability level.

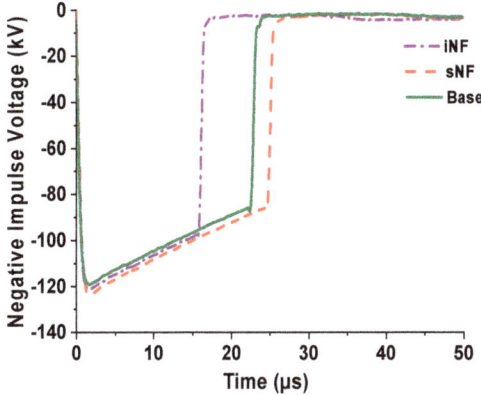

Figure 9. Oscillogram of the negative LI BDVs of the under-study samples at 50% breakdown probability level.

The LI strength of iNF under positive polarity is enhanced by 28.5% with respect to the base oil with the corresponding improvement of sNF being limited at 6%. The mean BD times of the samples after the application of positive LIV are almost steady, unlike with the corresponding ones under negative polarity. Under the application of negative LIV, the time to BD of iNF is restricted by 29.5% with respect to the one of the base oil, while sNF sample demonstrates almost the same BD time. In terms of their behavior under negative LIV, both NF samples show limited increase, as displayed in Table 6 and Figure 8. It is also evident that the addition of alumina NPs narrows the gap between positive and negative LI BDV.

3.4. Partial Discharge

In Figure 10, the change in apparent charge is depicted versus the applied AC voltage, while the PDIV and the value of apparent charge at the PDIV for each sample are presented in Table 7. It is evident that all the samples in question have similar behavior until about the level of 3 kV. Above this voltage level, only the paper impregnated to sNF demonstrates a stable resistance to PD activity for the whole range of AC stress, with the accumulation of apparent charge at the level of 5.5 kV (highest applied voltage) being improved by 97.5 and 95.8% in comparison to the corresponding one of the paper impregnated to base oil and iNF, respectively. Papers impregnated to sNF and iNF samples demonstrate improved resistance to PD appearance by 92 and 44% with respect to the paper impregnated to the matrix.

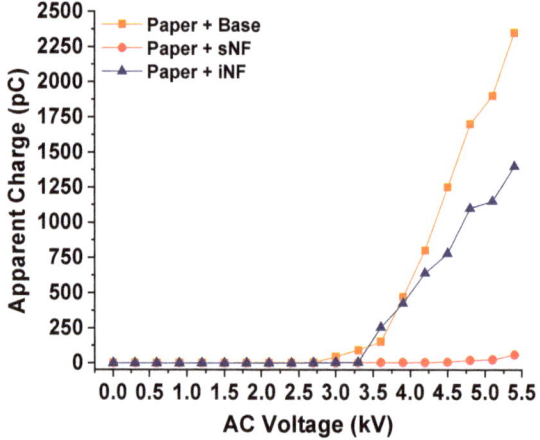

Figure 10. Change of the apparent charge as a function of the applied voltage for the three samples under investigation.

Table 7. PDIV and apparent charge at the PDIV.

Sample Name (n)	PDIV (kV)	Apparent Charge (pC)
Base	2.50	43.1
iNF	3.60	257.0
sNF	4.80	20.0

Figure 11 depicts the PD pulses related to the papers impregnated to sNF (Figure 11a) and iNF (Figure 11b) at the highest applied voltage. In the case of iNF impregnated insulating paper, the multiple pulses of irregular form indicate the presence of voids in the vicinity of the NF.

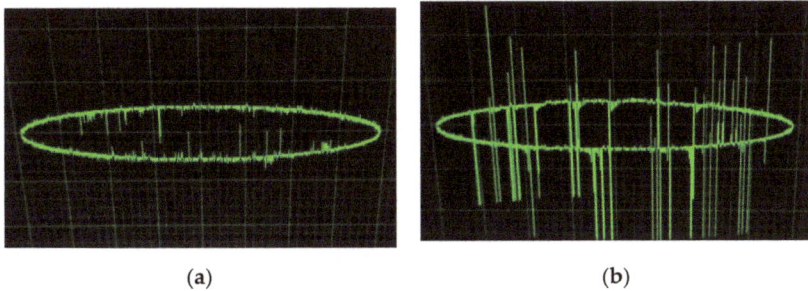

Figure 11. Waveforms of PD pulses at 5.5 kV applied voltage: (**a**) Paper impregnated to sNF; (**b**) Paper impregnated to iNF.

4. Discussion

Taking into account the $U_{50\%}$ values from Table 5 according to Weibull distribution, the mean AC dielectric strength for a gap of 2.5 mm is 24.9 kV·mm^{-1}, 25.9 kV·mm^{-1}, and 28.9 kV·mm^{-1} for base, iNF, and sNF samples, respectively. This finding could be attributed to the fact that the addition of semi-conducting SiC NPs for the selected weight fraction can lead to the delay of streamer propagation, by trapping and de-trapping the fast electrons at the tip of the streamer in shallow traps [27,28,32,38].

In an effort to explain the effect of the two types of NPs on BDV of the oil, Sima et al. [28] proposed a theory according to which the addition of NPs in the matrix results in change of the main electrodynamics in the dielectric liquid, regardless of their electrical properties. If considerable divergence exists in conductivity or permittivity between NP and base oil, then induced or polarized charges are generated at the interface between NP and matrix. These charges result in the production of a potential well that can trap the fast electrons at the tip of the streamer in shallow traps and slow down its propagation. As for the SiC NPs, they are characterized by much higher conductivity than that of the base oil, and due to their characterization as semi-conductors, charge induction involves also positive holes apart from electrons. These redistributed charges on the surface of SiC NPs produce a potential well, which is given by the formula [28] for the electric field direction ($\theta = 0$) and the opposite one ($\theta = \pi$):

$$\varphi_{SiC} = \begin{cases} \frac{\sigma_2-\sigma_1}{2\sigma_1+\sigma_2}R^3 E_0 \frac{1}{r^2}, & \theta = 0, \; r \geq R \\ -\frac{\sigma_2-\sigma_1}{2\sigma_1+\sigma_2}R^3 E_0 \frac{1}{r^2}, & \theta = \pi, \; r \geq R \end{cases} \quad (3)$$

where σ_1, σ_2 are the conductivities of matrix and NPs in S·m^{-1}, respectively, R is the radius of NP in m, E_0 is the mean dielectric strength of the NF in V·m^{-1} and r is the distance from the NP's surface in m.

On the other hand, surface charges, known as bound charges, are formed at the surface of dielectric alumina NPs, due to polarization under the influence of an external electric field E_0. This mechanism incorporates displacement and turning-direction polarizations. The displacement polarizations are generated very quickly in only 10^{-15} to 10^{-12} s, while turning-direction polarizations in time ranging from 10^{-10} s to 10^{-2} s [28]. Therefore, they can also produce a potential well, which is given by:

$$\varphi_{Al2O3} = \begin{cases} \frac{\varepsilon_2-\varepsilon_1}{2\varepsilon_1+\varepsilon_2}R^3 E_0 \frac{1}{r^2}, & \theta = 0, \; r \geq R \\ -\frac{\varepsilon_2-\varepsilon_1}{2\varepsilon_1+\varepsilon_2}R^3 E_0 \frac{1}{r^2}, & \theta = \pi, \; r \geq R \end{cases} \quad (4)$$

where ε_1, ε_2 are the permittivities of matrix and NPs in F·m^{-1}, respectively.

The potential well on each occasion is able to catch the fast-moving electrons and transform them into negatively charged NPs which are moving slowly due to their larger radius. This could lead to the delay of streamer propagation and thus higher required voltage level to bridge the gap.

Substituting the values of conductivity and permittivity of FR3™, SiC and Al$_2$O$_3$ NPs, as well as the corresponding values of mean dielectric strength and radius of NPs in Equations (3) and (4), the potential well of the colloidal suspensions of SiC and Al$_2$O$_3$ is given as:

$$\varphi_{SiC\,(r)} = 3.61 \times 10^{-15} \times \frac{1}{r^2}\,[V] \tag{5}$$

$$\varphi_{Al2O3\,(r)} = 1.32 \times 10^{-15} \times \frac{1}{r^2}\,[V] \tag{6}$$

Figure 12 demonstrates the higher potential well of the suspended SiC NPs in the oil, especially while moving closer to their surface. These types of NPs have the ability to capture more free electrons in shallow traps than the alumina NPs. This fact explains their increased dielectric performance. The total amount of charge that can be trapped by each of these two types of NPs is expressed by (7) and (8), respectively:

$$Q_{SiC} = -12\pi\varepsilon_1 E_0 R^2 \tag{7}$$

$$Q_{Al2O3} = -12\pi\varepsilon_1 E_0 R^2 \frac{\varepsilon_2}{2\varepsilon_1 + \varepsilon_2} \tag{8}$$

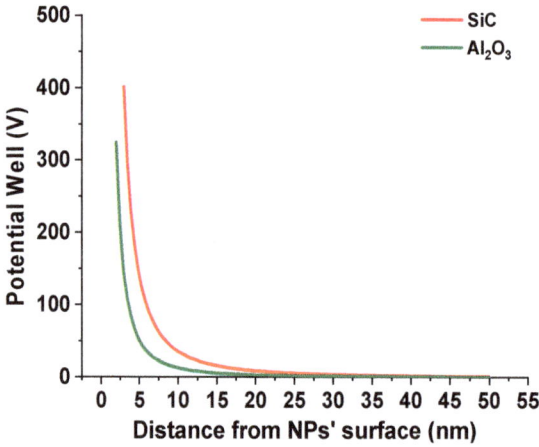

Figure 12. Potential well distribution versus the distance from the NPs' surface.

Additionally, with substitution of the corresponding values as indicated above, they give:

$$Q_{SiC} = -7.71 \times 10^{-17}\,C \tag{9}$$

$$Q_{Al2O3} = -4.18 \times 10^{-17}\,C \tag{10}$$

This means that each SiC and Al$_2$O$_3$ NP could potentially capture approximately 482 and 300 electrons, respectively, until they are saturated. The higher amount of trapped charge carriers in sNF is ought to the higher conductivity of SiC NPs and is a result of the mechanism described above. Streamers propagate in different modes, which are characterized by increasing velocity. Atiya et al. [32] indicated that the operation of NPs as electron traps hinders the first two slow modes, which are applicable under AC voltage.

Unlike AC, higher streamer propagation velocities are applicable under Lightning Impulse Voltage (LIV). Ionization of the oil molecules begins above a threshold value of electric field which includes positive ions and electrons. During positive LIV, the space charge field created by the positive ions increases towards the grounded sphere, facilitating the propagation of positive streamers. Under negative polarity, the electric field towards the grounded sphere is weakened by the space charge field, hindering the propagation of negative streamers. This explains why negative LI BDV is higher than the positive LI BDV. Both types of NFs demonstrate increased strength under positive and negative LIV, which indicates that the trapping of high mobility electrons at the tip of the streamer in shallow traps results in increased BD performance [2,7,39]. However, the better performance of iNF during positive LIV and its worse performance during negative LIV with respect to sNF could be attributed to the distribution of the space charge field [39,40]. The positive charges induced by the Al_2O_3 NPs can trap the fast electrons at the tip of the streamer, inhibiting its initiation and propagation towards the grounded sphere electrode. On the other hand, under negative polarity, the addition of Al_2O_3 NPs increases the electric field towards the grounded sphere, accelerating the propagation of negative streamers.

The improved resistance to PD activity of both NF samples, with respect to the base oil, can be attributed to the formation of the electrical double layer (EDL) between the surface of the NP and the oil [3,7,32]. The addition of these types of NPs can augment the interfacial EDL, which can capture the charge carriers and delay the initiation of PD activity. The better performance of sNF can be interpreted by the greater ability of SiC NPs to trap the charge carriers, foremost the high mobility electrons, and delay the appearance of PD in the vicinity of the liquid, as it was explained above.

5. Conclusions

In summary, two NF samples, containing Al_2O_3 (iNF) and SiC (sNF) NPs in the same concentration (0.004% w/w), were synthesized and characterized. Electron microscopy showed that primary particles have a narrow particle size distribution. DLS has revealed rather good stability of the dispersion towards agglomeration over a period of five days. Both dispersions exhibit similar agglomeration behavior. AC, positive and negative LI BDV, and PD activity of the NF samples were measured and compared to the corresponding dielectric properties of a pure base oil sample. AC BDV results indicated that sNF demonstrates the highest improvement in terms of dielectric strength by 16.3%. The statistical analysis also showed that the same sample demonstrates increased AC BDV by 15.6 and 13.7% at 10 and 1% probability levels, providing an indication of enhanced reliability for the uninterrupted operation of the power transformer. These findings have been attributed to the greater ability of SiC NPs to trap charge carriers in shallow traps, because of their higher conductivity.

On the other hand, the increase in positive LI BDV of iNF can be attributed to the positive charge induced by alumina NPs, which results in delaying the streamer propagation from the positive point. Finally, sNF impregnated paper also demonstrates the greatest resistance to the appearance of PD, delaying its initiation by 92% with respect to the paper impregnated to matrix. This finding could be attributed to the role of charge carriers' trapping from the augmented interfacial zone between SiC NPs and matrix. Our future research work includes an attempt to find an appropriate surface modification technique in order to achieve long-term stability of the Al_2O_3 and SiC dispersions, which will be followed by the conduction of measurements of dielectric and thermal properties to address the effect of surface modification on them.

Author Contributions: Conceptualization, K.N.K., I.A.N., E.C.P. and V.P.C.; methodology, K.N.K., I.A.N. and A.E.A.; validation, E.C.P., I.F.G. and S.N.Y.; formal analysis, K.N.K. and I.A.N.; investigation, K.N.K., I.A.N., E.C.P. and V.P.C.; resources, E.C.P.; data curation, I.F.G., A.E.A. and S.N.Y.; writing—original draft preparation, K.N.K. and V.P.C.; writing—review and editing, I.F.G. and S.N.Y.; supervision, E.C.P. and S.N.Y. All authors have read and agreed to the published version of the manuscript.

Funding: This research received no external funding.

Acknowledgments: Authors want to express their gratitude to Christina Polyzou of the Laboratory of Physical, Inorganic and Nuclear Chemistry, Department of Chemistry, University of Patras for the assistance in the synthesis of the NF samples.

Conflicts of Interest: The authors declare no conflict of interest.

References

1. Hussain, M.R.; Khan, Q.; Khan, A.A.; Refaat, S.S.; Abu-Rub, H. Dielectric Performance of Magneto-Nanofluids for Advancing Oil-Immersed Power Transformer. *IEEE Access* **2020**, *8*, 163316–163328. [CrossRef]
2. Du, Y.; Lv, Y.; Li, C.; Chen, M.; Zhong, Y.; Zhou, J.; Li, X.; Zhou, Y. Effect of semiconductive nanoparticles on insulating performances of transformer oil. *IEEE Trans. Dielectr. Electr. Insul.* **2012**, *19*, 770–776.
3. Thomas, P.; Hudedmani, N.; Prasath, R.T.A.R.; Roy, N.K.; Mahato, S.N. Synthetic Ester Oil Based High Permittivity $CaCu_3Ti_4O_{12}$ (CCTO) Nanofluids an Alternative Insulating Medium for Power Transformer. *IEEE Trans. Dielectr. Electr. Insul.* **2019**, *26*, 314–321. [CrossRef]
4. Peppas, G.D.; Charalampakos, V.P.; Pyrgioti, E.C.; Gonos, I.F. Electrical and optical measurements investigation of the pre-breakdown processes in natural ester oil under different impulse voltage waveforms. *IET Sci. Meas. Technol.* **2016**, *10*, 545–551. [CrossRef]
5. Farade, R.A.; Wahab, N.I.A.; Mansour, D.A.; Azis, N.B.; Jasni, J.; Soudagar, M.E.M.; Siddappa, V. Development of Graphene Oxide-Based Nonedible Cottonseed Nanofluids for Power Transformers. *Materials* **2020**, *13*, 2569. [CrossRef]
6. Katiyar, A.; Dhar, P.; Nandi, T.; Maganti, L.S.; Das, S.K. Enhanced Breakdown Performance of Anatase and Rutile Titania based Nanooils. *IEEE Trans. Dielectr. Electr. Insul.* **2016**, *23*, 3494–3503. [CrossRef]
7. Koutras, K.N.; Naxakis, I.A.; Antonelou, A.E.; Charalampakos, V.P.; Pyrgioti, E.C.; Yannopoulos, S.N. Dielectric strength and stability of natural ester oil based TiO_2 nanofluids. *J. Mol. Liq.* **2020**, *316*, 113901. [CrossRef]
8. Zhong, Y.; Lv, Y.; Li, C.; Du, Y.; Chen, M.; Zhang, S.; Zhou, Y.; Chen, L. Insulating properties and charge characteristics of natural ester fluid modified by TiO_2 semiconductive nanoparticles. *IEEE Trans. Dielectr. Electr. Insul.* **2013**, *20*, 135–140. [CrossRef]
9. Charalampakos, V.P.; Peppas, G.D.; Pyrgioti, E.C.; Bakandritsos, A.; Polykrati, A.D.; Gonos, I.F. Dielectric Insulation Characteristics of Natural Ester Fluid Modified by Colloidal Iron Oxide Ions and Silica Nanoparticles. *Energies* **2019**, *12*, 3259. [CrossRef]
10. McShane, C. New safety dielectric coolants for distribution and power transformers. *IEEE Ind. Appl. Mag.* **2000**, *6*, 24–32. [CrossRef]
11. Salama, M.M.M.; Mansour, D.A.; Daghrah, M.; Abdelkasouda, S.M.; Abbasa, A.A. Thermal performance of transformers filled with environmentally friendly oils under various loading conditions. *Int. J. Electr. Power Energy Syst.* **2020**, *118*, 105743. [CrossRef]
12. Daghrah, M.; Wang, Z.; Liu, Q.; Hilker, A.; Gyore, A. Experimental study of the influence of different liquids on the transformer cooling performance. *IEEE Trans. Power Deliv.* **2019**, *34*, 588–595. [CrossRef]
13. Ali, M.; Sahito, M.F.; Jha, N.K.; Arain, Z.U.A.; Memon, S.; Keshavarz, A.; Iglauer, S.; Saeedi, A.; Sarmadivaleh, M. Effect of nanofluid on CO_2-wettability reversal of sandstone formation; implications for CO_2 geo-storage. *J. Colloid Interface Sci.* **2020**, *559*, 304–312. [CrossRef] [PubMed]
14. Jha, N.K.; Lebedev, M.; Iglauer, S.; Ali, M.; Roshan, H.; Barifcani, A.; Sangwai, J.S.; Sarmadivaleh, M. Pore scale investigation of low salinity surfactant nanofluid injection into oil saturated sandstone via X-ray micro-tomography. *J. Colloid Interface Sci.* **2020**, *562*, 370–380. [CrossRef] [PubMed]
15. Al-Anssari, S.; Arain, Z.U.A.; Shanshool, H.A.; Ali, M.; Keshavarz, A.; Iglauer, S.; Sarmadivaleh, M. Effect of Nanoparticles on the Interfacial Tension of CO_2-Oil System at High Pressure and Temperature: An Experimental Approach. In Proceedings of the SPE Asia Pacific Oil & Gas Conference and Exhibition, Bali, Indonesia, 12–14 October 2020.
16. Aftab, A.; Ali, M.; Arif, M.; Panhwar, S.; Saady, N.M.C.; Al-Khdheeawi, E.A.; Mahmoud, O.; Ismail, A.R.; Kesharvaz, A.; Iglauer, S. Influence of tailor-made TiO_2/API bentonite nanocomposite on drilling mud performance: Towards enhanced drilling operations. *Appl. Clay Sci.* **2020**, *199*, 105862. [CrossRef]
17. Rafiq, M.; Shafique, M.; Azam, A.; Ateeq, M. Transformer oil-based nanofluid: The application of nanomaterials on thermal, electrical and physicochemical properties of liquid insulation—A review. *Ain Shams Eng. J.* **2020**. in Press. [CrossRef]
18. Choi, S.U.S.; Eastman, J.A. Enhancing Thermal Conductivity of Fluids with Nanoparticles. In Proceedings of the 1995 International Mechanical Engineering Congress Exposition, San Francisco, CA, USA, 12–17 November 1995; pp. 99–105.

19. Mohamad, N.A.; Azis, N.; Jasni, J.; Ab Kadir, M.Z.A.; Yunus, R.; Yaakub, Z. Impact of Fe_3O_4, CuO and Al_2O_3 on the AC Breakdown Voltage of Palm Oil and Coconut Oil in the Presence of CTAB. *Energies* **2019**, *12*, 1605. [CrossRef]
20. Wang, X.; Xu, X.; Choi, S.U.S. Thermal conductivity of nanoparticle-fluid mixture. *J. Thermophys. Heat Transf.* **1999**, *13*, 474–480. [CrossRef]
21. Du, B.X.; Li, X.L.; Xiao, M. High Thermal Conductivity Transformer Oil Filled with BN Nanoparticles. *IEEE Trans. Dielectr. Electr. Insul.* **2015**, *22*, 851–858. [CrossRef]
22. Du, Y.; Lv, Y.; Li, C.; Chen, M.; Zhou, J.; Li, X.; Zhou, Y.; Tu, Y. Effect of electron shallow trap on breakdown performance of transformer oil-based nanofluids. *J. Appl. Phys.* **2011**, *110*, 104104. [CrossRef]
23. Khaled, U.; Beroual, A. AC Dielectric Strength of Synthetic Ester Based Fe_3O_4, Al_2O_3 and SiO_2 Nanofluids—Conformity with Normal and Weibull Distributions. *IEEE Trans. Dielectr. Electr. Insul.* **2019**, *26*, 625–633. [CrossRef]
24. Rafiq, M.; Chengrong, L.; Lv, Y. Effect of Al_2O_3 nanorods on dielectric strength of aged transformer oil/paper insulation system. *J. Mol. Liq.* **2019**, *284*, 700–708. [CrossRef]
25. Fontes, D.H.; Ribatski, G.; Filho, E.P.B. Experimental evaluation of thermal conductivity, viscosity and breakdown voltage AC of nanofluids of carbon nanotubes and diamond in transformer oil. *Diam. Relat. Mater.* **2015**, *58*, 115–121. [CrossRef]
26. Ilyas, S.U.; Pendyala, R.; Narahari, M.; Susin, L. Stability, rheology and thermal analysis of functionalized alumina- thermal oil-based nanofluids for advanced cooling systems. *Energy Convers. Manag.* **2017**, *142*, 215–229. [CrossRef]
27. Hwang, J.G.; Zahn, M.; O'Sullivan, F.M.; Pettersson, L.A.A.; Hjortstam, O.; Liu, R. Effects of nanoparticle charging on streamer development in transformer oil-based nanofluids. *J. Appl. Phys.* **2010**, *107*, 014310. [CrossRef]
28. Sima, W.; Shi, J.; Yang, Q.; Huang, S.; Cao, X. Effects of Conductivity and Permittivity of Nanoparticle on Transformer Oil Insulation Performance: Experiment and Theory. *IEEE Trans. Dielectr. Electr. Insul.* **2015**, *22*, 380–390. [CrossRef]
29. Miao, J.; Dong, M.; Ren, M.; Wu, X.; Shen, L.; Wang, H. Effect of nanoparticle polarization on relative permittivity of transformer oil-based nanofluids. *J. Appl. Phys.* **2013**, *113*, 204103. [CrossRef]
30. Koutras, K.; Pyrgioti, E.; Naxakis, I.; Charalampakos, V.; Peppas, G. AC Breakdown Performance of Al_2O_3 and SiC Natural Ester Based Nanofluids. In Proceedings of the 2020 IEEE International Conference on Environment and Electrical Engineering and 2020 IEEE Industrial and Commercial Power Systems Europe (EEEIC/I&CPS Europe), Madrid, Spain, 9–12 June 2020; pp. 1–5.
31. Peppas, G.D.; Charalampakos, V.P.; Pyrgioti, E.C.; Danikas, M.G.; Bakandritsos, A.; Gonos, I.F. Statistical investigation of AC breakdown voltage of nanofluids compared with mineral and natural ester oil. *IET Sci. Meas. Technol.* **2016**, *10*, 644–652. [CrossRef]
32. Atiya, E.G.; Mansour, D.A.; Khattab, R.M. Dispersion behavior and breakdown strength of transformer oil filled with TiO_2 nanoparticles. *IEEE Trans. Dielectr. Electr. Insul.* **2015**, *22*, 2463–2472. [CrossRef]
33. Pyrgioti, E.C.; Koutras, K.N.; Charalampakos, V.P.; Peppas, G.D.; Bakandritsos, A.P.; Naxakis, I.A. An experimental study on the influence of surface modification of TiO_2 nanoparticles on breakdown voltage of natural ester nanofluid. *Lect. Notes Electr. Eng.* **2020**, *599*, 279–288.
34. Martin, D.; Wang, Z.D. Statistical analysis of the AC breakdown voltages of ester based transformer oils. *IEEE Trans. Dielectr. Electr. Insul.* **2008**, *15*, 1044–1050. [CrossRef]
35. Khaled, U.; Beroual, A. AC dielectric strength of mineral oil-based Fe_3O_4 and Al_2O_3 nanofluids. *Energies* **2018**, *11*, 3505. [CrossRef]
36. Dang, V.-H.; Beroual, A.; Perrier, C. Comparative study of statistical breakdown in mineral, synthetic and natural ester oils under AC voltage. *IEEE Trans. Dielectr. Electr. Insul.* **2012**, *19*, 1508–1513. [CrossRef]
37. Khaled, U.; Beroual, A. Statistical Investigation of AC Breakdown Voltage of Natural Ester with Electronic Scavenger Additives. *IEEE Trans. Dielectr. Electr. Insul.* **2019**, *26*, 2012–2018. [CrossRef]
38. Li, J.; Zhang, Z.; Zou, P.; Grzybowski, S.; Zahn, M. Preparation of a vegetable oil-based nanofluid and investigation of its breakdown and dielectric properties. *IEEE Electr. Insul. Mag.* **2012**, *28*, 43–50. [CrossRef]
39. Lv, Y.; Ge, Y.; Li, C.; Wang, Q.; Zhou, Y.; Qi, B.; Yi, K.; Chen, X.; Yuan, J. Effect of TiO_2 Nanoparticles on Streamer Propagation in Transformer Oil under Lightning Impulse Voltage. *IEEE Trans. Dielectr. Electr. Insul.* **2016**, *23*, 2110–2115. [CrossRef]

40. Massala, G.; Lesaint, O. Positive streamer propagation in large oil gaps: Electrical properties of streamers. *IEEE Trans. Dielectr. Electr. Insul.* **1998**, *5*, 371–381. [CrossRef]

Publisher's Note: MDPI stays neutral with regard to jurisdictional claims in published maps and institutional affiliations.

© 2020 by the authors. Licensee MDPI, Basel, Switzerland. This article is an open access article distributed under the terms and conditions of the Creative Commons Attribution (CC BY) license (http://creativecommons.org/licenses/by/4.0/).

Article

Polydopamine Coated CeO₂ as Radical Scavenger Filler for Aquivion Membranes with High Proton Conductivity

Roberto D'Amato [1,†], Anna Donnadio [2,3,*], Chiara Battocchio [4], Paola Sassi [1], Monica Pica [2,3], Alessandra Carbone [5], Irene Gatto [5] and Mario Casciola [1,3]

[1] Department of Chemistry, Biology and Biotechnologies, University of Perugia, via Elce di Sotto 8, 06123 Perugia, Italy; roberto.damato@inl.int (R.D.); paola.sassi@unipg.it (P.S.); mario.casciola@unipg.it (M.C.)
[2] Department of Pharmaceutical Sciences, University of Perugia, via del Liceo 1, 06123 Perugia, Italy; monica.pica@unipg.it
[3] CEMIN, Materiali Innovativi Nanostrutturali per Applicazioni Chimica Fisiche e Biomediche, University of Perugia, via Elce di Sotto 8, 06123 Perugia, Italy
[4] Department of Sciences, University of Roma Tre, via della Vasca Navale 79, 00146 Rome, Italy; chiara.battocchio@uniroma3.it
[5] CNR ITAE, via S. Lucia Sopra Contesse, 5, 98126 Messina, Italy; alessandra.carbone@itae.cnr.it (A.C.); irene.gatto@itae.cnr.it (I.G.)
* Correspondence: anna.donnadio@unipg.it
† Present address: International Iberian Nanotechnology Laboratory, 4715-330 Braga, Portugal.

Abstract: CeO₂ nanoparticles were coated with polydopamine (PDA) by dopamine polymerization in water dispersions of CeO₂ and characterized by Infrared and Near Edge X-ray Absorption Fine Structure spectroscopy, Transmission Electron Microscopy, Thermogravimetric analysis and X-ray diffraction. The resulting materials (PDAx@CeO₂, with x = PDA wt% = 10, 25, 50) were employed as fillers of composite proton exchange membranes with Aquivion 830 as ionomer, to reduce the ionomer chemical degradation due to hydroxyl and hydroperoxyl radicals. Membranes, loaded with 3 and 5 wt% PDAx@CeO₂, were prepared by solution casting and characterized by conductivity measurements at 80 and 110 °C, with relative humidity ranging from 50 to 90%, by accelerated ex situ degradation tests with the Fenton reagent, as well as by in situ open circuit voltage stress tests. In comparison with bare CeO₂, the PDA coated filler mitigates the conductivity drop occurring at increasing CeO₂ loading especially at 110 °C and 50% relative humidity but does not alter the radical scavenger efficiency of bare CeO₂ for loadings up to 4 wt%. Fluoride emission rate data arising from the composite membrane degradation are in agreement with the corresponding changes in membrane mass and conductivity.

Keywords: cerium oxide; polydopamine; radical scavenger; proton conductivity; chemical degradation

1. Introduction

In recent years, the need to arrest the effects of climate change is pushing governments worldwide to plan and coordinate efforts to achieve a dramatic reduction in CO₂ emissions. This requires a revolution in energy supply toward much more flexible renewable energy systems. Hydrogen offers several benefits for simultaneously decarbonizing transport, housing and industrial sectors. Among hydrogen-based technologies, proton exchange membrane (PEM) fuel cells have revealed promising for stationary and automotive applications. Because of the lifetime targets for large-scale stationary applications (≈40,000 h), as well as for automotive applications (>6000 h) [1], the proton exchange membrane durability represents a key element for the longevity of the device. However, several factors limit the membrane's long-term stability. A source of membrane degradation is due to chemical degradation caused by radical species such as H•, OH• and HOO• [2–6], which gives rise to a thinning of the membrane leading to short the lifetime of PEM fuel cells. A strategy to mitigate such radical attacks consists of the incorporation of radical scavengers.

For example, the introduction of metal cations, such as Ce^{4+} and Mn^{2+}, or their oxides, including CeO_2 and MnO_2, revealed to be effective in mitigating the chemical degradation of PFSA (PerFluoroSulfonic Acid) polymers [7–15] because, due to the multivalent oxidation state of the metals, they can act as catalysts for the decomposition of hydroxyl and hydroperoxyl radicals.

In a recent paper, it was reported that CeO_2 nanoparticles dispersed in an Aquivion matrix undergo partial solubilization at relative humidity in the range 50–90%, when the temperature is increased from 80 to 110 °C [16]. As a consequence, for CeO_2 loadings greater than 2 wt%, a decrease in the composite membrane conductivity was observed with increasing temperature, in such a way that the larger the CeO_2 loading, the more severe the conductivity drop. It was also found that the formation of a protective shell on the oxide surface, made of fluorophosphonates bonded to cerium ions through the $-PO_3$ groups, partially avoided the conductivity drop. However, the organically modified CeO_2 nanoparticles show reduced radical scavenger activity in comparison with the pristine nanoparticles. A reasonable compromise between stable conductivity and improved membrane stability towards radical was reached by bonding a fluorobenzyl phosphonate (hereafter Bz) to the CeO_2 surface. Based on these results and taking into account that the phosphonate can be hydrolyzed after long-term operation under conditions of high membrane hydration, it was of interest to coat the oxide surface with a polymeric film that could be hydrolytically more stable than the phosphonate coating.

To this aim, polydopamine (PDA, Scheme 1) was chosen for its strong and universal adhesion ability and the simple deposition process through self-polymerization in an alkaline aqueous solution [17–33].

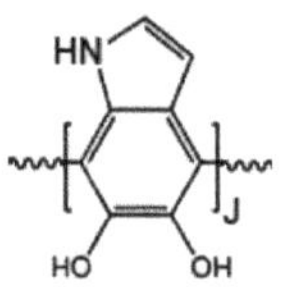

Scheme 1. Structure of polydopamine.

The preparation of PDA-based materials has rapidly advanced in recent years with a significant expansion in their applications [21–34], becoming one of the most attractive areas within the materials field including surface modification, biosensing [35,36], nanomedicine [37] and systems for energy applications [38–43]. In particular, PDA allows obtainment of a beneficial and advantageous interface between CeO_2 and PFSA improving the lifetime of PEM fuel cells [44].

This paper reports the formation of a PDA film on the surface of CeO_2 nanoparticles by dopamine polymerization in a water suspension of CeO_2, and the use of this composite material (PDA@CeO_2) as a filler of membranes made of Aquivion.

Membranes containing 3 and 5 wt% filler loadings were characterized by conductivity measurements at 80 and 110 °C, in the RH range 50–90%, to test their stability at increasing temperature. These membranes were also subjected to accelerated ageing by using the Fenton reagent to assess the radical scavenger efficiency of the filler based on the fluoride emission rate (FER). Both conductivity and FER data collected in the present work are compared with the corresponding literature data for composite Aquivion membranes filled with pristine CeO_2 and Bz@CeO_2. The most stable membrane was also characterized by Open Circuit Voltage (OCV) stress tests coupled with hydrogen crossover determinations.

2. Materials and Methods

2.1. Materials

Cerium (III) nitrate hexahydrate ($Ce(NO_3)_3 \cdot 6H_2O$) was from Carlo Erba. A 20 wt% Aquivion dispersion in water (D83-6A, ionomer equivalent weight = 830 g/equiv.) was kindly provided by Solvay Specialty Polymers, Italy. The citric acid ($C_6H_8O_7 \cdot H_2O$), dopamine and all other reagents were purchased from Sigma-Aldrich and used without purification.

2.2. Synthesis of CeO_2 and $PDA@CeO_2$

Nanopolyhedral CeO_2 was synthesized by sol-gel followed by thermal decomposition according to the procedure reported in 15. Three composite materials made of cerium dioxide nanoparticles coated with PDA were prepared by reacting, under stirring, a weighed amount of cerium dioxide nanoparticles with 10 mL of a dopamine hydrochloride solution: specifically, 100 mg CeO_2 with 0.01 M dopamine (hereafter Sample 1), 150 mg CeO_2 with 0.05 M dopamine (Sample 2) or with 0.1 M dopamine (Sample 3). A suitable amount of NaOH 0.1 M was added to keep pH at 8.5 to achieve a polydopamine film on the nanoparticles. Subsequently, the dispersion was stirred for 24 h at room temperature in dark conditions. The powder was then recovered by centrifugation, washed with water several times and then dried at 80 °C overnight. These composite materials will hereafter be indicated as PDAx@CeO_2, where x is the PDA weight percentage in the composite.

2.3. Membrane Preparation

The Aquivion dispersion in water was cast on a Petri dish and dried at 80 °C in an oven. The resulting membrane was dissolved in propanol (1 g in 20 mL) at 80 °C. A weighted amount of pristine CeO_2 or PDA coated CeO_2 nanoparticles was added to 20 mL of the Aquivion dispersion in propanol. The mixtures were treated with ultrasounds for 10 min, stirred for 2 h, cast by an Elcometer Doctor Blade Film Applicator on a glass support and dried in an oven at 80 °C. After that, all membranes were treated according to the following procedure: 2 h in HCl 1 M, 1 h in H_2O at room temperature, 2 h at 90 °C and 1 h at 160 °C. Composite membranes, 20–25 μm thick, containing 3% and 5% wt% of PDAx@CeO_2 were prepared. The same procedure was used to prepare the neat Aquivion membrane.

2.4. Ex Situ Accelerated Ageing

Accelerated ex situ ageing tests were performed by treating a membrane sample (ca. 60 mg) with 20 mL of the Fenton reagent (20 ppm iron sulfate, $FeSO_4 \cdot 7H_2O$, in 30 wt% hydrogen peroxide solution) for 4 h, at 75 °C. The membrane was then washed with deionized water and dried at room temperature. The concentration of fluoride ions in the Fenton's solution was determined using a Mettler Toledo fluoride ion-selective electrode. The pH was kept in the range of 5–7 using an electrolyte solution (TISAB) with the appropriate total ionic strength adjustment buffer [16].

2.5. Conductivity Measurements

The in-plane conductivity was determined according to the four-point impedance technique on 5 cm ± 0.5 cm membrane strips at frequencies ranging from 10 Hz to 100 kHz, with 100 mV, signal amplitude using an Autolab, PGSTAT30 potentiostat/galvanostat equipped with a frequency response analyzer module, as described in ref. [45].

2.6. Transmission Electron Microscopy

Transmission electron microscopy (TEM) images were collected on powders previously dispersed in ethanol by using a sonicator and then supported and dried on copper grids (200 mesh) coated with Formvar carbon film. A Philips 208 transmission electron microscope, operating at an accelerating voltage of 100 kV, was used.

2.7. X-ray Diffraction

X-ray diffraction (XRD) patterns of powders were collected with a Philips X'Pert PRO MPD diffractometer as described in [46].

2.8. Ionic Exchange Capacity Determination

Membrane samples (~250 mg) were dried at 120 °C for 3 h, weighed and then equilibrated in 20 mL of 0.1 M NaCl overnight to exchange the membrane protons with Na^+ ions. The solution was titrated, in the presence of the membrane, with 0.01 M NaOH through a Radiometer automatic titrimeter (TIM900 TitraLab, Radiometer Copenhagen, Denmark), according to the equilibrium point method. The reported Ionic Exchange Capacity (IEC) values are the average of five replicate titrations [16].

2.9. ATR-FTIR

IR spectra were collected by means of Bruker Optics Alpha FTIR instrument equipped with a Platinum-ATR accessory (Bruker Optics, Karlsruhe, Germany). The samples were deposited on the diamond ATR (attenuated total reflection) crystal and their spectrum was recorded at room temperature over the range 5000–400 cm^{-1} with a 2 cm^{-1} resolution.

2.10. Near Edge X-ray Absorption Fine Structure (NEXAFS)

NEXAFS spectra were acquired at the ELETTRA storage ring using the BEAR (bending magnet for emission absorption and reflectivity) beamline, installed at the left exit of the 8.1 bending magnet exit. The BEAR beamline has a bending magnet as a source, and beamline optics deliver photons from 5 eV up to 1600 eV; the degree of ellipticity of the beam is selectable. The experimental station is in UHV, and it is equipped with a movable hemispherical electron analyzer and a set of photodiodes to collect angle-resolved photoemission spectra, optical reflectivity and fluorescence yield. In the here reported experiments ammeters to measure drain current from the sample were used. We collected C K-edge and O K-edges spectra at a magic-incidence angle (54.7°) of the linearly polarized photon beam with respect to the sample surface. Both photon energy and spectral resolution were calibrated and experimentally tested using the absorption K-edges of Ar, N_2 and Ne. The acquired spectra were normalized by subtracting a straight line that fits the part of the spectrum below the edge and imposing an Absorption Intensity value of 1 at 320.00 eV for C K-edge and 560.00 eV for O K-edge.

2.11. In Situ Accelerated Stress Tests

Gas diffusion electrodes (GDEs) were prepared by a spray technique. Sigracet 25- BC Gas Diffusion Layer (SGL), was used as a GDL, and the catalytic ink was deposited onto its surface, as reported elsewhere [47]. The same Pt loading of 0.2 mg cm^{-2} was used for cathode and anode. The GDEs were hot-pressed, at a pressure of 20 $kgcm^{-2}$ for 5 min at 125 °C, onto Aquivion membrane to realize the Membrane-Electrode Assemblies (MEAs). The Accelerated stress tests (AST) in a H_2/air 25 cm^2 single cell, at the Open Circuit Voltage and steady-state conditions, were carried out in the following operative conditions: 80 °C, 50% RH, 1.5 bar_{abs}, flow rate 1.5 and 2 times the stoichiometry for H_2 and air, respectively [48]. The tests were performed by connecting the single cell with a commercial test station (Fuel Cell Technologies Inc.), and an AUTOLAB Metrohm Potentiostat/Galvanostat with a 20 A current booster to carry out the electrochemical diagnostics measurements. Linear Sweep Voltammetry (LSV) was carried out by feeding the anode and cathode with hydrogen and nitrogen, respectively, to determine the H_2 crossover. A potential scan, ranging from 0 to 0.8 V with a scan rate of 4 mVs^{-1}, was used to perform the LSV.

3. Results and Discussion

3.1. Filler Materials

The composite samples obtained by reacting dopamine with an aqueous dispersion of CeO_2 nanoparticles were first characterized by ATR-FTIR spectroscopy to prove the PDA formation. In Figure 1 the ATR spectrum of Sample 3 (see Section 2.2) is compared with the spectra of CeO_2 and PDA. The main bands of PDA are recognized in the spectrum of the composite. Specifically, the bands centered at 1596 and 1510 cm^{-1} can be attributed to (C=C) and (C–N) stretching modes, respectively, and confirm the presence of aromatic amine species in the coating. The band at ca. 1600 cm^{-1}, as well as the feature at 1723 cm^{-1}, are assigned to C=O quinone groups. All these peaks increase in intensity as dopamine concentration increases in the reacting mixture, which indicates the increasing PDA concentration in the coatings [49–52].

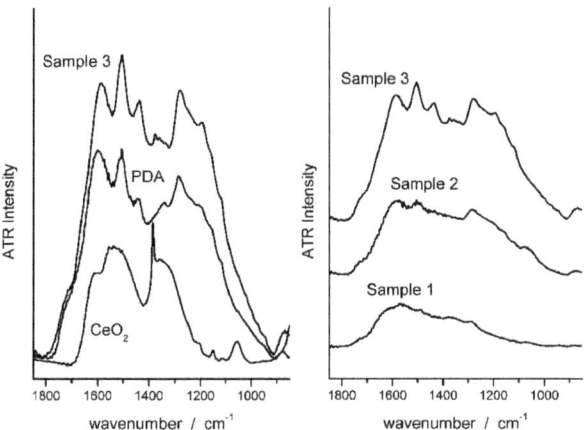

Figure 1. ATR spectra of CeO$_2$, PDA and PDAx@ CeO$_2$ samples.

PDAx@CeO$_2$ materials were further characterized by NEXAFS to get information about the PDA oxidation state. All samples show similar features in C K-edge spectra (Figure 2). The main feature appears in the π* region, at about 286 eV, and is attributed to C1s –π*C=O transitions [53]. A couple of features around 289 eV are indicative of the N–containing ring (C=C π* and C–N σ* excitations), confirming the molecular structure integrity. The large and broad feature at about 300 eV in σ* spectral region is associated with C1s – σ*C=O excitations. C K-edge spectra suggest an abundance of C=O functional groups in the examined samples.

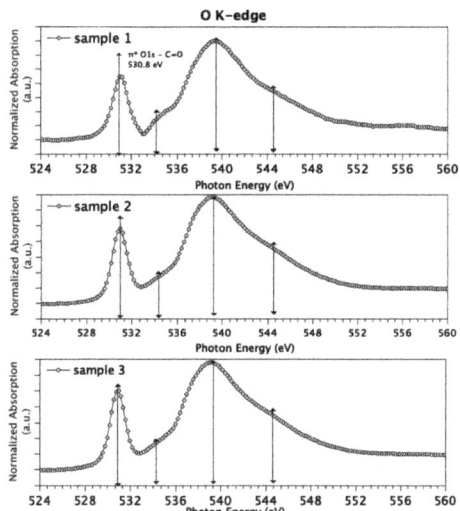

Figure 2. C K-edge spectra measured for Sample 1 (**top**), Sample 2 (**middle**) and Sample 3 (**bottom**).

As with C K-edge spectra, O K-edge spectra (Figure 3) are similar for the three measured samples. The energies of the features in O K-edge spectra and proposed assignments are summarized as follows: the sharp and intense peak centered at 530.8 eV is attributed to the transition of 1 s electrons of C=O groups to antibonding molecular orbitals π*C=O, while the small feature around 534 eV is indicative for transitions of 1 s electrons of hydroxyl-like O atoms to π*O–C and 3 s/σ*O–H [54]. As for the σ* region, features around 540 and 544 eV are associated with O1s C–O and C=O σ* transitions, respectively.

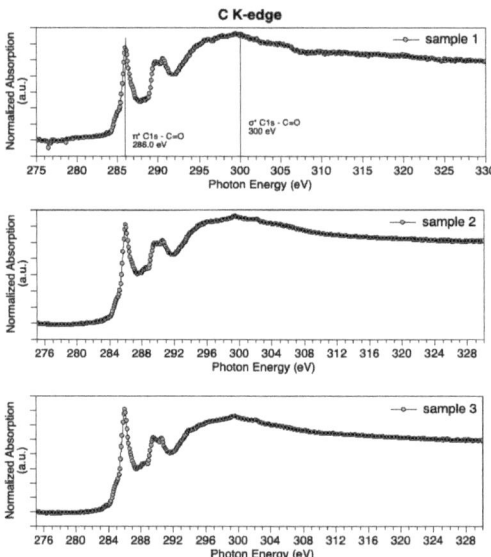

Figure 3. O K-edge spectra measured for Sample 1 (**top**), Sample 2 (**middle**) and Sample 3 (**bottom**).

Since in NEXAFS data analysis the so-called *building block approach* can be successfully applied [55] (i.e., the NEXAFS spectrum of a complex molecule or sample can be built by summing up the contribution arising by the different functional groups, weighted for their abundance in the sample), the presence of strong features diagnostic for carbonyl groups, and only weak contributions arising by hydroxyls, suggests that the polymer is mainly in the oxidized state.

The PDA content in the PDAx@CeO$_2$ composite samples was determined by thermogravimetric analysis. The weight loss curves for Samples 1, 2 and 3, as well as for bare CeO$_2$, are displayed in Figure 4. While CeO$_2$ does not present any appreciable loss, the curves of the composites show a small weight loss up to 100 °C, due to the water loss, and a second loss above 200 °C arising from PDA decomposition, which increases with increasing the dopamine concentration used for the polymerization reaction. Based on the second weight loss, the PDA content in anhydrous PDAx@CeO$_2$ turned out to be 10.2 wt% (Sample 1), 24.7 wt% (Sample 2) and 49.5 wt% (Sample 3); these samples will be hereafter indicated as PDA10@CeO$_2$, PDA25@CeO$_2$ and PDA50@CeO$_2$, respectively.

The morphology of CeO$_2$ and the PDAx@CeO$_2$ composites was investigated by TEM. The pictures of Figure 5 reveal that PDA can coat the cerium oxide surface forming an irregular layer without affecting the shape and the dimension of the pristine CeO$_2$ particles, which in all cases lies around 10 nm. In particular, as the amount of PDA in the composite increases, the thickness of the coating becomes more evident reaching a thickness of some nanometers for the highest PDA content.

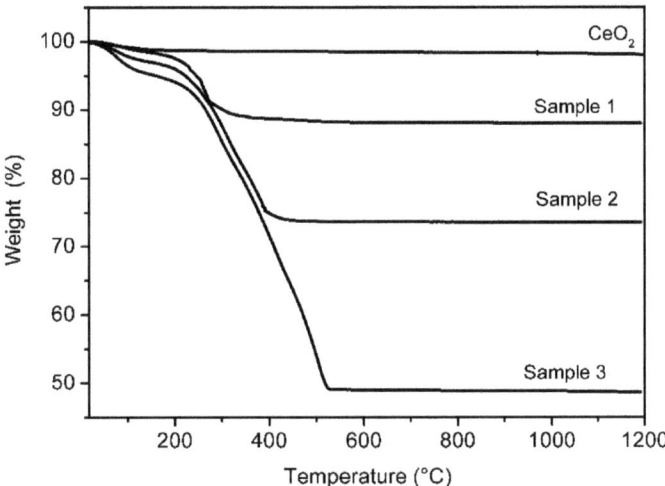

Figure 4. Weight loss curves for bare CeO_2 and for PDAx@CeO_2 samples.

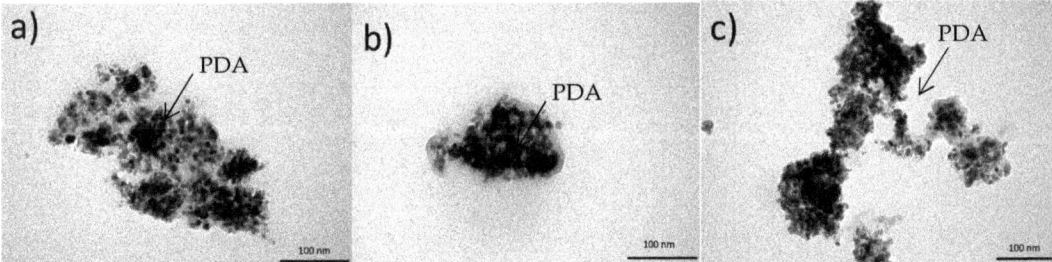

Figure 5. TEM images of (**a**) PDA10@CeO_2, (**b**) PDA25@CeO_2 and (**c**) PDA50@CeO_2.

To use the PDAx@CeO_2 materials as fillers of Aquivion composite membranes, we checked that the PDA coating is not soluble in the solvent (propanol) used for membrane preparation. To this aim, 0.05 g of PDA was dispersed in 20 mL of propanol and the mixture was kept under stirring at room temperature for 2 h and then at 80 °C in a closed bottle for 2 h. After centrifugation, the solid was dried at 80 °C. The weight loss curve of the starting material (PDA13@CeO_2) is coincident, within the experimental error, with the curve of the treated material (PDA13@CeO_2 PrOH 80) thus indicating that PDA is not soluble under the conditions of membrane preparation (data not shown).

X-ray diffraction (XRD) patterns were collected to reveal possible structural modifications or changes in crystallinity induced by the PDA formation. Figure 6 shows that the position and the intensity of the peaks of bare CeO_2 do not change in the PDA coated samples suggesting that the presence of the PDA coating does not affect the CeO_2 crystal structure.

Moreover, in agreement with the TEM images, the particle size calculated using the Scherrer equation lies in the range from 9.9 to 11.2 nm.

The PDA coating of CeO_2 turned out to be insoluble in propanol at 80 °C, which is the solvent for Aquivion 830: this allowed the PDAx@CeO_2 materials to be used as fillers of Aquivion based composite membranes.

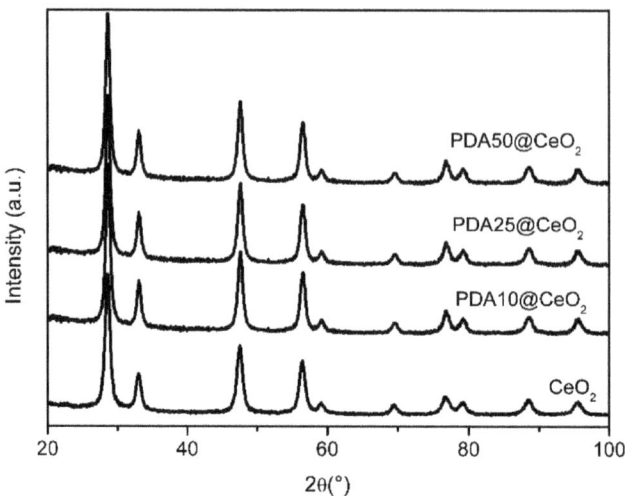

Figure 6. XRD patterns for pristine and coated CeO_2.

3.2. Composite Membranes

The IEC values (in milliequivalents per gram) of the composite membranes with 3 and 5 wt% PDAx@CeO_2 loadings (Table 1) are at most by 4% lower than those calculated based on the ionomer weight percentage. In principle, this could be due to the protonation of nitrogen atoms of PDA and/or of surface oxide ions of CeO_2. The fact that, for the same PDAx@CeO_2 loading, the IECs show the sequence:

$$IEC(PDA50) > IEC(PDA25) > IEC(PDA10)$$

indicates that the protonation of the oxide ions is mainly responsible for the IEC decrease because the amount of cerium oxide in the filler is minimum for PDA50 and maximum for PDA10.

Table 1. IEC values (meq g^{-1}) of PDAx@CeO_2 composite membranes. The IEC values calculated based on the Aquivion wt% (AQ_{calc}) are also reported.

wt%	PDA10	PDA25	PDA50	AQ_{calc}
3	1.13	1.15	1.16	1.16
5	1.10	1.12	1.15	1.14

Standard deviation = 0.0032.

The conductivity (σ) of membranes containing 3 and 5 wt% PDAx@CeO_2 (with x = 10, 25 or 50), as well as the conductivity of a membrane containing 6 wt% PDA10@CeO_2, was determined for RH increasing in the range 50–90%, first at 80 °C and then at 110 °C. In all cases, the plot of logσ as a function of RH is linear. As an example, Figure 7 displays the conductivity of composite membranes containing 5 wt% filler together with that of bare Aquivion. At both temperature and for each RH value, the following conductivity sequence is observed:

$$\sigma(PDA50) > \sigma(PDA25) > \sigma(PDA10)$$

which indicates that the conductivity increases with decreasing the CeO_2 content in the filler. Moreover, going from 80 to 110 °C, the conductivity evolution depends also on the CeO_2 mass fraction in the filler in such a way that it increases in the presence of PDA50@CeO_2 but keeps nearly constant in the presence of PDA10@CeO_2.

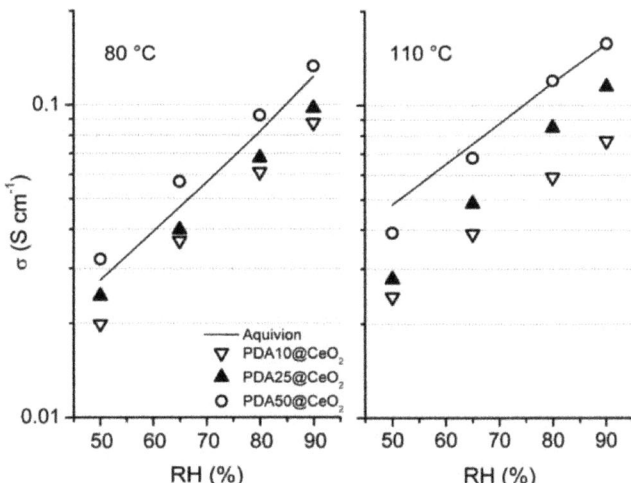

Figure 7. Conductivity as a function of relative humidity, at 80 and 110 °C, for composite Aquivion membranes containing 5 wt% PDAx@CeO_2. The conductivity of bare Aquivion is also reported.

A similar trend was already reported for Aquivion/CeO_2 composite membranes in a recent work [16] where it was shown that the increase in temperature favors the acid-base reaction between cerium oxide and ionomer protons thus causing an IEC decrease which, depending on CeO_2 content, offsets the expected increase in conductivity.

To get insight into the dependence of the conductivity on CeO_2 loading, the conductivity of the PDAx@CeO_2 composite membranes is plotted in Figures 8 and 9 as a function of CeO_2 wt% in the membrane at constant temperature (80 and 110 °C) and RH (50 and 90%). For comparison, the conductivity of composite Aquivion membranes filled with bare CeO_2 and Bz@CeO_2 is also reported [16]. At 80 °C, the conductivity of the composite membranes filled with PDAx@CeO_2 is weakly dependent on CeO_2 loading and is close to the conductivity of Aquivion. On the other hand, at 110 °C, the composite membranes become progressively less conductive with increasing of the CeO_2 loading so that the membrane with 5.4 wt% CeO_2 is by a factor of about 2.5 less conductive than Aquivion both at 50 and 90% RH.

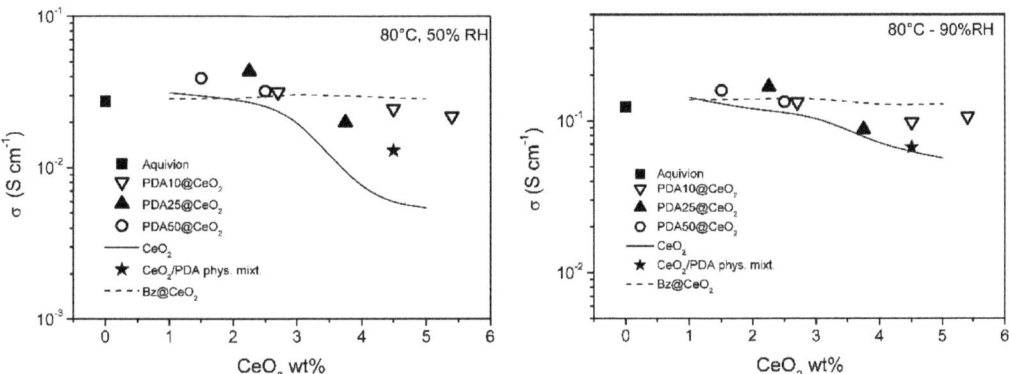

Figure 8. Conductivity as a function of CeO_2 loading, at 80 °C, for composite Aquivion membranes filled with PDAx@CeO_2 (x = 10, 25 and 50), as well as with the physical mixture PDA/CeO_2 (see text). The conductivity of composite Aquivion membranes containing bare CeO_2 and Bz@CeO_2 (redrawn from ref. [16]) is reported for comparison.

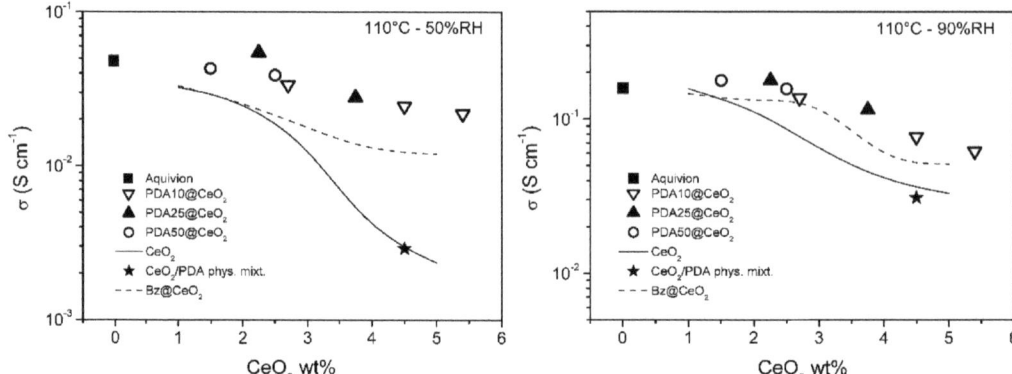

Figure 9. Conductivity as a function of CeO_2 loading, at 110 °C, for composite Aquivion membranes filled with PDAx@CeO_2 (x = 10, 25 and 50), as well as with the physical mixture PDA/CeO_2 (see text). The conductivity of composite Aquivion membranes containing bare CeO_2 and Bz@CeO_2 (redrawn from ref. [16]) is reported for comparison.

The CeO_2 loading being the same, the conductivity of the PDAx@CeO_2 membranes is similar to the conductivity of the Bz@CeO_2 membranes except for 110 °C and 50% RH, where the PDAx@CeO_2 membranes are more conductive by a factor of ~2 at the highest CeO_2 loadings.

Moreover, the PDAx@CeO_2 membranes are always more conductive than the corresponding membranes filled with bare CeO_2 and the difference in conductivity increases with decreasing RH and with increasing filler loading and temperature. As a consequence, at 110 °C and 50% RH, the conductivity of the membrane with PDA10@CeO_2 containing 4.5 wt% CeO_2 is by one order of magnitude higher than that of the corresponding membrane containing bare CeO_2.

It was of interest to prove that the better conductivity of the PDAx@CeO_2 membranes, in comparison with the corresponding membranes loaded with bare CeO_2, is indeed due to the presence of the PDA shell on the CeO_2 surface. To this end, a composite Aquivion membrane containing the same amount of PDA and CeO_2 as the membrane loaded with 5 wt% PDA10@CeO_2 (i.e., 0.5 wt% PDA and 4.5 wt% CeO_2) was prepared by mixing the Aquivion dispersion with a physical mixture of PDA and bare CeO_2. The conductivity of this membrane, determined at 50 and 90% RH, first at 80 °C and then at 110 °C (the asterisk in Figures 8 and 9), was always lower than the conductivity of the membrane loaded with 5 wt% PDA10@CeO_2, being in three cases even coincident with the conductivity of the membrane containing 4.5 wt% bare CeO_2. These results show that it is the PDA coating that efficiently protects the cerium oxide particles against the acidic sulfonic groups of the ionomer, thus avoiding to a large extent the severe conductivity drop occurring with bare CeO_2.

To evaluate the membrane resistance towards radical species generated by the decomposition of hydrogen peroxide, ex situ degradation tests were performed by treating the composite membranes with the Fenton solution (see Experimental section). The results of these tests are expressed in terms of fluoride emission rate, FER, defined as the ratio between the mass of released fluoride ions and the initial mass of the anhydrous membranes. Figure 10 shows the FER values of the PDAx@CeO_2 composite membranes as a function of CeO_2 loading and, for comparison, the FER values of composite Aquivion membranes filled with bare CeO_2 and Bz@CeO_2. The FER values obtained with PDAx@CeO_2 are significantly lower than those of Bz@CeO_2. Like Bz, the PDA coating prevents to a large extent the decrease in conductivity, but, unlike Bz, it does not compromise the radical scavenger activity of CeO_2. The radical scavenger efficiency of PDAx@CeO_2 is nearly coincident with that of membranes loaded with bare CeO_2, for CeO_2 percentage up to 4 wt%, and similar to that for higher loadings: a FER value of 10^{-3} is indeed obtained with

5.4 wt% of PDA coated CeO_2 and with 4.7 wt% of bare CeO_2. It can also be observed that the membranes loaded with 5 wt% PDA50@CeO_2 and with 3 wt% PDA10@CeO_2 have the same FER and close CeO_2 content (2.5 and 2.7 wt%, respectively) but very different PDA content (2.5 and 0.3 wt%, respectively). Thus, the radical scavenger properties of PDAx@CeO_2 are mainly dependent on the CeO_2 weight percentage, while the PDA coating does not shield significantly the radical scavenger activity of CeO_2.

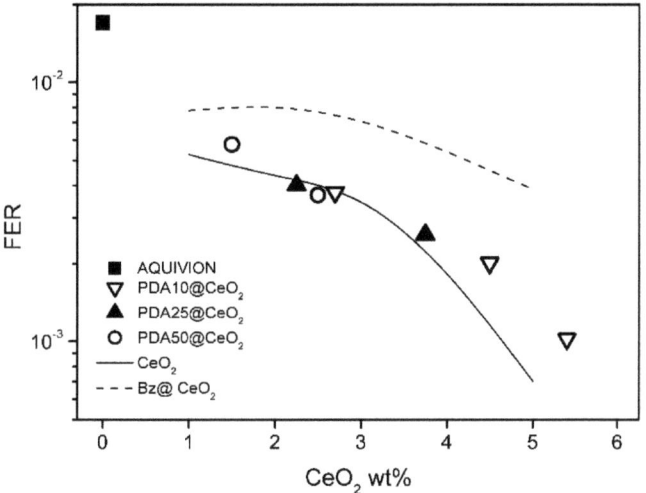

Figure 10. Fluoride emission rate (FER) as a function of CeO_2 content for composite Aquivion membranes filled with PDA10@CeO_2, PDA25@CeO_2 and PDA50@CeO_2. The FER of composite Aquivion membranes containing bare CeO_2 and Bz@CeO_2 (redrawn from ref. [16]) is reported for comparison.

After the Fenton test, the membranes with 3 wt% PDAx@CeO_2 and the membrane with 6 wt% PDA10@CeO_2 were washed with 1 M HCl and water, dried at 120 °C and weighed. The percentage weight loss concerning the initial weight of the anhydrous membrane (Table 2) decreases with increasing the CeO_2 loading in the composite membrane, going from 25.4% for the membrane with 3 wt% PDA50@CeO_2 to 5.9% for the membrane with 6 wt% PDA10@CeO_2.

Table 2. Dry weight and conductivity percentage changes (%Δw and %$\Delta\sigma$) for bare Aquivion and Aquivion composite membranes loaded with PDAx@CeO_2 after the Fenton test.

Filler	Filler wt%	CeO_2 wt%	%Δw	%$\Delta\sigma$
Aquivion	0	0	−22	−25
PDA50@CeO_2	3	1.5	−25	−25
PDA25@CeO_2	3	2.3	−14	−22
PDA10@CeO_2	3	2.7	−12	−8
PDA10@CeO_2	6	5.4	−6	−1

The conductivity of the aged membranes was also determined at 80 °C and 90% RH (Table 2). The percentage decrease in conductivity with respect to the initial membrane conductivity reflects qualitatively the trend of the weight changes. Thus, both the weight and conductivity of the aged membranes are consistent with the FER data.

Based on the results of ex situ characterization, the membrane loaded with 6 wt% PDA10@CeO_2 (hereafter AQ-PDA@CeO_2) was selected for MEA realization and in situ characterized by OCV stress tests. Hydrogen crossover measurements were carried out before and during the stress tests after 24 and 47 h from the beginning to check the stability of the membrane. Figure 11 shows the OCV vs. time curves for the bare Aquivion membrane (AQ) and a composite membrane containing 5 wt% CeO_2 (AQ-CeO_2). In all cases, after a non-linear drop during the first 3 h, the cell potential decays linearly as a function of time. The same trend is also observed in the time interval between hours 24 and 47 for AQ-PDA@CeO_2 and AQ after the second hydrogen crossover determination (AQ-CeO_2 stopped working).

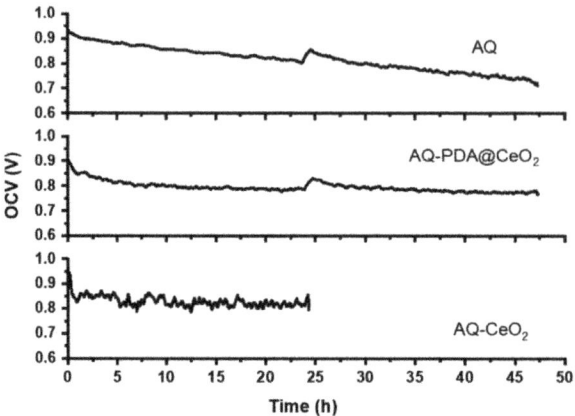

Figure 11. OCV vs. time curves for AQ, AQ-PDA@CeO_2 and AQ-CeO_2 membranes.

The OCV decay rate in the linear regions (Table 3) is lower for AQ-PDA@CeO_2 than for AQ both in the first and in the second time interval of the stress test; as a consequence, the overall OCV decay is ~15% for AQ-PDA@CeO_2 and ~25% for AQ. Moreover, during the first interval, the decay rate of AQ-PDA@CeO_2 and AQ@CeO_2 is similar, thus confirming that the PDA coating does not compromise the radical scavenger activity of CeO_2.

Table 3. OCV decay rate for the indicated membranes at 80 °C and 50% RH.

Membrane	OCV Decay Rate (mV h^{-1})		Overall OCV Percentage Loss
	Hour 4–24	Hour 28–47	
AQ-PDA@CeO_2	1.6	1.5	14.7
AQ-CeO_2	1.6		15.8
AQ	4.0	4.2	24.6

Consistently with the evolution of the OCV decay rate, the increase in the hydrogen crossover during the stress test (Table 4) is much larger for AQ (~23 times) than for AQ-PDA@CeO_2 (~2 times).

Table 4. Hydrogen crossover for the indicated membranes at 80 °C and 50% RH.

Membrane	H_2 Crossover at 400 mV (mA cm^{-2})		
	Beginning	24 h	47 h
AQ-PDA@CeO_2	5.5	12	129
AQ-CeO_2	1.4	2.3	
AQ	3.4	5.4	6.6

4. Conclusions

Polymerization of dopamine in water suspensions of CeO_2 allowed preparation of three types of PDA coated CeO_2 nanoparticles containing 10, 25 and 50 wt% PDA. These materials were used as fillers of Aquivion membranes to improve their chemical stability towards hydroxyl and hydroperoxyl radicals while hindering as much as possible the reaction between the oxide surface and the sulfonic groups of the ionomer.

Composite membranes, loaded with 3 and 5 wt% of each filler, were characterized by conductivity measurements and accelerated degradation tests (based on the Fenton reaction) to assess the radical scavenger efficiency of the filler, and the results of these investigations were compared with those obtained, under the same conditions, for Aquivion membranes containing bare CeO_2 or Bz@ CeO_2. It was found that the decrease in conductivity associated with the increase in the CeO_2 loading is appreciably less severe for membranes containing PDAx@CeO_2 than for those containing bare CeO_2, especially at 110 °C and 50% RH. Moreover, the FER of membranes filled with PDAx@CeO_2 is significantly lower than that of membranes containing Bz@CeO_2 and, for CeO_2 percentage up to 4 wt%, it is coincident with that of membranes loaded with bare CeO_2. Therefore, the PDA coating prevents to a large extent the decrease in conductivity without compromising the radical scavenger activity of CeO_2. Interestingly, the radical scavenger efficiency of PDAx@CeO_2 is nearly unaffected by the PDA content as also proved by the results of the OCV stress tests.

Author Contributions: Conceptualization, A.D.; Data curation, R.D., A.D., C.B., P.S., M.P., A.C., I.G. and M.C.; Investigation, R.D., C.B., P.S., M.P., A.C. and I.G.; Methodology, A.D.; Supervision, A.D. and M.C.; Writing—original draft, A.D. and M.C. All authors have read and agreed to the published version of the manuscript.

Funding: This research was funded by MIUR grants AMIS and DELPHI through the program "Dipartimenti di Eccellenza 2018–2022".

Institutional Review Board Statement: Not applicable.

Informed Consent Statement: Not applicable.

Data Availability Statement: The data underlying this article will be shared on reasonable request from the corresponding author.

Conflicts of Interest: The authors declare that they have no known competing financial interests or personal relationships that could have appeared to influence the work reported in this paper.

References

1. Wu, J.; Yuan, X.Z.; Martin, J.J.; Wang, H.; Zhang, J.; Shen, J.; Wu, S.; Merida, W. A review of PEM fuel cell durability: Degradation mechanisms and mitigation strategies. *J. Power Sources* **2008**, *184*, 104–119. [CrossRef]
2. Luo, X.; Ghassemzadeh, L.; Holdcroft, S. Effect of free radical-induced degradation on water permeation through PFSA ionomer membranes. *Int. J. Hydrog. Energy* **2015**, *40*, 16714–16723. [CrossRef]
3. Danilczuk, M.; Coms, F.D.; Schlick, S. Visualizing chemical reactions and crossover processes in a fuel cell inserted in the ESR resonator: Detection by spin trapping of oxygen radicals, nafion-derived fragments, and hydrogen and deuterium atoms. *J. Phys. Chem. B* **2009**, *113*, 8031–8042. [CrossRef] [PubMed]
4. Danilczuk, M.; Coms, F.D.; Schlick, S. Fragmentation of fluorinated model compounds exposed to oxygen radicals: Spin trapping ESR experiments and implications for the behaviour of proton exchange membranes used in fuel cells. *Fuel Cells* **2008**, *8*, 436–452. [CrossRef]
5. Zatoń, M.; Rozière, J.; Jones, D.J. Current understanding of chemical degradation mechanisms of perfluorosulfonic acid membranes and their mitigation strategies: A review. *Sustain. Energ. Fuels* **2017**, *1*, 409–438. [CrossRef]
6. Collier, A.; Wang, H.; Zi, Y.X.; Zhang, J.; Wilkinson, D.P. Degradation of polymer electrolyte membranes. *Int. J. Hydrog. Energy* **2006**, *31*, 1838–1854. [CrossRef]
7. Zatoń, M.; Rozière, J.; Jones, D.J. Mitigation of PFSA membrane chemical degradation using composite cerium oxide–PFSA nanofibers. *J. Mater. Chem.* **2017**, *A5*, 5390–5401. [CrossRef]
8. Coms, F.D.; Liu, H.; Owejan, J.E. Mitigation of perfluorosulfonic acid membrane chemical degradation using cerium and manganese Ions. *ECS Trans.* **2008**, *16*, 1735–1747. [CrossRef]
9. Trogadas, P.; Parrondo, J.; Ramani, V. Degradation mitigation in polymer electrolyte membranes using cerium oxide as a regenerative free-radical scavenger. *Electrochem. Solid-State Lett.* **2008**, *11*, B113–B116. [CrossRef]
10. Wang, L.; Advani, S.G.; Prasad, A.K. Degradation reduction of polymer electrolyte membranes using CeO_2 as a free-radical scavenger in catalyst layer. *Electrochim. Acta* **2013**, *109*, 775–780. [CrossRef]

11. Lei, M.; Yang, T.Z.; Wang, W.J.; Huang, K.; Zhang, Y.C.; Zhang, R.; Jiao, R.Z.; Fu, X.L.; Yang, H.J.; Wang, Y.G.; et al. One-dimensional manganese oxide nanostructures as radical scavenger to improve membrane electrolyte assembly durability of proton exchange membrane fuel cells. *J. Power Sources* **2013**, *230*, 96–100. [CrossRef]
12. D'Urso, C.; Oldani, C.; Baglio, V.; Merlo, L.; Aricò, A.S. Fuel cell performance and durability investigation of bimetallic radical scavengers in Aquivion® perfluorosulfonic acid membranes. *Int. J. Hydrog. Energy* **2017**, *42*, 27987–27994. [CrossRef]
13. Wong, C.Y.; Wong, W.Y.; Ramya, K.; Khalid, M.; Loh, K.S.; Daud, W.R.W.; Lim, K.L.; Walvekar, R.; Kadhum, A.A.H. Additives in proton exchange membranes for low- and high-temperature fuel cell applications: A review. *Int. J. Hydrog. Energy* **2019**, *12*, 6116–6135. [CrossRef]
14. Saccà, A.; Gatto, I.; Carbone, A.; Pedicini, R.; Maisano, S.; Stassi, A.; Passalacqua, E. Influence of doping level in Yttria-Stabilised-Zirconia (YSZ) based-fillers as degradation inhibitors for proton exchange membranes fuel cells (PEMFCs) in drastic conditions. *Int. J. Hydrog. Energy* **2019**, *44*, 31445–31457. [CrossRef]
15. Xing, Y.; Li, H.; Avgouropoulos, G. Research progress of proton exchange membrane failure and mitigation strategies. *Materials* **2021**, *14*, 2591. [CrossRef] [PubMed]
16. Donnadio, A.; D'Amato, R.; Marmottini, F.; Panzetta, G.; Pica, M.; Battocchio, C.; Capitani, D.; Ziarelli, F.; Casciola, M. On the evolution of proton conductivity of Aquivion membranes loaded with CeO_2 based nanofillers: Effect of temperature and relative humidity. *J. Membr. Sci.* **2019**, *574*, 17–23. [CrossRef]
17. Ball, V. Polydopamine nanomaterials: Recent advances in synthesis methods and applications. *Front. Bioeng. Biotechnol.* **2018**, *6*, 109–121. [CrossRef] [PubMed]
18. Dreyer, D.R.; Miller, D.J.; Freeman, B.D.; Paul, D.R.; Bielawski, C.W. Perspectives on poly(dopamine). *Chem. Sci.* **2013**, *4*, 3796–3802. [CrossRef]
19. Sedó, J.; Saiz-Poseu, J.; Busqué, F.; Ruiz-Molina, D. Catechol-based biomimetic functional materials. *Adv. Mater.* **2013**, *25*, 653–701. [CrossRef]
20. D'Ischia, M.; Napolitano, A.; Ball, V.; Chen, C.T.; Buehler., M.J. Polydopamine and eumelanin: From structure-property relationships to a unified tailoring strategy. *Acc. Chem. Res.* **2014**, *47*, 3541–3550. [CrossRef]
21. Ryu, J.H.; Messersmith, P.B.; Lee, H. Polydopamine surface chemistry: A decade of discovery. *ACS Appl. Mater. Interfaces* **2018**, *10*, 7523–7540. [CrossRef]
22. Liebscher, J.; Mrówczyński, R.; Scheidt, H.A.; Filip., C.; Hădade, N.D.; Turcu, R.; Bende, A.; Beck, S. Structure of polydopamine: A never-ending story? *Langmuir* **2013**, *29*, 10539–10548. [CrossRef]
23. Ding, Y.; Weng, L.T.; Yang, M.; Yang, Z.; Lu, X.; Huang, N.; Leng, Y. Insights into the aggregation/deposition and structure of a polydopamine film. *Langmuir* **2014**, *30*, 12258–12269. [CrossRef] [PubMed]
24. Alfieri, M.L.; Panzella, L.; Oscurato, S.L.; Salvatore, M.; Avolio, R.; Errico, M.E.; Maddalena, P.; Napolitano, A.; d'Ischia, M. The chemistry of polydopamine film formation: The amine-quinone interplay. *Biomimetics* **2018**, *3*, 26. [CrossRef] [PubMed]
25. Tamakloe, W.; AdjeiAgyeman, D.; Park, M.; Yang, J.; Kang, Y.M. Polydopamine-induced surface functionalization of carbon nanofibers for Pd deposition enabling enhanced catalytic activity for the oxygen reduction and evolution reactions. *J. Mater. Chem. A* **2019**, *7*, 7396–7405. [CrossRef]
26. Oh, K.H.; Choo, M.J.; Lee, H.; Park, K.H.; Park, J.K.; Choi, J.W. Mussel-inspired polydopamine-treated composite electrolytes for long-term operations of polymer electrolyte membrane fuel cells. *J. Mater. Chem. A* **2013**, *1*, 14484–14490. [CrossRef]
27. Lee, H.; Dellatore, S.M.; Miller, W.M.; Messersmith, P.B. Mussel-inspired surface chemistry for multifunctional coatings. *Science* **2007**, *318*, 426–430. [CrossRef]
28. Hong, S.; Lee, J.S.; Ryu, J.; Lee, S.H.; Lee, D.Y.; Kim, D.P.; Park, C.B.; Lee, H. Bio-inspired strategy for on-surface synthesis of silver nanoparticles for metal/organic hybrid nanomaterials and LDI-MS substrates. *Nanotechnology* **2011**, *22*, 494020–494027. [CrossRef]
29. Lee, H.; Lee, Y.H.; Statz, A.R.; Rho, J.; Park, T.G.; Messersmith, P.B. Substrate-independent layer-by-layer assembly by using mussel-adhesive-inspired polymers. *Adv. Mater.* **2008**, *20*, 1619–1623. [CrossRef]
30. Ho, C.C.; Ding, S.J. Structure, properties and applications of mussel-inspired polydopamine. *J. Biomed. Nanotechnol.* **2014**, *10*, 3063–3084. [CrossRef]
31. Kang, S.M.; Hwang, N.S.; Yeom, J.; Park, S.Y.; Messersmith, P.B.; Choi, I.S.; Langer, R.; Anderson, D.G.; Lee, H. One-step multipurpose surface functionalization by adhesive catecholamine. *Adv. Funct. Mater.* **2012**, *22*, 2949–2955. [CrossRef]
32. Lynge, M.E.; van der Westen, R.; Postma, A.; Stadler, B. Polydopamine—A nature-inspired polymer coating for biomedical science. *Nanoscale* **2011**, *3*, 4916–4928. [CrossRef]
33. Ju, K.Y.; Lee, Y.; Lee, S.; Park, S.B.; Lee, J.K. Bioinspired polymerization of dopamine to generate melanin-like nanoparticles having an excellent free-radical-scavenging property. *Biomacromolecules* **2011**, *12*, 625–632. [CrossRef] [PubMed]
34. Liu, Y.; Ai, K.; Lu, L. Polydopamine and its derivative materials: Synthesis and promising applications in energy, environmental, and biomedical fields. *Chem. Rev.* **2014**, *114*, 5057–5115. [CrossRef] [PubMed]
35. Martín, M.; Salazar, P.; Villalonga, R.; Campuzano, S.; Pingarrond, J.M.; González-Mora, J.L. Preparation of core–shell Fe_3O_4@poly(dopamine) magnetic nanoparticles for biosensor construction. *J. Mater. Chem. B* **2014**, *2*, 739–746. [CrossRef]
36. Fedorenko, V.; Damberga, D.; Grundsteins, K.; Ramanavicius, A.; Ramanavicius, S.; Emerson Coy, E.; Iatsunskyi, I.; Viter, R. Application of polydopamine functionalized zinc oxide for glucose biosensor design. *Polymers* **2021**, *13*, 2918. [CrossRef] [PubMed]

37. Black, K.C.; Yi, J.; Rivera, J.G.; Zelasko-Leon, D.C.; Messersmith, P.B. Polydopamine-enabled surface functionalization of gold nanorods for cancer cell-targeted imaging and photothermal therapy. *Nanomedicine* **2013**, *8*, 17–28. [CrossRef]
38. Liu, T.; Kim, K.C.; Lee, B.; Chen, Z.; Noda, S.; Jang, S.S.; Lee, S.W. Self-polymerized dopamine as an organic cathode for Li- and Na-Ion batteries. *Energy Environ. Sci.* **2017**, *10*, 205–215. [CrossRef]
39. He, Y.; Wang, J.; Zhang, H.; Zhang, T.; Zhang, B.; Cao, S.; Liu, J. Polydopamine-modified graphene oxide nanocomposite membrane for proton exchange membrane fuel cell under anhydrous conditions. *J. Mater. Chem. A* **2014**, *2*, 9548–9558. [CrossRef]
40. Wang, J.; Bai, H.; Zhang, H.; Zhao, L.; Chen, H.; Li, Y. Anhydrous proton exchange membrane of sulfonated poly(ether ether ketone) enabled by polydopamine-modified silica nanoparticles. *Electrochim. Acta* **2015**, *152*, 443–455. [CrossRef]
41. Wang, J.; Gong, C.; Wen, S.; Liu, H.; Qin, C.; Xiong, C.; Dong, L. A facile approach of fabricating proton exchange membranes by incorporating polydopamine-functionalized carbon nanotubes into chitosan. *Int. J. Hydrog. Energy* **2019**, *44*, 6909–6918. [CrossRef]
42. Zhang, H.; Zhang, T.; Wang, J.; Pei, F.; He, Y.; Liu, J. Enhanced proton conductivity of sulfonated poly(ether ether ketone) membrane embedded by dopamine-modified nanotubes for proton exchange membrane fuel cell. *Fuel Cells* **2013**, *13*, 1155–1165. [CrossRef]
43. Zhou, J.; Duan, B.; Fang, Z.; Song, J.; Wang, C.; Messersmith, P.B.; Duan, H. Interfacial assembly of mussel-inspired Au@Ag@polydopamine core-shell nanoparticles for recyclable nanocatalysts. *Adv. Mater.* **2014**, *26*, 701–705. [CrossRef] [PubMed]
44. Yoon, K.R.; Lee, K.A.; Jo, S.; Yook, S.H.; Lee, K.Y.; Kim, L.D.; Kim, J.Y. Mussel-inspired polydopamine-treated reinforced composite membranes with self-supported CeO$_x$ radical scavengers for highly stable PEM fuel cells. *Adv. Funct. Mater.* **2019**, *29*, 1806929. [CrossRef]
45. Casciola, M.; Donnadio, A.; Sassi, P. A critical investigation of the effect of hygrothermal cycling on hydration and in-plane/through-plane proton conductivity of Nafion 117 at medium temperature (70–130 °C). *J. Power Sour.* **2013**, *235*, 129–134. [CrossRef]
46. Capitani, D.; Casciola, M.; Donnadio, A.; Vivani, R. High yield precipitation of crystalline α-zirconium phosphate from oxalic acid solutions. *Inorg. Chem.* **2010**, *49*, 9409–9415. [CrossRef]
47. Gatto, I.; Saccà, A.; Baglio, V.; Aricò, A.S.; Oldani, C.; Merlo, L.; Carbone, A. Evaluation of hot-pressing parameters on the electrochemical performance of MEAs based on Aquivion® PFSA membranes. *J. Energy Chem.* **2019**, *35*, 168–173. [CrossRef]
48. Gatto, I.; Carbone, A.; Saccà, A.; Passalacqua, E.; Oldani, C.; Merlo, L.; Sebastiána, D.; Aricò, A.S.; Baglio, V. Increasing the stability of membrane-electrode assemblies based on Aquivion® membranes under automotive fuel cell conditions by using proper catalysts and ionomers. *J. Electroanal. Chem.* **2019**, *842*, 59–65. [CrossRef]
49. Luo, H.; Gu, C.; Zheng, W.; Dai, F.; Wang, X.; Zheng, Z. Facile synthesis of novel size-controlled antibacterial hybrid spheres using silver nanoparticles loaded with poly-dopamine spheres. *RSC Adv.* **2015**, *5*, 13470–13477. [CrossRef]
50. Silverstein, R.; Bassler, G.; Morrill, R. *Spectrometric Identification of Organic Compounds*; John Wiley & Sons: New York, NY, USA, 1981.
51. Zangmeister, R.A.; Morris, T.A.; Tarlov, M.J. Characterization of polydopamine thin films deposited at short times by autoxidation of dopamine. *Langmuir* **2013**, *29*, 8619–8628. [CrossRef]
52. Knorr, D.B., Jr.; Tran, N.T.; Gaskell, K.J.; Orlicki, J.A.; Woicik, J.C.; Fischer, D.A.; Lenhart, J.L. Synthesis and characterization of aminopropyltriethoxysilane-polydopamine coatings. *Langmuir* **2016**, *32*, 4370–4381. [CrossRef] [PubMed]
53. Stöhr, J. *NEXAFS Spectroscopy*; Springer-Verlag: Berlin/Heidelberg, Germany, 1992.
54. Feyer, V.; Plekan, O.; Richter, R.; Coreno, M.; Prince, K.C.; Carravetta, V. Photoemission and photo absorption spectroscopy of glycyl-glycine in the gas phase. *J. Phys. Chem. A* **2009**, *113*, 10726–10733. [CrossRef] [PubMed]
55. Stewart-Ornstein, J.; Hitchcock, A.P.; Hernández Cruz, D.; Henklein, P.; Overhage, J.; Hilpert, K.; Hale, J.D.; Hancock, R.E. Using intrinsic X-ray absorption spectral differences to identify and map peptides and proteins. *J. Phys. Chem. B* **2007**, *111*, 7691–7699. [CrossRef] [PubMed]

Article

The Effect of In Situ Synthesis of MgO Nanoparticles on the Thermal Properties of Ternary Nitrate

Zhiyu Tong [1], Linfeng Li [1], Yuanyuan Li [1], Qingmeng Wang [2] and Xiaomin Cheng [1,2,*]

[1] School of Materials Science and Engineering, Wuhan University of Technology, Wuhan 470070, China; tongzy@whut.edu.cn (Z.T.); lilinfeng0514@whut.edu.cn (L.L.); yyli@whut.edu.cn (Y.L.)
[2] School of Electromechanical and Automobile Engineering, Huanggang Normal University, Huanggang 438000, China; wangqingmeng@whut.edu.cn
* Correspondence: chengxm@whut.edu.cn; Tel.: +86-135-0711-7513

Abstract: The multiple eutectic nitrates with a low melting point are widely used in the field of solar thermal utilization due to their good thermophysical properties. The addition of nanoparticles can improve the heat transfer and heat storage performance of nitrate. This article explored the effect of MgO nanoparticles on the thermal properties of ternary eutectic nitrates. As a result of the decomposition reaction of the $Mg(OH)_2$ precursor at high temperature, MgO nanoparticles were synthesized in situ in the $LiNO_3$–$NaNO_3$–KNO_3 ternary eutectic nitrate system. XRD and Raman results showed that MgO nanoparticles were successfully synthesized in situ in the ternary nitrate system. SEM and EDS results showed no obvious agglomeration. The specific heat capacity of the modified salt is significantly increased. When the content of MgO nanoparticles is 2 wt %, the specific heat of the modified salt in the solid phase and the specific heat in the liquid phase increased by 51.54% and 44.50%, respectively. The heat transfer performance of the modified salt is also significantly improved. When the content of MgO nanoparticles is 5 wt %, the thermal diffusion coefficient of the modified salt is increased by 39.3%. This study also discussed the enhancement mechanism of the specific heat capacity of the molten salt by the nanoparticles mainly due to the higher specific surface energy of MgO and the semi-solid layer that formed between the MgO nanoparticles and the molten salt.

Keywords: MgO nanoparticles; eutectic nitrates; in situ; specific heat capacity; thermal diffusion coefficient

1. Introduction

The exploitation of solar energy is essential to sustainable development. To solve the problem of intermittent solar energy in solar thermal utilization, it is necessary to use the thermal energy storage (TES) system to store and release heat when solar radiation is weak or absent [1,2]. As an excellent heat storage carrier, molten salt heat storage material has the advantages of sizeable latent heat, high energy storage density, low subcooling, good thermal stability and low cost, which is widely used as a heat storage medium for solar heat. Currently, solar salt is widely used in TES technology. The components of solar salt are $NaNO_3$ and KNO_3. The melting point of solar salt is 220 °C. Due to its high melting point, the pipeline needs to be heated to a higher temperature to prevent the pipeline from freezing, resulting in additional energy input and power generation costs. The higher melting point of molten salt limits the application of molten salt in the field of heat storage. Therefore, the ideal heat transfer fluid is supposed to have a low melting point, reducing the risk of freezing and heating energy consumption of the pipeline. Studies have shown that mixing several molten salts in a certain proportion forms a eutectic salt that can reduce the melting point while ensuring the thermal stability of the molten salt. The main melting salts are nitrates, carbonates, and sulfates [3,4]. In recent years, the development of low-melting, high-stability multi-element molten salt systems has become a research hotspot in molten salt modification. Wu et al. [5] formulate 19 kinds of binary mixed molten salts in different proportions, the main component of which is KNO_3-$Ca(NO_3)_2 \cdot 4H_2O$. The results showed that the thermodynamic properties of these molten salts performed well. Ren et al. [6] further explored the $Ca(NO_3)_2$–$NaNO_3$ binary salt and modified it with expanded graphite, which effectively improved the thermophysical properties of the molten salt. In recent years, more and more ternary and quaternary molten salts have been developed [7–13]. The main research systems are $LiNO_3$–$NaNO_3$–KNO_3, $NaNO_3$–$NaNO_2$–KNO_3, $Ca(NO_3)_2$–$NaNO_3$–KNO_3, and $LiNO_3$–$NaNO_3$–KNO_3–$Ca(NO_3)_2$. These multi-molten salts have lower melting points and higher stability. The use of relevant phase diagrams to create ternary or higher salt mixtures can obtain low melting point molten salts. The ideal freezing temperature for Hitec and Hitec XL is 120–140 °C, and they can withstand temperatures exceeding 500 °C. The $LiNO_3$–$NaNO_3$–KNO_3 ternary mixture is considered as a promising heat transfer and storage medium, with a low melting point (120 °C) and high thermal stability (550 °C). Multi-element eutectic molten salt has a wide operating temperature range (low melting point and high decomposition point), which is very suitable as a heat transfer fluid and heat storage carrier in the TES system of a concentrating solar power plant to store solar energy.

Nanomaterials have special physical and chemical properties due to their unique structure, so they have important applications in heat storage [14–18]. For example, nano-SiC and nano-MgO have not only higher specific heat capacity but also better heat transfer efficiency, and they are very good heat storage materials. Therefore, the research of nanomaterials is of great significance to the development of heat storage materials. Researchers tried to add nanoparticles to molten salt to increase the specific heat capacity of molten salt. Among the research of using nanoparticles to modify molten salt, the most common materials are SiO_2 and Al_2O_3 nanoparticles, most of which have been observed to have an increase in the specific heat and thermal conductivity [19–21]. Dudda et al. [22] and Seo et al. [23] explored the effect of nanoparticle size on the specific heat capacity of the nanoparticle/molten salt eutectic mixture. It was observed that the salt compounds around the nanoparticles formed a large number of nano-sized structures, which may be the main reason for the increase in specific heat. From the view of structure, one reason for the increased specific heat is the thermal resistance of the interface between the nanoparticles and the molten salt. Another reason is that a semi-solid layer is formed between the nanoparticles and the molten salt. From the perspective of energy, the high surface energy of nanoparticles can also store part of the thermal energy. Hu et al. [24] performed molecular dynamics simulations on Al_2O_3 nanoparticles doped in solar salt and explored the reason for the specific heat enhancement from the view of energy. The result shows that the change of Coulomb energy is the reason for the change of specific heat capacity.

The addition of nanoparticles can also improve the heat transfer performance of molten salt to a certain extent. Gupta et al. [25] added different types of nanoparticles (TiO_2, ZnO, Fe_2O_3, and SiO_2) to the phase change material (PCM) of $Mg(NO_3)_2 \cdot 6H_2O$ and formed the PCM–metal oxide nanocomposite material through the melting and mixing technology. The PCM–metal oxide nanocomposite with a 0.5 wt % nanoparticle addition increased the thermal conductivity by 147.5% (TiO_2), 62.5% (ZnO), 55% (Fe_2O_3), and 45% (SiO_2), respectively. Ho et al. [26] discussed the effect of nanoparticle concentration on the convective heat transfer performance of molten nano-HITEC fluid laminar flow in microtubes. The heat transfer performance of HITEC fluid with Al_2O_3 nanoparticle concentration as high as 0.25 wt % has been improved. The study of Yu et al. [27] observed that SiO_2 and TiO_2 nanoparticles can improve the thermal conductivity of molten salt. Under normal circumstances, the thermal conductivity of molten salt is about 0.2–2.0 W/(m·K), and the specific heat is about 1.35 J/(g·°C). The specific heat of the molten salt added with these two kinds of nanoparticles increased by 28.1%, and the thermal conductivity increased by 53.7%. Studies have shown that there are heat transfer channels in high-density nanostructures, which can contribute to the enhancement of thermal conductivity. D. Shin et al. [28]

found that in traditional nanofluids, nanoparticles can form fractal fluid nanostructures to enhance thermal conductivity.

Poor particle dispersion can reduce the specific heat capacity of the molten salt [29]. Therefore, to achieve the particular heat enhancement of nanomaterials, the preparation method must be carefully controlled. There are many methods of using nanoparticles to modify molten salts, such as the high-temperature melting method, aqueous solution method, combustion method, and in situ synthesis method. The high-temperature melting method is to directly melt and stir molten salt and nanoparticles at high temperature to form a uniform eutectic system. The aqueous solution method is to dissolve the molten salt in water, then add nanomaterials to form a stable suspension, and finally, by heating, precipitation to obtain the eutectic salt. The combustion method is to mix the precursor, molten salt, and fuel together, then ignite the fuel and generate a lot of heat through violent combustion, so that the molten salt forms a eutectic system. The in situ synthesis method is to mix the precursor and molten salt, and then, the precursor reacts in the molten salt at a certain temperature to generate nanoparticles. Li et al. [20] and Zhang et al. [30] used SiO_2 and Al_2O_3 nanoparticles as additives and added the nanoparticles to the molten salt by the high-temperature melting method, and they successfully prepared the modified salt. Xiong et al. [21] used the aqueous solution method to prepare the SiO_2/molten salt nanofluid successfully. Lasfargues et al. [31,32] used copper sulfate pentahydrate and titanium sulfate as precursors to synthesize CuO and TiO_2 nanoparticles in situ in solar salt. The specific heat of solar salt was observed to increase. In our previous research, we successfully synthesized MgO nanoparticles in situ in solar salt, which significantly increased the specific heat capacity of solar salt [33].

At present, there are relatively few studies on the performance improvement of multi-element molten salt by nanoparticles [19]. In the research of nanoparticle modification of molten salt, MgO nanoparticles are an excellent modified particle. MgO has several types of bulk intrinsic defects, including oxygen and magnesium vacancies, interstitials, their agglomerates, etc. [34,35]. This has aroused the interest of many researchers. In this work, we prepared a ternary eutectic nitrate and applied the in situ generation method to generate MgO nanoparticles in molten salt. By testing the specific heat capacity, latent heat of phase change, and thermal conductivity of the prepared nitrate-based composite materials, the influence of MgO nanoparticles on the heat transfer and heat storage performance of ternary nitrate was studied.

2. Materials and Methods

2.1. Materials

$Mg(OH)_2$ was obtained from Aladdin Chemical Co., Ltd., Shanghai, China. KNO_3, $NaNO_3$ and $LiNO_3$ were commercially supplied by Sinopharm Chemical Reagent Co., Ltd., Shanghai, China. All chemicals are of analytical grade.

2.2. Preparation

The first step is the preparation of lithium nitrate–potassium nitrate–sodium nitrate eutectic nitrate. Many scholars have studied the ternary eutectic point of the $LiNO_3$–$NaNO_3$–KNO_3 ternary system. Zhong et al. [36] predicted the phase diagram of the $LiNO_3$–$NaNO_3$–KNO_3 ternary system and experimentally verified the predicted ternary invariant points. This result is similar to the report by Coscia et al. [37]. The mass ratio of the $LiNO_3$–$NaNO_3$–KNO_3 ternary system is 29:58:13 (wt %). The ternary nitrate with the mass balance was ground in a mortar for 1 h; then, it was transferred to a crucible and placed in a resistance furnace at 300 °C for 5 h. Then, we took out the crucible, cooled it, and ground it to obtain ternary eutectic nitrate.

According to the ratio in Table 1, the magnesium hydroxide precursor was added to the ternary eutectic nitrate. Then, we ground it in an agate mortar for 30 min. After that, the mixture was transferred to a ceramic crucible and placed in a resistance furnace at 400 °C for 2 h to ensure complete decomposition of the magnesium hydroxide precursor.

After taking it out, it was quickly cooled in air and ground to obtain a sample. The process is shown in Figure 1.

Table 1. The content of each ingredient in the preparation.

Sample	Percentage of MgO (wt %)	Percentage of Precursor (wt %)	Mg(OH)$_2$ Precursor (g)	LiNO$_3$-NaNO$_3$-KNO$_3$ (g)
S$_1$	0.5	0.7	0.037	4.975
S$_2$	1.0	1.4	0.073	4.950
S$_3$	1.5	2.1	0.109	4.925
S$_4$	2.0	2.8	0.145	4.900
S$_5$	2.5	3.5	0.182	4.875
S$_6$	3.0	4.1	0.217	4.850
S$_7$	5.0	7.0	0.370	4.750

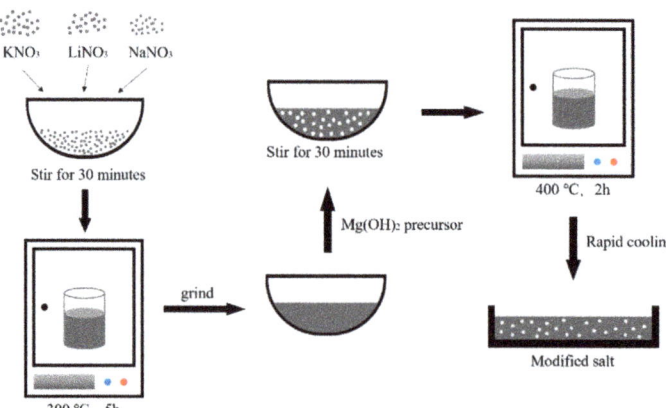

Figure 1. Preparation of modified salt.

To further test the material, it is necessary to obtain the nanoparticles synthesized in situ. The modified salt was washed with deionized water and then centrifuged. After the nitrate is washed away, the nanoparticles are dispersed by ultrasound. Finally, the resulting nanoparticles are dried.

2.3. Characterization

The crystal phases of all samples were characterized by X-ray diffraction (XRD, Empyrean, PANalytical B.V., Amsterdem, Netherlands). RENISHAW Raman microscope (Raman, In Via, RENISHAW, England) was used to measure modified salts and magnesium oxide nanoparticles. We observed the microstructure of the sample with a field emission scanning electron microscope (SEM, S-4800, HITACHI, Tokyo, Japan). The component analysis of the prepared modified nitrate was examined by X-ray energy-dispersive spectroscopy (EDS, AMETEK, Berwyn, PA, USA) combined with scanning electron microscopy under a constant nitrogen flow from 30 to 200 °C at a heating rate of 10 °C/min. The specific heat capacity and latent heat of phase change of the samples were measured with a differential scanning calorimeter (DSC8500, PERKINELMER, Waltham, MA, USA). Then, we used an infrared thermal imager (TESTO-872, Testo SE&Co. KGaA, Titisee-Neustadt, Germany) to characterize the heat transfer performance of ternary nitrate modified with different percentages of MgO nanoparticles. Thermal diffusivity was obtained by a laser thermal conductivity meter (LFA457, NETZSCH, Selb, Germany).

3. Results and Discussions

3.1. Components of Modified Salt

Figure 2 shows the XRD spectra of samples S_0, S_4, and the product after centrifugation. There are characteristic peaks at 19.03°, 23.52°, 33.06°, 33.84°, 41.15°, and 46.61° in sample S_0, corresponding to the (110), (111), (200), (112), (221), and (113) crystal planes of KNO_3. Two peaks at 29.41° and 38.99° represent the (104) and (113) crystal planes of $NaNO_3$, respectively. In addition, the crystal plane (104) of $LiNO_3$ could be found according to the characteristic peak at 32.21°. It can be seen that the ternary nitrate was successfully prepared. After centrifugation, characteristic peaks can be seen at 36.89°, 42.86°, 62.22°, 74.58°, and 78.51°, corresponding to (111), (200), (220), (311), and (222) crystal planes of MgO. Sample S_4 possesses several characteristic peaks of MgO at 42.96° and 62.22°. It can be seen that MgO is formed in situ in the ternary nitrate system. The characteristic peaks at 18.59°, 32.84°, 38.02°, 50.85°, 58.64°, 68.87°, and 72.03° mark to (001), (100), (101), (102), (110), (200), and (201) of $Mg(OH)_2$ planes. However, no characteristic peaks of $Mg(OH)_2$ were found in samples S_0 and S_4, indicating that the magnesium hydroxide precursor was completely decomposed when the nanoparticles were generated in situ. The reason why the characteristic peaks of $Mg(OH)_2$ can be seen in the product after centrifugation might be because MgO combines with water during the centrifugation process to form a trace amount of $Mg(OH)_2$.

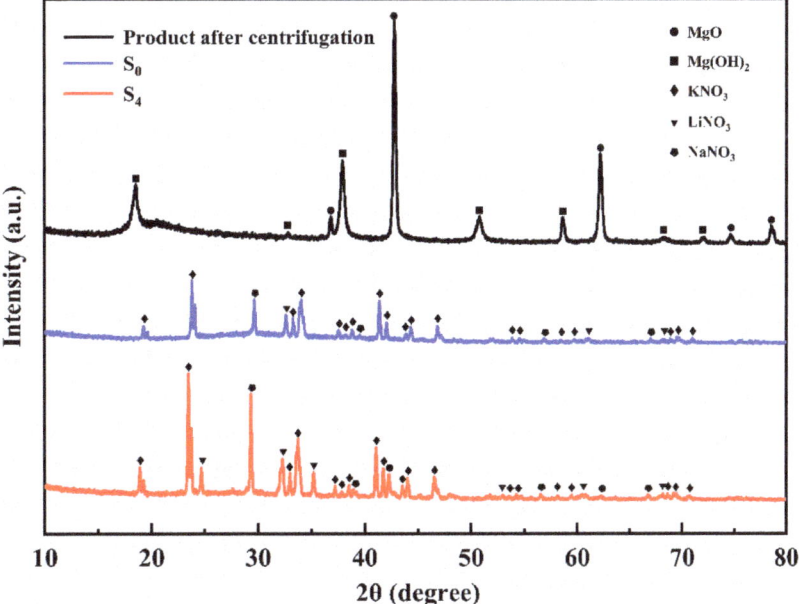

Figure 2. The XRD patterns of samples S_0, S_4, and product after centrifugation.

Figure 3 shows the Raman spectra of samples S_0, S_4, and the product after centrifugation. In samples S_0 and S_4, the peak at 713 cm^{-1} corresponds to the NO_3^- in-plane bending vibration (710–740 cm^{-1}), and the peak at the frequency of 1049 cm^{-1} corresponds to the NO_3^- symmetric stretching vibration (1020–1060 cm^{-1}). In the Raman spectrum of sample S_4 and the MgO nanoparticles obtained after centrifugation, peaks with frequencies of 1499 cm^{-1} and 1936 cm^{-1} can be observed. Combined with XRD analysis, it can be further known that MgO nanoparticles were successfully generated in situ in the ternary nitrate system [38].

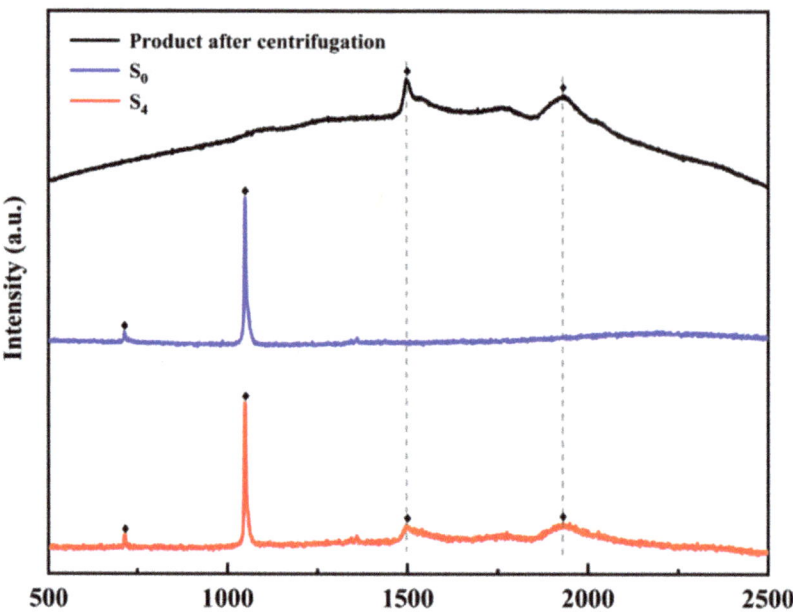

Figure 3. The Raman spectra of samples S_0, S_4, and product after centrifugation.

3.2. The Structure of Modified Salt

Figure 4 is SEM photos of samples S_0, S_2, S_4, S_6, and S_7. Figure 4a is the SEM diagram of $LiNO_3$–$NaNO_3$–KNO_3 ternary nitrate. The surface of the nitrate is relatively flat, and the material is uniform. The mass fractions of MgO nanoparticles in Figure 4b–d are 1 wt %, 2 wt %, and 3 wt %, respectively. It can be seen that the MgO nanoparticles are relatively evenly dispersed among the nitrates, which shows that the nanoparticles generated in situ have good dispersibility. Figure 4e,f are SEM of modified nitrate with 5 wt % MgO nanoparticles at different magnifications. When the content reached 5 wt %, the nanoparticles had obvious agglomeration. Due to the high surface energy of nanoparticles, they will agglomerate together and deposit in the nanofluid, resulting in poor system stability. In the modified salt, the size of the nanoparticles synthesized in situ is concentrated in the range of 50–200 nm.

The element distribution of the sample was determined by the area scanning method of the energy spectrum. Figure 5 is the element distribution of modified nitrate. Figure 5c is the distribution of the Mg element. Figure 5d shows the distribution of the N element, which represents the distribution of nitrate. It can be seen from the element distribution diagram that MgO nanoparticles are evenly dispersed among the ternary nitrates.

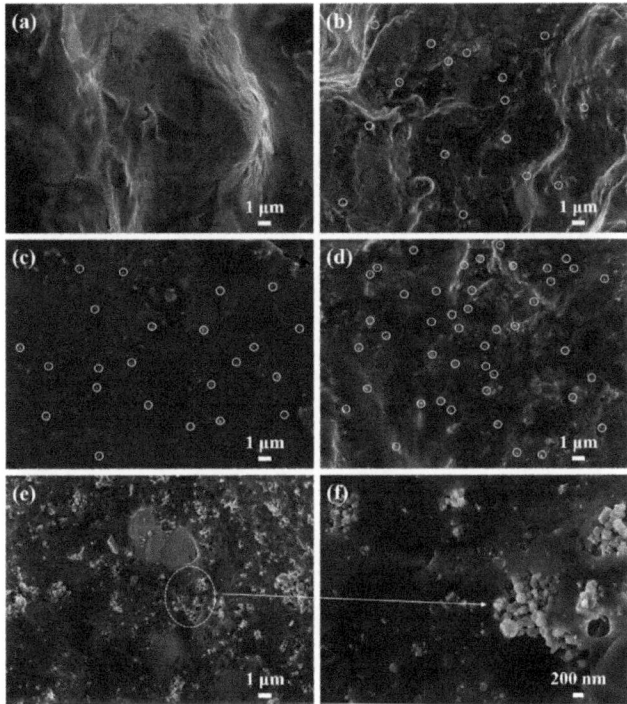

Figure 4. SEM micrographs of (**a**) sample S_0, (**b**) sample S_2, (**c**) sample S_4, (**d**) sample S_6, and (**e**,**f**) sample S_7.

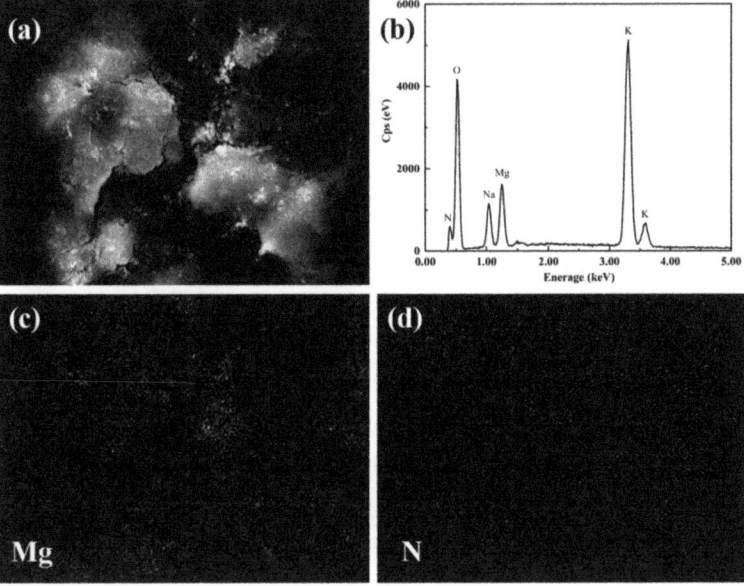

Figure 5. The elemental distribution of modified nitrate.

3.3. Specific Heat Capacity

Figure 6a,b are the specific heat capacity curves at 50–80 °C (solid phase) and 150–200 °C (liquid phase), respectively. The overall sensible heat capacity can be judged by selecting the specific heat capacities at 60 °C (solid phase) and 170 °C (liquid phase). Table 2 shows the specific heat capacities of ternary nitrate and modified salt in the solid phase and liquid phase. For the same sample, the specific heat in the liquid state is improved compared with that in the solid state. The reason for this phenomenon is that in the molten state, the ions in the molten salt perform randomly free movement, which can carry more energy. The specific heat in the solid state increases first and then decreases with the growth of nanoparticles. When the content of nanoparticles is 2%, the solid specific heat is the largest, which is 1.479 J/(g·°C). Compared with $LiNO_3$–$NaNO_3$–KNO_3 ternary nitrate, it has increased by 51.54%. With the increase of nanoparticles, the specific heat in the liquid state shows an increasing trend at the first stage and then a decreasing tendency at the following stage. When the content of nanoparticles is 2%, the specific heat of the material at the liquid state reaches the peak, which is 1.878 J/(g·°C): an increase by 44.50%.

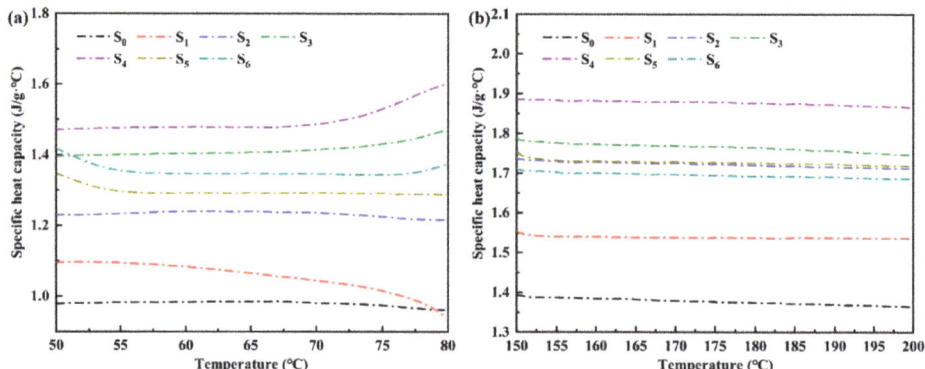

Figure 6. Specific heat capacity of ternary nitrate and modified nitrate: (**a**) solid phase, (**b**) liquid phase.

Table 2. Specific heat capacity of ternary nitrate and modified nitrate.

Sample	Cp (J/g·°C)	
	Solid	Liquid
S_0	0.976 (0%)	1.301 (0%)
S_1	1.084 (11.07%)	1.538 (18.22%)
S_2	1.240 (27.05%)	1.726 (32.67%)
S_3	1.406 (44.06%)	1.764 (35.59%)
S_4	1.479 (51.54%)	1.880 (44.50%)
S_5	1.291 (32.27%)	1.727 (32.74%)
S_6	1.348 (38.11%)	1.695 (30.28%)
Solar Salt	1.290	1.350

For MgO nanoparticles to increase the specific heat capacity of nitric acid ternary salt, the first possible reason is that MgO nanoparticles have higher specific surface energy. The second reason is the interface thermal resistance between MgO nanoparticles and nitrate, which can store and release additional energy. Another reason is that the MgO nanoparticles and the surrounding molten salt form a semi-solid layer. Hu et al. [24] explained the specific heat enhancement from the perspective of Coulomb energy. Molecular dynamics simulation is used to analyze the influence of nanoparticles on the energy composition of each atom type. The results show that the change in the Coulomb energy of each atom contributes the most to the enhanced specific heat capacity.

3.4. Latent Heat

The DSC endothermic and exothermic curves of ternary nitrate and modified nitrate are shown in Figure 7. The endothermic peak of the melting process and the exothermic peak of the solidification process can be seen in the figure. Table 3 shows the latent heat of phase change, the onset temperature, and the melting temperature of the ternary nitrate and the modified nitrate. The phase transition temperature of different samples is unchanged basically, indicating that the addition of trace nanoparticles has little effect on the phase transition process of the material. Compared with ternary nitrate, the latent heat of nitrate doped with nanoparticles is slightly lower. On the whole, there is little difference in the latent heat of phase change. In the field of medium and low-temperature energy storage, the latent heat of phase change materials is about 130 J/g, so this modified salt can meet the requirements.

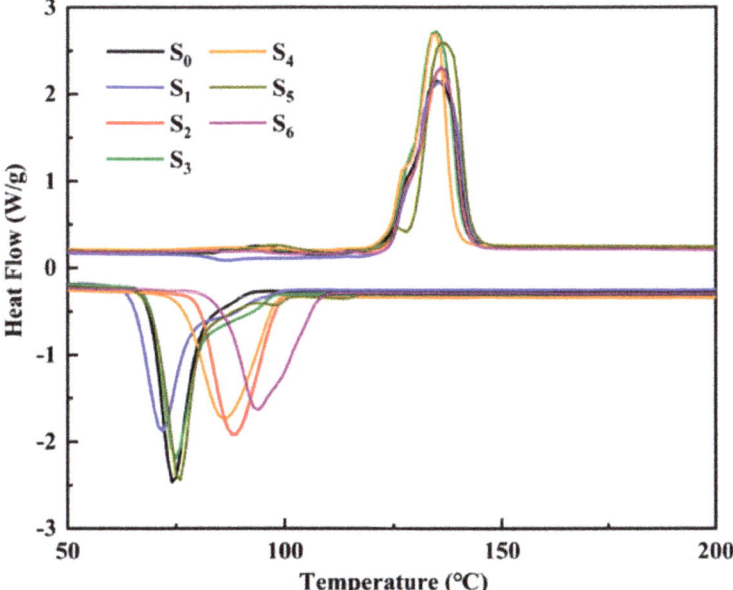

Figure 7. The DSC endothermic and exothermic curves of ternary nitrate and modified nitrate.

Table 3. Result of latent heat, onset temperature, and melting temperature.

Sample	Latent Heat (J/g)	Onset Temperature (°C)	Melting Temperature (°C)
S_1	135.8	110	150
S_2	142.5	112	148
S_3	132.3	114	152
S_4	131.2	112	152
S_5	138.8	116	151
S_6	128.2	110	151

3.5. Heat Transfer Characteristics

Ternary nitrate has good heat transfer performance and can be used as a heat transfer fluid. Figure 8 shows the thermal diffusivity of samples S_0, S_2, S_4, S_6, and S_7. Obviously, with the increase of MgO nanoparticles, the thermal diffusion coefficient increases significantly. When the mass fraction of MgO nanoparticles is 5%, the thermal diffusion coefficient is 0.425 mm$^2 \cdot$s^{-1}, which is 39.3% higher than that of the eutectic salt without MgO nanoparticles. It can be seen that the interface thermal resistance effect between nanoparticles and nitrate does not reduce the overall heat transfer performance of the material. The main reason for improving the heat transfer performance of the modified salt might be due to the high thermal conductivity of the MgO nanoparticles.

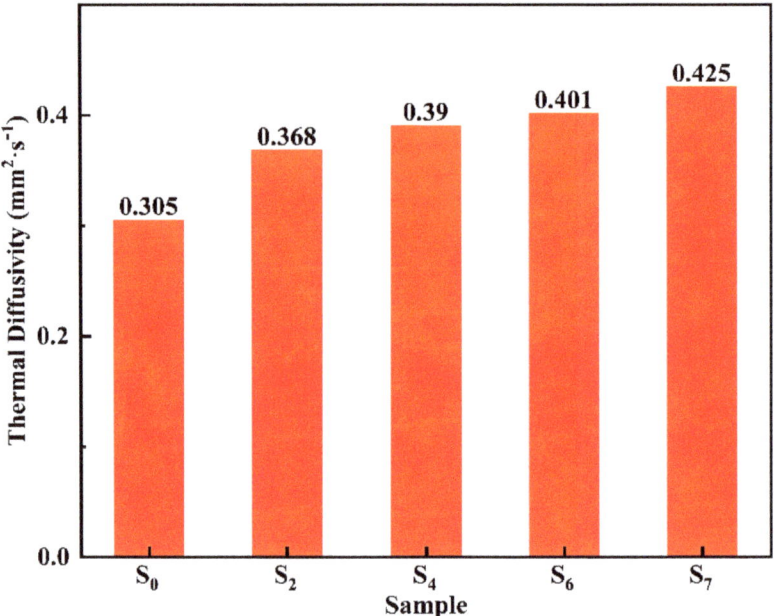

Figure 8. Thermal diffusivity of S_0, S_2, S_4, S_6, and S_7.

To explore the heat transfer performance of the modified nitrate, the samples S_0, S_2, S_4, and S_6 were placed on a heating plate, and an infrared thermal imaging instrument was used to characterize the heat transfer performance of the material. Figure 9 shows the infrared thermal images of samples S_0, S_2, S_4, and S_6 heated on the heating plate for different times. The heat is transferred upward from the bottom end. The heat transfer rate from sample S_0 to sample S_6 shows an increasing trend, and the transfer rate of sample S_6 is the fastest at the same time. Plot the temperature of samples S_0, S_2, S_4, and S_6 at the same geometric position with heating time, as shown in Figure 10. It can be seen that the heat transfer performance of the material increases with the increase of MgO nanoparticles. The doping of nanoparticles enables the material to transfer heat in the solid state quickly, accelerating the heat transfer process of the material, thus shortening the time required for the phase change of the material, and more efficiently storing thermal energy. At the same time, when the material acts as a nanofluid in a liquid state, the heat transfer performance is greatly improved, and the heat transfer process is accelerated. The results show that modified nitrate has high sensible heat and latent heat, a suitable phase transition temperature, and high thermal conductivity, which can improve the energy storage efficiency and can be used in thermal energy storage systems.

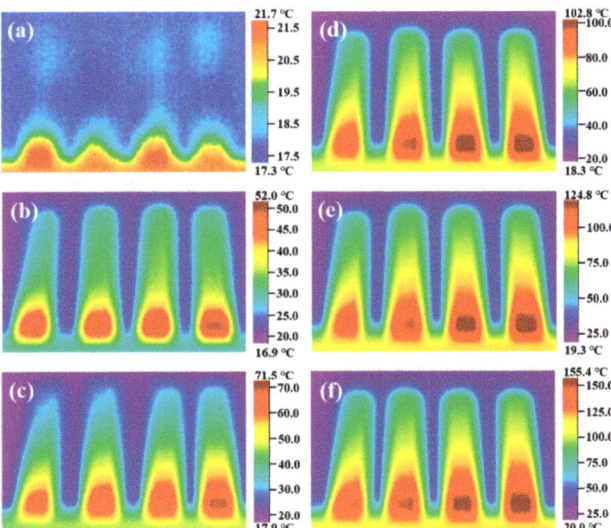

Figure 9. Infrared thermography images of samples S_0, S_2, S_4, and S_6 heated on the heating plate for different times: (**a**) 0 s, (**b**) 60 s, (**c**) 120 s, (**d**) 180 s, (**e**) 240 s, and (**f**) 300 s.

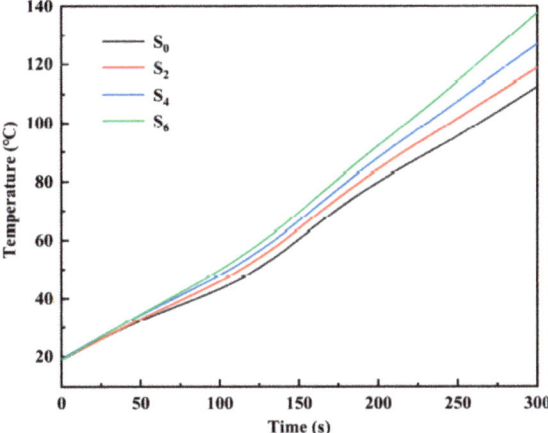

Figure 10. The temperature variation curve of the same geometric position of samples S_0, S_2, S_4, and S_6 with heating time.

4. Conclusions

1. We successfully synthesized magnesium oxide nanoparticles in situ in the ternary nitrate system through the high-temperature decomposition of the Mg(OH)$_2$ precursor. When the content of MgO nanoparticles does not exceed 3 wt %, the dispersion performance is more appropriate, and there is no agglomeration. The size of the nanoparticles synthesized in situ is concentrated in the range of 50–200 nm.

2. Nano-magnesium oxide can improve the heat storage performance of nitrate. When the content of MgO nanoparticles is 2 wt %, the specific heat capacity of the solid increases by 51.54%, and the specific heat capacity of the liquid increases by 44.50%. The increase in specific heat capacity is related to the interface thermal resistance and the semi-solid layer.

3. MgO nanoparticles synthesized in situ significantly improve the heat transfer performance of ternary nitrate. When the content of MgO nanoparticles is 5 wt %, the

thermal diffusion coefficient of the modified salt increased by 39.3%. In the molten state, MgO nanoparticles and ternary nitrate form a heat transfer fluid. Molten salt nanofluid is an excellent heat exchange medium in the field of heat storage, with a high heat exchange capacity.

Author Contributions: Conceptualization, X.C. and Z.T.; methodology, Z.T.; software, L.L.; validation, L.L., Y.L. and Q.W.; formal analysis, Z.T.; investigation, Y.L.; resources, Q.W.; data curation, Z.T.; writing—original draft preparation, Z.T.; writing—review and editing, X.C.; visualization, L.L.; supervision, Y.L.; project administration, X.C.; funding acquisition, X.C. All authors have read and agreed to the published version of the manuscript.

Funding: This research was funded by the Technology and Innovation Major Project of Hubei, grant number 2021BGE023 and 2020BED002.

Institutional Review Board Statement: Not applicable.

Informed Consent Statement: Not applicable.

Data Availability Statement: All the data are available within the manuscript.

Conflicts of Interest: The authors declare no conflict of interest.

References

1. Pelay, U.; Luo, L.; Fan, Y.; Stitou, D.; Rood, M. Thermal energy storage systems for concentrated solar power plants. *Renew. Sustain. Energy Rev.* **2017**, *79*, 82–100. [CrossRef]
2. Li, X.; Dai, Y.J.; Wang, R.Z. Performance investigation on solar thermal conversion of a conical cavity receiver employing a beam-down solar tower concentrator. *Sol. Energy* **2015**, *114*, 134–151. [CrossRef]
3. Shin, D.; Banerjee, D. Specific heat of nanofluids synthesized by dispersing alumina nanoparticles in alkali salt eutectic. *Int. J. Heat Mass Transf.* **2014**, *74*, 210–214. [CrossRef]
4. Vaka, M.; Walvekar, R.; Khalid, M.; Jagadish, P. Low-melting-temperature binary molten nitrate salt mixtures for solar energy storage. *J. Therm. Anal. Calorim.* **2020**, *141*, 2657–2664. [CrossRef]
5. Wu, Y.-T.; Li, Y.; Lu, Y.-W.; Wang, H.-F.; Ma, C.-F. Novel low melting point binary nitrates for thermal energy storage applications. *Sol. Energy Mater. Sol. Cells* **2017**, *164*, 114–121. [CrossRef]
6. Ren, Y.; Li, P.; Yuan, M.; Ye, F.; Xu, C.; Liu, Z. Effect of the fabrication process on the thermophysical properties of $Ca(NO_3)_2$–$NaNO_3$/expanded graphite phase change material composites. *Sol. Energy Mater. Sol. Cells* **2019**, *200*, 110005. [CrossRef]
7. Serrano-López, R.; Fradera, J.; Cuesta-López, S. Molten salts database for energy applications. *Chem. Eng. Process.* **2013**, *73*, 87–102. [CrossRef]
8. Vignarooban, K.; Xu, X.; Arvay, A.; Hsu, K.; Kannan, A.M. Heat transfer fluids for concentrating solar power systems—A review. *Appl. Energy* **2015**, *146*, 383–396. [CrossRef]
9. Fernández, A.G.; Ushak, S.; Galleguillos, H.; Pérez, F.J. Development of new molten salts with $LiNO_3$ and $Ca(NO_3)_2$ for energy storage in CSP plants. *Appl. Energy* **2014**, *119*, 131–140. [CrossRef]
10. Bonk, A.; Sau, S.; Uranga, N.; Hernaiz, M.; Bauer, T. Advanced heat transfer fluids for direct molten salt line-focusing CSP plants. *Prog. Energy Combust. Sci.* **2018**, *67*, 69–87. [CrossRef]
11. Li, C.-J.; Li, P.; Wang, K.; Emir Molina, E. Survey of Properties of Key Single and Mixture Halide Salts for Potential Application as High Temperature Heat Transfer Fluids for Concentrated Solar Thermal Power Systems. *AIMS Energy* **2014**, *2*, 133–157. [CrossRef]
12. Fernández, A.G.; Galleguillos, H.; Pérez, F.J. Corrosion Ability of a Novel Heat Transfer Fluid for Energy Storage in CSP Plants. *Oxid. Met.* **2014**, *82*, 331–345. [CrossRef]
13. Fernández, A.G.; Ushak, S.; Galleguillos, H.; Pérez, F.J. Thermal characterisation of an innovative quaternary molten nitrate mixture for energy storage in CSP plants. *Sol. Energy Mater. Sol. Cells* **2015**, *132*, 172–177. [CrossRef]
14. Akhmetov, B.; Navarro, M.E.; Seitov, A.; Kaltayev, A.; Bakenov, Z.; Ding, Y. Numerical study of integrated latent heat thermal energy storage devices using nanoparticle-enhanced phase change materials. *Sol. Energy* **2019**, *194*, 724–741. [CrossRef]
15. Wong-Pinto, L.-S.; Milian, Y.; Ushak, S. Progress on use of nanoparticles in salt hydrates as phase change materials. *Renew. Sustain. Energy Rev.* **2020**, *122*, 109727. [CrossRef]
16. Hajizadeh, M.R.; Alsabery, A.I.; Sheremet, M.A.; Kumar, R.; Li, Z.; Bach, Q.-V. Nanoparticle impact on discharging of PCM through a thermal storage involving numerical modeling for heat transfer and irreversibility. *Powder Technol.* **2020**, *376*, 424–437. [CrossRef]
17. Qi, G.-Q.; Yang, J.; Bao, R.-Y.; Liu, Z.-Y.; Yang, W.; Xie, B.-H.; Yang, M.-B. Enhanced comprehensive performance of polyethylene glycol based phase change material with hybrid graphene nanomaterials for thermal energy storage. *Carbon* **2015**, *88*, 196–205. [CrossRef]
18. Xie, B.; Li, C.; Chen, J.; Wang, N. Exfoliated 2D hexagonal boron nitride nanosheet stabilized stearic acid as composite phase change materials for thermal energy storage. *Sol. Energy* **2020**, *204*, 624–634. [CrossRef]

19. Seo, J.; Shin, D. Enhancement of specific heat of ternary nitrate (LiNO$_3$–NaNO$_3$–KNO$_3$) salt by doping with SiO$_2$ nanoparticles for solar thermal energy storage. *Micro Nano Lett.* **2014**, *9*, 817–820. [CrossRef]
20. Li, Y.; Chen, X.; Wu, Y.; Lu, Y.; Zhi, R.; Wang, X.; Ma, C. Experimental study on the effect of SiO$_2$ nanoparticle dispersion on the thermophysical properties of binary nitrate molten salt. *Sol. Energy* **2019**, *183*, 776–781. [CrossRef]
21. Xiong, Y.; Wang, Z.; Sun, M.; Wu, Y.; Xu, P.; Qian, X.; Li, C.; Ding, Y.; Ma, C. Enhanced thermal energy storage of nitrate salts by silica nanoparticles for concentrating solar power. *Int. J. Energy Res.* **2020**, *45*, 5248–5262. [CrossRef]
22. Dudda, B.; Shin, D. Effect of nanoparticle dispersion on specific heat capacity of a binary nitrate salt eutectic for concentrated solar power applications. *Int. J. Therm. Sci.* **2013**, *69*, 37–42. [CrossRef]
23. Seo, J.; Shin, D. Size effect of nanoparticle on specific heat in a ternary nitrate (LiNO$_3$–NaNO$_3$–KNO$_3$) salt eutectic for thermal energy storage. *Appl. Therm. Eng.* **2016**, *102*, 144–148. [CrossRef]
24. Hu, Y.; He, Y.; Zhang, Z.; Wen, D. Effect of Al$_2$O$_3$ nanoparticle dispersion on the specific heat capacity of a eutectic binary nitrate salt for solar power applications. *Energy Convers. Manag.* **2017**, *142*, 366–373. [CrossRef]
25. Gupta, N.; Kumar, A.; Dhasmana, H.; Kumar, V.; Kumar, A.; Shukla, P.; Verma, A.; Nutan, G.V.; Dhawan, S.K.; Jain, V.K. Enhanced thermophysical properties of Metal oxide nanoparticles embedded magnesium nitrate hexahydrate based nanocomposite for thermal energy storage applications. *J. Energy Storage* **2020**, *32*, 101773. [CrossRef]
26. Ho, M.X.; Pan, C. Experimental investigation of heat transfer performance of molten HITEC salt flow with alumina nanoparticles. *Int. J. Heat Mass Transf.* **2017**, *107*, 1094–1103. [CrossRef]
27. Yu, Q.; Lu, Y.; Zhang, X.; Yang, Y.; Zhang, C.; Wu, Y. Comprehensive thermal properties of molten salt nanocomposite materials base on mixed nitrate salts with SiO$_2$/TiO$_2$ nanoparticles for thermal energy storage. *Sol. Energy Mater. Sol. Cells* **2021**, *230*, 111215. [CrossRef]
28. Shin, D.; Tiznobaik, H.; Banerjee, D. Specific heat mechanism of molten salt nanofluids. *Appl. Phys. Lett.* **2014**, *104*, 121914. [CrossRef]
29. Riazi, H.; Mesgari, S.; Ahmed, N.A.; Taylor, R.A. The effect of nanoparticle morphology on the specific heat of nanosalts. *Int. J. Heat Mass Transf.* **2016**, *94*, 254–261. [CrossRef]
30. Zhang, Y.; Li, J.; Gao, L.; Wang, M. Nitrate based nanocomposite thermal storage materials: Understanding the enhancement of thermophysical properties in thermal energy storage. *Sol. Energy Mater. Sol. Cells* **2020**, *216*, 110727. [CrossRef]
31. Lasfargues, M.; Bell, A.; Ding, Y. In situ production of titanium dioxide nanoparticles in molten salt phase for thermal energy storage and heat-transfer fluid applications. *J. Nanopart. Res.* **2016**, *18*, 150. [CrossRef]
32. Lasfargues, M.; Stead, G.; Amjad, M.; Ding, Y.; Wen, D. In Situ Production of Copper Oxide Nanoparticles in a Binary Molten Salt for Concentrated Solar Power Plant Applications. *Materials* **2017**, *10*, 537. [CrossRef] [PubMed]
33. Huang, Y.; Cheng, X.; Li, Y.; Yu, G.; Xu, K.; Li, G. Effect of in-situ synthesized nano-MgO on thermal properties of NaNO$_3$-KNO$_3$. *Sol. Energy* **2018**, *160*, 208–215. [CrossRef]
34. Monge, M.A.; González, R.; Muñoz Santiuste, J.E.; Pareja, R.; Chen, Y.; Kotomin, E.A.; Popov, A.I. Photoconversion and dynamic hole recycling process in anion vacancies in neutron-irradiated MgO crystals. *Phys. Rev. B* **1999**, *60*, 3787–3791. [CrossRef]
35. Popov, A.I.; Shirmane, L.; Pankratov, V.; Lushchik, A.; Kotlov, A.; Serga, V.E.; Kulikova, L.D.; Chikvaidze, G.; Zimmermann, J. Comparative study of the luminescence properties of macro-and nanocrystalline MgO using synchrotron radiation. *Nucl. Instrum. Methods Phys. Res. Sect. B-Beam Interact. Mater. Atoms* **2013**, *310*, 23–26. [CrossRef]
36. Zhong, Y.; Yang, H.; Wang, M. Thermodynamic evaluation and optimization of LiNO$_3$–KNO$_3$–NaNO$_3$ ternary system. *Calphad* **2020**, *71*, 102202. [CrossRef]
37. Coscia, K.; Elliott, T.; Mohapatra, S.; Oztekin, A.; Neti, S. Binary and Ternary Nitrate Solar Heat Transfer Fluids. *J. Sol. Energy Eng.* **2013**, *135*, 1–6. [CrossRef]
38. Karbovnyk, I.; Bolesta, I.; Rovetskyi, I.; Lesivtsiv, V.; Shmygelsky, Y.; Velgosh, S.; Popov, A.I. Long-term evolution of luminescent properties in CdI$_2$ crystals. *Low Temp. Phys.* **2016**, *42*, 594–596. [CrossRef]

Article

Design of Distributed Bragg Reflectors for Green Light-Emitting Devices Based on Quantum Dots as Emission Layer

Iman E. Shaaban [1], Ahmed S. Samra [1], Shabbir Muhammad [2,3] and Swelm Wageh [4,5,6,*]

[1] Electronics and Communications Department, Faculty of Engineering, Mansoura University, Mansoura 35516, Egypt; iman_mshaban@students.mans.edu.eg (I.E.S.); shmed@mans.edu.eg (A.S.S.)
[2] Research Center for Advanced Materials Science (RCAMS), King Khalid University, P.O. Box 9004, Abha 61413, Saudi Arabia; mshabbir@kku.edu.sa
[3] Department of Chemistry, College of Science, King Khalid University, P.O. Box 9004, Abha 61441, Saudi Arabia
[4] Department of Physics, Faculty of Science, King Abdulaziz University, P.O. Box 80200, Jeddah 21589, Saudi Arabia
[5] K.A.CARE Energy Research and Innovation Center, King Abdulaziz University, P.O. Box 80200, Jeddah 21589, Saudi Arabia
[6] Physics and Engineering Mathematics Department, Faculty of Electronic Engineering, Menoufia University, Menouf 32952, Egypt
* Correspondence: wswelm@kau.edu.sa; Tel.: +966-505-277-380

Abstract: Light-emitting diodes based on quantum dots as an active emission can be considered as a promising next generation for application in displays and lighting. We report a theoretical investigation of green emission at 550 nm of microcavity inorganic–organic light-emitting devices based on Zn (Te, Se) alloy quantum dots as an active layer. Distributed Bragg Reflector (DBR) has been applied as a bottom mirror. The realization of high-quality DBR consisting of both high and low refractive index structures is investigated. The structures applied for high refractive index layers are (ZrO_2, SiN_x, ZnS), while those applied for low index layers are (Zr, SiO_2, CaF_2). DBR of ZnS/CaF_2 consisting of three pairs with a high refractive index step of ($\Delta n = 0.95$) revealed a broad stop bandwidth (178 nm) and achieved a high reflectivity of 0.914.

Keywords: Zn (Te, Se) alloy quantum dot; organic light-emitting devices; green emission; distributed Bragg reflector

1. Introduction

Quantum dots (QDs) have a unique property that originates almost individually due to the size regime in which they exist. The unique optical properties of quantum dots take place because of the quantum confinement effect [1]. The current study focused on wide bandgap II–VI semiconductors quantum dots because of their fundamental structural, electrical, and distinguished optical properties [2].

Zn (Te, Se) ternary alloy QDs are considered one of the essential types of II–VI semiconductors that possess favorable optical properties, as they displayed tunable and narrow-band photoluminescence (PL) emission [3,4].

Zn (Te, Se) is considered a good candidate for Cd-free and green emission materials with a long lifetime. The existence of the green band emission of Zn (Te, Se) partially results from the combination of ZnTe QDs and Te isoelectronic centers in ZnSe. The fact that ZnTe quantum dots as a partial source of green emission was confirmed by the power-dependent PL spectra. In addition, a long lifetime of the PL emission arises from the alignment of the type-II band between ZnSe and ZnTe [5]. The electrons from the conduction band in ZnSe and holes from the valence band in ZnTe were involved in the recombination process, leading to a smaller energy bandgap than that of either ZnTe or ZnSe.

The energy of the bandgap of an alloyed system such as Zn (Te, Se), which consists of materials with considerably numerous chemical features and lattice constants, exhibited a significant negative deviation from the mean of the bandgap mole fraction weighted. Therefore, the Zn (Te, Se) system is named a highly mismatched semiconductor alloy [6]. The band gaps of this alloy possess much smaller energies than their constituents of unalloyed materials. The band gaps of ZnTe and ZnSe, are, respectively, 2.25 and 2.72 [7], while the minimal energy of the bandgap of the Zn (Te, Se) alloy is considered to be 2.03 eV at Zn (Te$_{0.63}$Se$_{0.37}$) [8]. Thus, the green emission of Zn (Te$_{1-x}$Se$_x$) QDs can be realized by controlling their particle size and composition [3].

It is worth mentioning the importance of semiconductor nanocrystals generated from their unique size-dependent optical and electrical properties, which can be utilized in constructing optoelectronic devices [9]. Therefore, semiconductor nanostructures are promising for application as a pure source of monochrome light-emitting diodes [10]. Light-emitting devices based on organic and inorganic structures with electrically excited quantum dots have experienced a large development. This type of configuration becomes competitive to the organic light-emitting devices for the application in displays because of their unique outstanding features of simple solution processability, tunable emission color with high saturation, and high brightness [11,12].

This paper demonstrates the realization of bottom green emission of QD-organic light emitting devices (QD-OLED) by using Distributed Bragg Reflector (DBR) as an optical reflector. The DBR is characterized by having a periodic structure with alternating dielectric layers. Therefore, it can be utilized to afford a high degree of reflection in a certain range of wavelengths by manipulating the thickness and differences in the refractive indexes of the dielectric layers [13]. DBR mirrors have high reflectivity and a small intrinsic absorption coefficient. The high reflectivity of light comes from the constructive interference between the incident light and the reflected light due to Fresnel reflection. The utilization of high-purity dielectric material leads to the production of highly efficient DBR mirrors. It suppresses absorption and controls the thickness of layers during fabrication, leading to obtaining the desired reflection wavelength [14]. According to these features, DBRs can be used in photovoltaic devices [15], vertical-cavity surface-emitting lasers (VCSELs) [16], light-emitting diodes (LEDs) [17], and solar cell actuators [18]. Furthermore, DBR is sensitive toward electric, magnetic, mechanic, and chemical stimuli, giving various characters to be used in different applications [19].

Kitabayashi and co-workers fabricated OLED with ZnS/CaF$_2$ DBR with different pairs and evaluated their reflectance [20]. High-efficiency white OLED with ZrO$_2$/Zr DBR investigated by Yonghua et al. [17]. In addition, Zhang and co-workers fabricated green OLEDs with ZrO$_2$/Zr DBR by atomic layer deposition, and it was found that the ZrO$_2$/Zr DBR structure significantly improves the light purity of green OLEDs without interfering with intrinsic electroluminescence properties [21].

In this work, we have attempted theoretically to compare the performance of green QD-OLED based on different types of DBRs. We have chosen the materials of DBRs with different index contrast to show the effect of this parameter on the performance of the devices. In addition, different pair numbers of DBRs were used to demonstrate the effect of this parameter on the performance of the devices. DBRs have been used as a bottom mirror for light-emitting devices based on organic and inorganic QD structures with green emission to investigate their microcavity effects on device performance. Zn (Te, Se) alloy QD with green emission at a wavelength of 550 nm has been applied as an active layer. The schematic structure of the designed device of QD-OLED with multilayered film at the bottom side are shown in Figure 1a. As shown in the figure, we used indium tin oxide (ITO)/DBR as a reflector anode. N,N′-Di(1-naphthyl)-N,N′-diphenyl-(1,1′-biphenyl)-4,4′-diamine (NPB) was used as the hole transport layer and ZnTeSe alloy QD as an emissive layer. Bis[2-(2-hydroxyphenyl)-pyridine]-beryllium (Bepp2) was used as the electron transport layer. Finally, Lithium quniolate (Liq) was used as an electron injection layer and Al as a cathode.

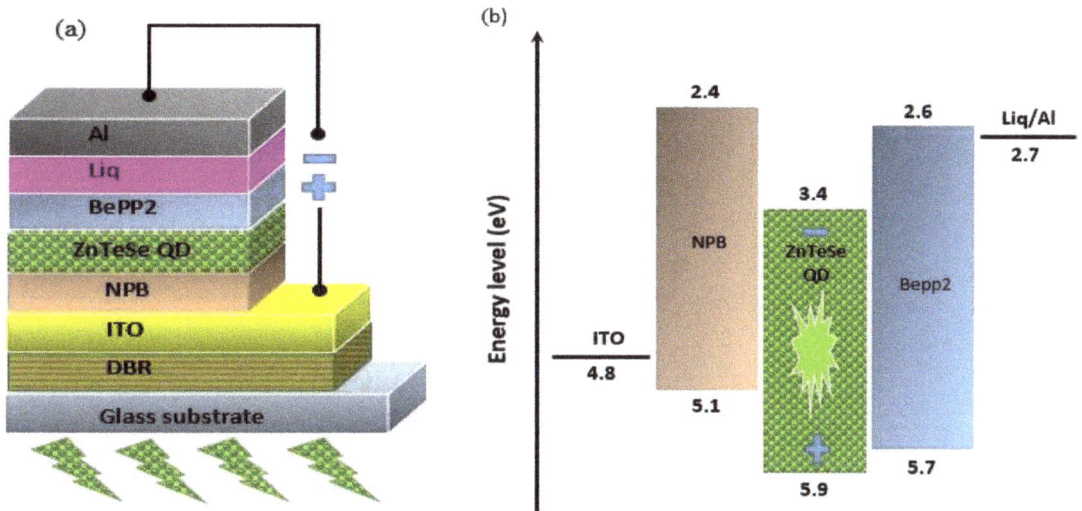

Figure 1. (**a**) Schematic structure of device, (**b**) energy level diagram.

The investigation of light-emitting devices with green emission is very important for many applications, one of them that can be used as a source of illumination for oxygen saturation measurements in blood. Additionally, the green color has an important application in non pharmacological therapy. Recent investigations proved that green light acts as a potential therapy in patients with episodic or chronic migraines with no side effects. It amended the number of headache days/months and improved the quality of life in both episodic and chronic migraine. In addition, the green light provides an additional therapy for the prevention of episodic and chronic migraine [22].

The energy-level diagram of the device is shown in Figure 1b. According to the energy-level diagram, the emission zone is confined to the QD layer due to good energy-level alignments at interfaces of adjacent layers.

2. Theoretical Analysis

2.1. Distributed Bragg Reflector

The most common OLED microcavity architectures incorporate two similar metal mirrors with variant thicknesses; one of them is fractional reflective. Another design includes one mirror with extreme reflection consisting of a dense dielectric distributed Bragg reflector (DBR) and other metal mirrors with low work function [17].

DBRs have tremendous relevance and acceptance in optoelectronic and photonic devices due to their high reflectance and wavelength selectivity compared to metallic mirrors [19]. DBR mirror contained alternating high and low refractive index layers of semiconductor compounds. The thickness of each of the layers is one-quarter wavelength ($\lambda/4$). The reflectivity of DBR is determined by the refractive index contrast and a number of periods [23].

A DBR mirror optical principle is based on successive Fresnel reflection at normal incidence at interfaces between two alternating layers with high and low refractive indices n_h and n_l respectively: $r = \frac{n_h - n_l}{n_h + n_l}$. When each layer's quarter wavelength ($\lambda/4$) optical thickness is maintained, the path difference between reflections from successive interfaces equals half of the wavelength ($\lambda/2$), or 180° out of phase. Despite the reflections (r) at successive interfaces having alternating signs, the 180° compensation phase shift can be obtained through the differences in path length, and all the reflected components interfere constructively. In this case, the cumulative reflection can be enhanced by changing the design. The transmission matrix theory has been used in calculating the reflectance spectrum of a DBR [24].

Reflected light from multiple films is due to the interference from the numerous lights reflected from each of the different surfaces. The interference of light reflected and transmitted by other contact surfaces of multiple film layers is depicted in the scheme shown in Figure 2 [25]. The structure is designed so that all reflected components from the interfaces interfere effectively, leading to a strong reflection. The range of wavelengths that are reflected is termed a photonic stopband. The stopband is controlled mainly by the index contrast of the two materials [19] that can be calculated by [24]:

$$\Delta \lambda_{max} = \frac{4\lambda_B}{\pi} \sin^{-1}\left(\frac{\Delta n}{n_h + n_l}\right) \quad (1)$$

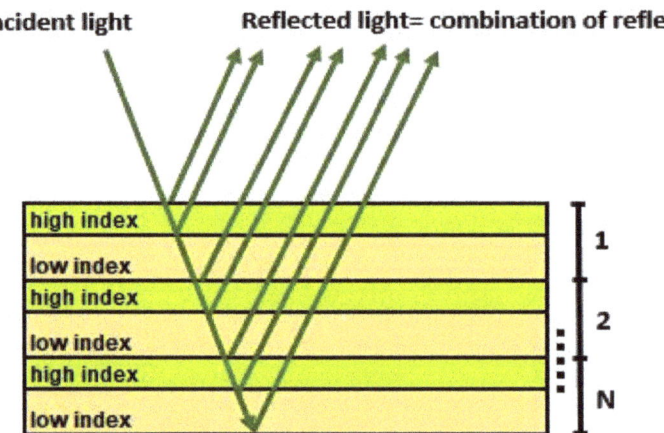

Figure 2. Reflectance of DBR structure.

$\Delta \lambda_{max}$ is proportional to the Bragg wavelength λ_B and is sensitively affected by the index contrast ($\Delta n = n_h - n_l$). Thus, the elevation in Δn is very desirable in DBR fabrication for both a high peak reflectance and a wide stopband width.

2.2. Cavity Emission Characteristics

Based on the concept of interfaces, the most straightforward formula for the emission of light from the thin-film structure, including an emissive layer, can be developed into a method that explains the transmission of a Fabry–Perot resonator structure [26]. In this study, the microcavity was designed as shown in Figure 3 of a top mirror composed of (Bepp2, Liq, and Al) and a bottom mirror is consisting of (NPB, ITO, and DBR), with using ZnTeSe alloy QD as an active emission layer. The theoretical spectrum for external emission normal to the plane of the layers of the device can be calculated based on classical optics by the following equation [27]:

$$|E_{cav}(\lambda)|^2 = \frac{\frac{(1-R_b)}{i}\sum_i\left[1 + R_t + 2(R_t)^{0.5}\cos\left(\frac{4\pi x_i \cos\theta_0}{\lambda} + \Phi_t\right)\right]}{1 + R_t R_b - (R_t R_b)^{0.5}\cos\left(\frac{4\pi L \cos\theta_0}{\lambda} + \Phi_t + \Phi_b\right)} \times |E_{nc}(\lambda)|^2 \quad (2)$$

where R_t and R_b are the reflectivities of the top and bottom mirrors, respectively, L is the optical thickness of the cavity, $|E_{nc}(\lambda)|^2$ is the free space emission intensity at wavelength λ and x_i is the optical thickness between the emitting sublayer and top mirror. θ_0 is the internal observation angle from the surface normal to the microcavity; Φ_t and Φ_b are the phase changes upon reflection to the effective reflectivities of R_t and R_b. According to Equation (3), the resonance condition to maximize luminance from a cavity is given by [26]:

$$\frac{\Phi_t + \Phi_b}{2} - \frac{2\pi L \cos\theta_0}{\lambda} = m\pi \qquad (3)$$

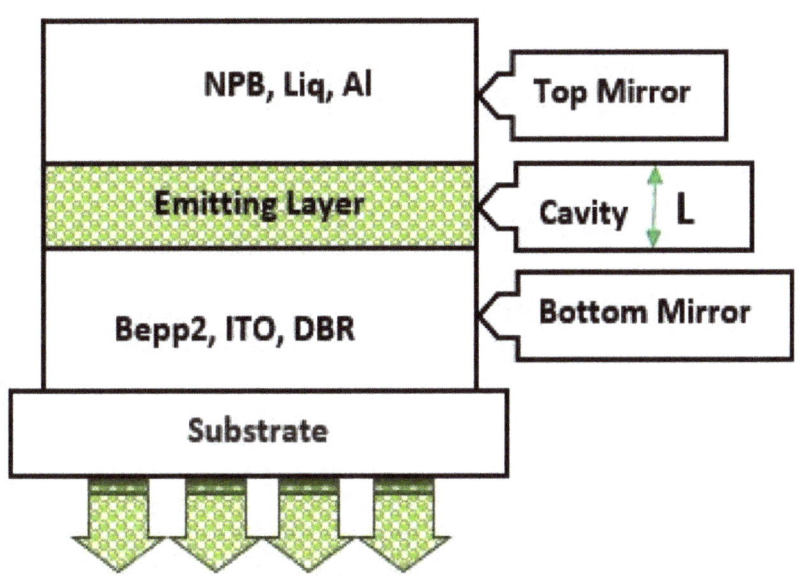

Figure 3. Microcavity structure used for an optical analysis.

A measure of the quality of the resonance of the cavity is given by the finesse [28]. Finesse is described as the number of light oscillations between two mirrors at the free space wavelength (λ) before its energy decays by a factor of $e^{-2\pi}$. A higher finesse exists as a result of a higher average number of times a photon is reflected back and forth within the cavity [29]. The finesse can be written in terms of reflectivity as follows [30]:

$$F = \frac{\pi \sqrt{R_t R_b}}{1 - R_t R_b} \qquad (4)$$

A cavity quality factor Q, which is defined as the reciprocal of the energy loss per cycle per energy stored in the cavity and may also be interpreted as the number of oscillations observed before decay below $e^{-2\pi}$, can be expressed as [31]:

$$Q = \frac{4 R_t R_b}{(1 - R_t R_b)^2} \qquad (5)$$

Cavity photon lifetime τ_p is a time constant that represents the rate at which photons are lost from the cavity and is given by [32],

$$\tau_p = \frac{\frac{2nL}{c}}{1 - R_t R_b} \qquad (6)$$

where c is the speed of light.

Multilayer Calculation

The transfer matrix method (TMM) has been applied to calculate the reflection spectra for our devices; this method inherently included a standing wave enhancement effect and a multi-reflection with the optical cavity [33]. The electric fields, together with the magnetic field inside and outside multilayer structures, have been calculated using TMM, giving us the transmittance and reflectance of this kind of structure.

For homogeneous and isotropic multilayer structures, each layer is represented by a 2 × 2 matrix M_j of the form:

$$M_j = \begin{bmatrix} \cos \delta_j & \frac{(i \sin \delta_j)}{\mu_j} \\ i\mu_j \sin \delta_j & \cos \delta_j \end{bmatrix} \qquad (7)$$

where μ_j is the optical admittance of j layer and δ_j is the phase change of the electromagnetic radiation traversing on the j layer and can be written as follows:

$$\delta_j = \frac{(2\pi n_j d_j \cos \theta_j)}{\lambda} \qquad (8)$$

where n_j is the refractive index of j layer, d_j is the physical thickness of j layer and θ_j is the incidence angle at j layer; herein, a quarter wavelength is considered as the thickness of each layer to obtain the highest reflection [34].

$$d_j = \frac{\lambda}{4n_j} \qquad (9)$$

The matrix relation defining the electric field (B) and magnetic field (C) of the multilayer structure adopted from [35] is as follows:

$$\begin{pmatrix} B \\ C \end{pmatrix} = \prod_{j=1}^{K} M_j \begin{pmatrix} 1 \\ \mu_{sub} \end{pmatrix} \qquad (10)$$

Using Equation (10) and considering the admittance introduced by the interfaces, which is indistinguishable from the reflectance, this idea has been applied for the reflectance calculation through the assembly of thin films. Then, the transmittance has been deduced from the reflectance R through $T = 1 - R$ [36]. The reflectance R, transmittance T and the phase change on reflection Ψ are given by

$$R = \left| \frac{\mu_0 B - C}{\mu_0 B + C} \right|^2 \qquad (11)$$

$$T = \frac{4\mu_0 \mu_{sub}}{|\mu_0 B + C|^2} \qquad (12)$$

$$\Psi = arg \left| \frac{\mu_0 B - C}{\mu_0 B + C} \right| \qquad (13)$$

where μ_0 and μ_{sub} represent the optical admittance of the emission layer and substrate layer, respectively. Mathcad software has been used to compute the data needed for the calculations.

2.3. Quantum Dot Emission

In this work, we have investigated the effect of various structures of DBR bottom mirrors on the performance of emission at 550 nm wavelength. We have adopted an emission profile based on the published experimental results for ZnTeSe quantum dots for internal emission. The emission peak wavelength of the ZnTeSe alloy QD varied in the range from 530 to 579 nm by controlling the particle size in the range from 3.8 to 6.0 nm [3], the PL emission peak wavelength for of $Zn(Te_{1-x}Se_x)$ QDs with x = 0.24 ± 0.04 QDs with various diameters of 4.0, 4.2, 4.9 and 6.0 nm are 535, 540, 557 and 576 nm, respectively. The dependence of the emission peaks on the size of the quantum dots is presented in

Figure 4. From these experimental data, an empirical fitting equation has been reproduced to express the relation between the diameter of QD (d) and the position of the emission peak wavelength $P(d)$:

$$P(d) = -3.584d^2 + 56.433d + 366.446 \tag{14}$$

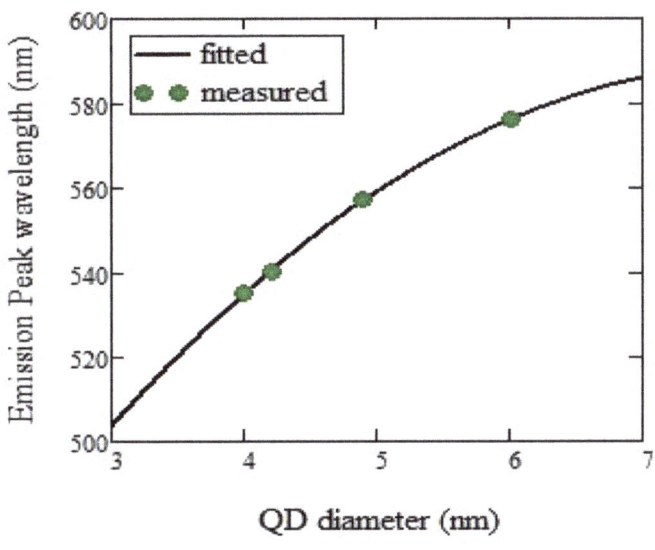

Figure 4. Emission peak positions against QD size.

The QDs spontaneous emission, $E_{int}(\lambda)$, is reproduced by adopting simulation using Gaussian distribution function, as follows:

$$E_{int}(\lambda, d) = \exp\left(\frac{-(\lambda - P(d))^2}{2v^2}\right) \tag{15}$$

where P is the position center of the peak and v was given by:

$$v = \frac{\frac{\Delta\lambda}{2}}{\sqrt{2\ln(2)}} \tag{16}$$

where $\Delta\lambda$ is the full width at half maximum (FWHM) of QD emission. The FWHM of QD emission was assumed to be 30 nm [7]. According to Equation (14), the emission at 550 nm wavelength of ZnTeSe alloy QD is compatible with quantum dots with a diameter of 4.6 nm.

The refractive index and band gap of QD are the main properties that are changed by the size of QD. The refractive index of QD n_{QD} is calculated as follows [37,38]:

$$n_{QD} = \sqrt{1 + \frac{\left[(n_{bulk})^2 - 1\right]}{\left[1 + \left(\frac{0.75}{d}\right)^{1.2}\right]}} \tag{17}$$

where n_{bulk} is the refractive index of bulk material, the refractive index of bulk ZnTe and ZnSe at 550 nm are 3.1 and 2.7, respectively [39]. The calculated refractive index of ZnTeSe QD with diameter 4.6 nm for the emission at 550 nm is 2.748.

The shift of the optical band gap of QD due to quantum confinement has a quantitative form. According to an early effective mass model calculation by Brus, the magnitude of this confinement energy can be modeled as a particle in a box, as seen in Equation (18) [1].

$$E_{confinemet} = \frac{\hbar^2 \pi^2}{2 a^2}\left(\frac{1}{m_e} + \frac{1}{m_h}\right) = \frac{\hbar^2 \pi^2}{2 a^2 \mu^2} \tag{18}$$

where m_e is the effective mass of the electron, m_h is the effective mass of the hole, μ is the reduced mass of the exciton system, and a is the radius of the quantum dot. For Zn (Te$_{0.76}$Se$_{0.24}$), m_e is 0.11 m_0 and m_h is 0.64 m_0 (where m_0 denotes the electron rest mass) [3]. The calculated optical band gap of Zn (Te, Se) alloy QD at 550 nm emission is approximately 2.5 eV, and this value is within the origin of the green band.

3. Results and Discussion

This study focuses on the use of a DBR as a bottom mirror; three DBRs labeled DBR1(ZrO$_2$/Zr), DBR2 (SiN$_x$/SiO$_2$), and DBR3 (ZnS/CaF$_2$) have been applied as bottom mirrors. (Zr, SiO$_2$, CaF$_2$) played as low refractive index materials, while (ZrO$_2$, SiN$_x$, ZnS) were used as high refractive index materials. Schematic-layer structures of ZnTeSe QD-OLED with three periods of DBR1, DBR2, and DBR3 are shown in Figure 5a–c, respectively. The thicknesses of all layers are also displayed in Figure 5. For designing the thickness of each layer in the DBRs, we have applied $L = \frac{\lambda_{Bragg}}{4n}$. According to the Bragg wavelength of 550 nm, and by using the refractive indices shown in Table 1, the thickness of multilayer films can be deduced as follows: the thicknesses of the ZrO$_2$ and Zr of DBR1 are:

$$L(ZrO_2) = \frac{550}{4 \times 2.17} = 63.4 \text{ (nm)} \tag{19}$$

$$L(Zr) = \frac{550}{4 \times 1.62} = 84.8 \text{ (nm)} \tag{20}$$

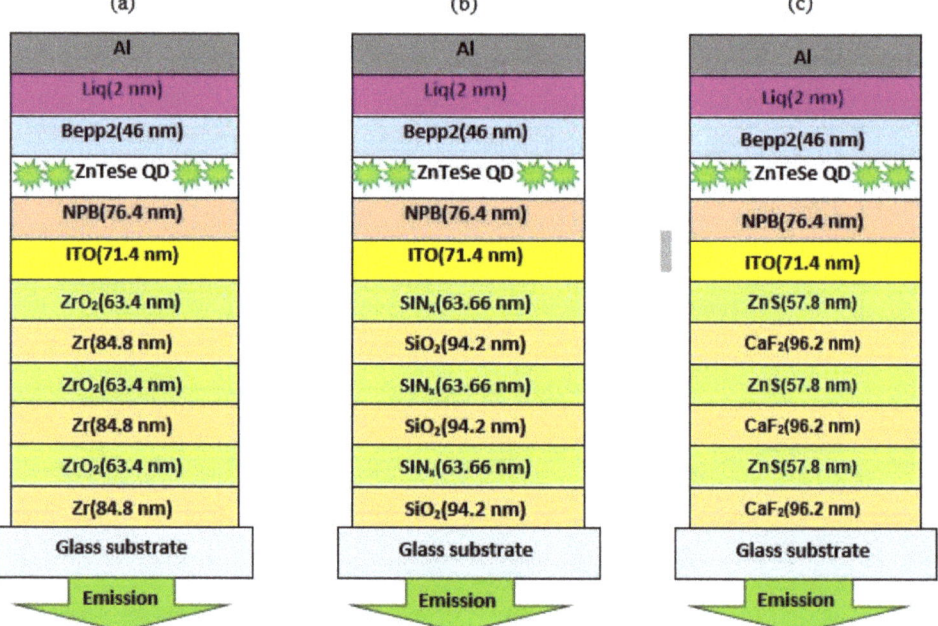

Figure 5. Schematic layer structure of three devices at 550 nm based on (**a**) DBR1 (**b**) DBR2 (**c**) DBR3.

Table 1. Refractive indexes of multilayer films at 550 nm.

Material	Zr	ZrO_2	SiO_2	SiN_x	CaF_2	ZnS
Refractive index (n)	1.62 [21]	2.17 [40]	1.46 [41]	2.16 [42]	1.43 [20]	2.38 [20]

The thicknesses of the SiN_x and SiO_2 of DBR2 are:

$$L(SiN_x) = \frac{550}{4 \times 2.16} = 63.7 \text{ (nm)} \qquad (21)$$

$$L(SiO_2) = \frac{550}{4 \times 1.46} = 94.2 \text{ (nm)} \qquad (22)$$

The thicknesses of the ZnS and CaF_2 of DBR3 are:

$$L(ZnS) = \frac{550}{4 \times 2.38} = 57.8 \text{ (nm)} \qquad (23)$$

$$L(CaF_2) = \frac{550}{4 \times 1.43} = 96.2 \text{ (nm)} \qquad (24)$$

Three structures were designed with a variation of a number of periods of DBR from one to three. The reflectivities calculated by the transfer matrix model and their reflectance are displayed in Figure 6. The refractive index at 550 nm for ITO is 1.9251 + i0.0021684 [43], Al is 0.6 + i5.2745 [44], and the refractive index of organic materials is assumed to be 1.8. The main parameters that affect the performance of DBR are the number of periods (N) and index contrast (Δn). Therefore, the comparison of the performance of devices with emission at 550 nm based on different DBRs by changing (N) and (Δn) is estimated and tabulated in Table 2.

Table 2. Comparison of the main characteristics for devices based on different designs of DBRs at 550 nm.

DBR Design	Index Contrast Δn	No. of Periods (N)	Stop Band Width (nm)	Peak Reflectivity	Finesse	Quality Factor	EL Intensity
DBR1 (ZrO_2/Zr)	0.55	1	101	0.336	2.461	2.459	5.319
		2		0.550	4.338	7.637	8.261
		3		0.718	7.04	20.124	11.711
DBR2 (SiN_x/SiO_2)	0.7	1	136.008	0.412	3.018	3.701	6.228
		2		0.67	6.077	14.965	10.601
		3		0.833	10.67	46.054	14.648
DBR3 (ZnS/CaF_2)	0.95	1	178	0.498	3.784	5.803	7.43
		2		0.779	8.67	30.417	13.279
		3		0.914	15.613	98.816	15.409

The main feature for the effect of variation of DBR periods can be observed in Figure 6. Clearly, the calculated reflectance seems to achieve higher values with the increasing number of DBR periods. In addition, the maximum reflectivity increased for DBR based on ZnS/CaF_2, which has the highest (Δn) relative to DBRs consisting of ZrO_2/Zr and SiN_x/SiO_2.

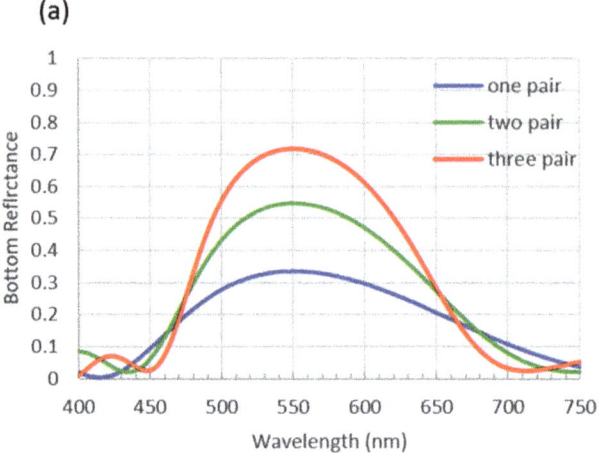
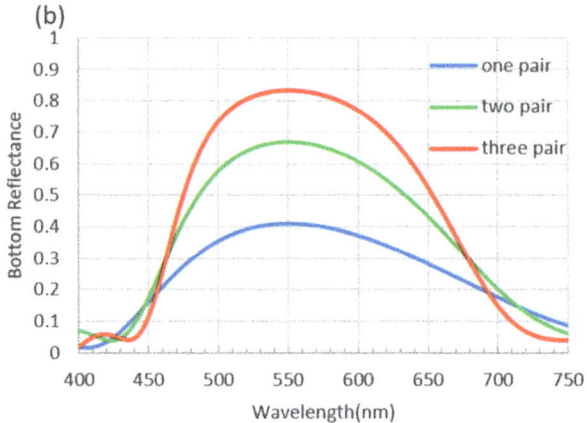
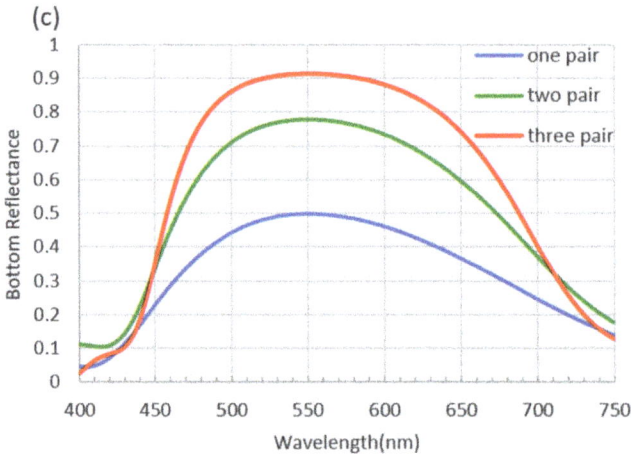

Figure 6. Reflectance spectra by the number of pairs for devices based on (**a**) DBR1 (**b**) DBR2 (**c**) DBR3.

It is worth mentioning that the reflectivity of DBR3 with a number of periods equal to 2 is larger than that for the DBR1 with $n = 3$; this is because of the large differences between the values of index contrast. The values of Δn for DBR1 and DBR3 are 0.55 and 0.95, respectively. Consequently, by using DBR3 consisting of ZnS/CaF$_2$, we can obtain high reflectance and a wide stopband by applying a small number of periods. The decreasing number of periods has many scientific and economic benefits, such as avoiding scattering losses and saving time and materials.

The calculated finesses of DBR structures with a varying number of periods are presented in Table 2. Figure 7 shows the finesse as a function of a number of DBR periods for three configurations applied in this work; the finesse increases with an increasing number of periods for the three devices with various DBR structures; additionally, the highest finesse (15.613) was found for DBR3 with three periods due it is the highest reflectivity of (0.914).

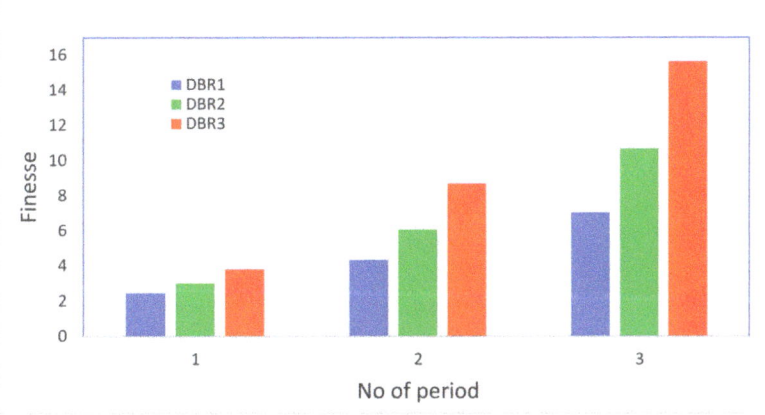

Figure 7. Calculated finesse versus number of DBR periods for devices based on (DBR1, DBR2, and DBR3).

The graph of stopband against index contrast is shown in Figure 8. Obviously, the increase in index contrast leads to an increase in the stopband width, reaching 178. CaF$_2$ clearly exhibits a smaller refractive index than ZnS, leading to a high index contrast ($\Delta n = 0.95$) compared to SiN$_x$/SiO$_2$ ($\Delta n = 0.7$) and ZrO$_2$/Zr ($\Delta n = 0.55$), as reported in Table 2. The enhancement of the stopband width of DBR3 arises from the highest refractive index contrast between high and low refractive index materials ($\Delta n = 0.95$).

The characteristics of the multimode cavity of the device with emissions at 550 nm based on three periods of DBR are discussed. We have investigated different devices based on the three types of DBR with varying cavity modes through changing the cavity length. The cavity lengths of several resonance modes are given by Equation (3); by varying mode index ($m = 1$, $m = 2$ and $m = 3$), the cavity length can be adjusted for resonance mode. Table 3 summarizes the main characteristic parameters of the device with emission at 550 nm for three cavity modes. Clearly, the highest cavity enhancement factor and photon lifetime is obtained for the device with mode index $m = 3$ and attributed to the increase in the cavity length at this mode.

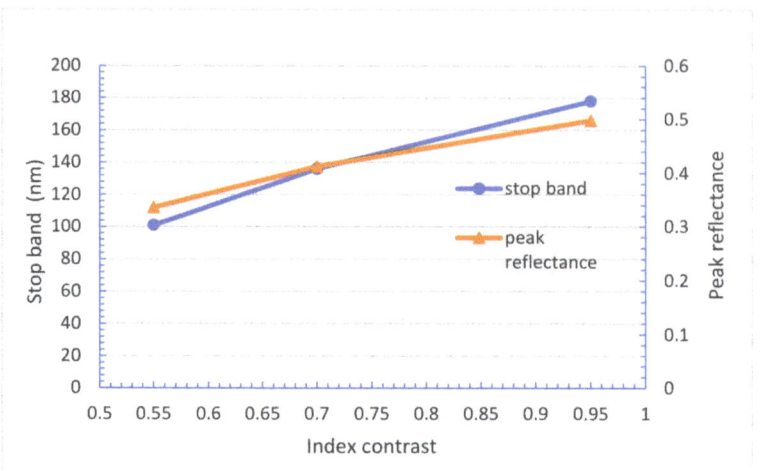

Figure 8. Plots of stopband width and bottom mirror peak reflectance versus refractive index contrast.

Table 3. Cavity characteristics of the device with three periods of DBR3 at 550 nm.

Cavity Mode	Optical Cavity Length	Cavity Enhancement Factor	Photon Life Time (ns)
1st	236	15.4	8.667
2nd	511	16.451	18.74
3rd	786	16.997	28.82

The external emission intensity spectrum of the devices was simulated using Equation (2) with an increasing number of periods of DBR1, DBR2, and DBR3. The obtained intensities of the emission peaks are presented in Table 2. The emission spectra of all devices with different structures of DBR are shown in Figure 9a–c. Clearly, the number of periods causes a pronounced effect on the emission characteristics. The emission spectrum's full width at half maximum (FWHM) decreases, and the peak intensity increases at the resonance wavelength as the number of periods increases. The improvement in the output light intensity of devices with increasing period number is ascribed to the increase in reflectivity. It is worth mentioning that the performance of the device based on DBR3 consisting of ZnS/CaF_2 revealed better performance relative to other devices. Consequently, we selected the device based on DBR3 consisting of ZnS/CaF_2 to investigate the effect of resonance mode on the emission spectra. Furthermore, ZnS/CaF_2 possesses low absorption and a high index of contrast for emission at 550 nm. In addition, future fabrication of ZnS/CaF_2 can be carried out on the substrate at room temperature, which benefits consuming time and avoiding damage to soft materials. The electroluminescence spectra of the device with three cavity modes are shown in Figure 10. Increasing the mode index of the cavity leads to a pronounced improvement in emission intensity. The increase in emission intensity is attributed to the increase in cavity length. Furthermore, the increase in cavity mode leads to the enhancement of the central wavelength at the expense of the other wavelengths and consequently causes a pronounced decrease in the line width.

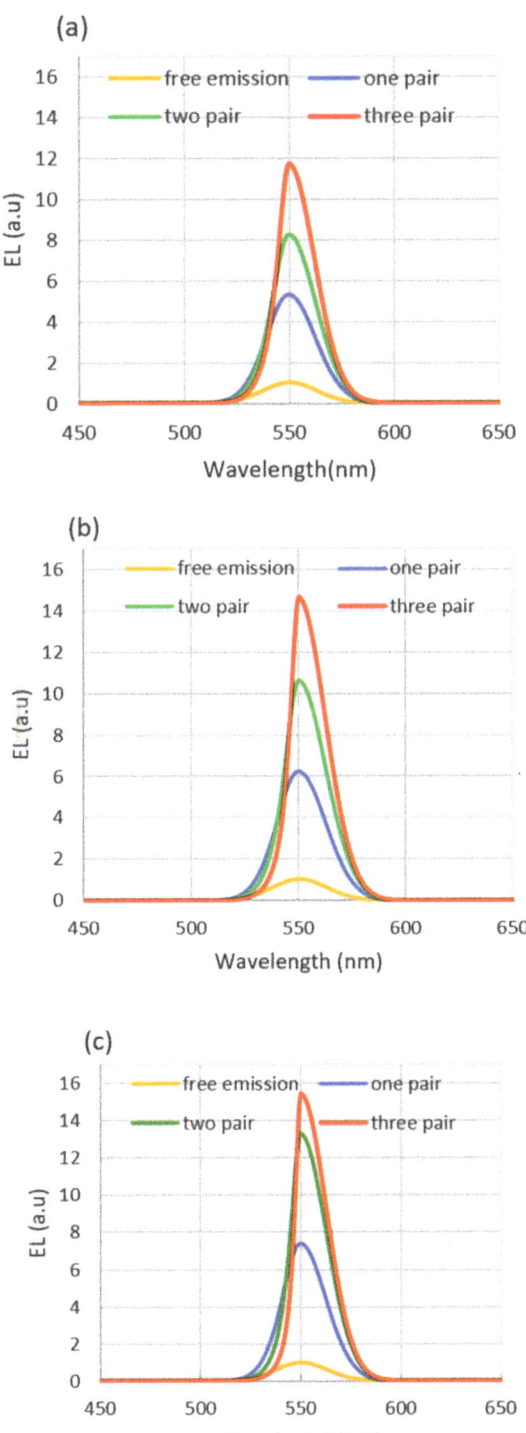

Figure 9. Electroluminance spectra for devices based on (**a**) DBR1 (**b**) DBR2 (**c**) DBR3.

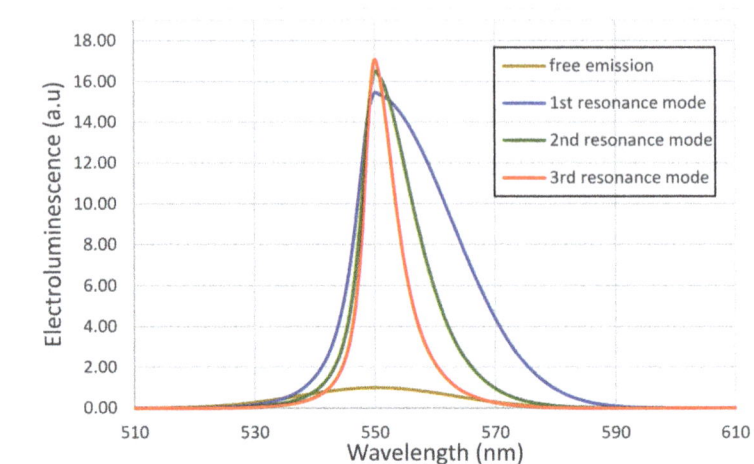

Figure 10. Electroluminescence spectra of device based on DBR3 at three-resonance mode.

To verify the validity of the methodology (TMM) that we used in this work, we applied this method to calculate the bottom reflectance of the device fabricated by Kitabayashi et al. [20]. In this calculation, we have considered the same values of layer thickness in the published work. Figure 11 shows the calculated bottom reflectance spectra of the device fabricated by Kitabayashi et al. [20] with 3, 5, and 7 pairs of ZnS/CaF$_2$ DBR. The comparison between the calculated bottom reflectance and the measured values at a wavelength of 550 nm of the device fabricated by Kitabayashi et al. [20] is shown in Table 4. The results indicated that the calculated bottom reflectances are compatible with the measured experimental results.

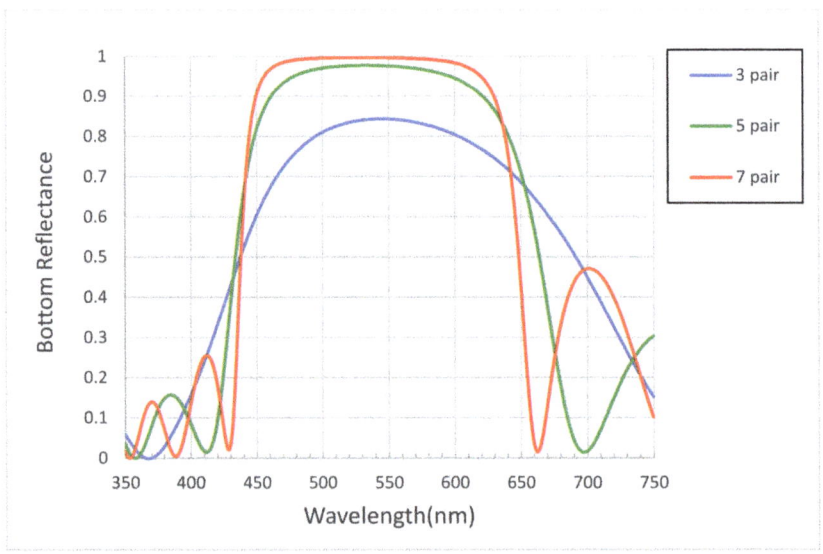

Figure 11. Calculated bottom reflection spectra of the device fabricated by Kitabayashi et al. [20].

Table 4. The measured and calculated bottom reflectance at 550 nm of the device fabricated by Kitabayashi et al. [20] with different pairs of ZnS/CaF$_2$ DBR.

Calculated Bottom Reflectance	Measured Bottom Reflectance [20]	Number of Pairs
0.844	0.817	3
0.976	0.974	5
0.997	0.992	7

4. Conclusions

The external emissions of microcavity light-emitting devices based on Zn (Te, Se) alloy quantum dots as an active layer and organic structures as electron and hole-transporting layers were investigated. The mirrors of the microcavity consist of one mirror with extreme reflection, consisting of a dense dielectric distributed Bragg reflector (DBR), and other metal mirrors with low work functions. The emission characteristics based on the variation in the number of periods of DBR and cavity length were investigated. The increase in the number of periods of DBR or refractive index contrast causes a pronounced increase in reflectivity, which in turn improves the eternal emission of the light-emitting devices. Additionally, the increase in the mode index of the cavity with adjusting cavity length leads to a pronounced improvement in emission intensity.

Author Contributions: Conceptualization, I.E.S., A.S.S. and S.W.; methodology, I.E.S. and S.W.; software, I.E.S. and A.S.S.; formal analysis, I.E.S., A.S.S., S.M. and S.W.; investigation, I.E.S., A.S.S., S.M. and S.W.; data curation, I.E.S., A.S.S., S.M. and S.W.; writing—original draft preparation, I.E.S.; writing—review and editing, I.E.S., A.S.S., S.M. and S.W.; supervision, A.S.S. and S.W.; project administration, S.M.; funding acquisition, S.M. All authors have read and agreed to the published version of the manuscript.

Funding: The authors acknowledge the support and funding of Research Center for Advanced Material Science (RCAMS) at King Khalid University through Grant No. RCAMS/KKU/009-21. Authors acknowledge the support provided by King Abdullah City for Atomic and Renewable Energy (K.A.CARE) under K.A.CARE-King Abdulaziz University Collaboration Program.

Informed Consent Statement: Not applicable.

Data Availability Statement: All data generated or analyzed during this study are included in this article.

Conflicts of Interest: The authors declare no conflict of interest.

References

1. Melville, J.; Kapelewski, M. *Optical Properties of Quantum Dots*; UC Berkeley College of Chemistry: Berkeley, CA, USA, 2015; pp. 2–3.
2. Ibrahim, A.; El-Sayed, N.; Kaid, M.; Ashour, A. Structural and electrical properties of evaporated ZnTe thin films. *Vacuum* **2004**, *75*, 189–194. [CrossRef]
3. Tsukuda, S.; Kita, M.; Omata, T. Zn (Te1 − xSex) quantum dots synthesized through a facile route and their band-edge and surface state driven visible-light emission. *J. Lumin.* **2021**, *231*, 117829. [CrossRef]
4. Wageh, S. Ternary ZnS: Te nanoparticles capped with 3-mercaptopropionic acid prepared in aqueous media. *J. Mater. Sci. Mater. Electron.* **2016**, *27*, 10877–10887. [CrossRef]
5. Chang, L.; Cheng, J.; Hsu, C.; Chao, H.; Li, W.; Chang, Y.; Chen, K.; Chen, Y.; Laing, C.-T. Isoelectronic centers and type-II quantum dots: Mechanisms for the green band emission in ZnSeTe alloy. *J. Appl. Phys.* **2009**, *105*, 113511. [CrossRef]
6. Asano, H.; Arai, K.; Kita, M.; Omata, T. Synthesis of colloidal Zn (Te, Se) alloy quantum dots. *Mater. Res. Express.* **2017**, *4*, 106501. [CrossRef]
7. Asano, H.; Tsukuda, S.; Kita, M.; Fujimoto, S.; Omata, T. Colloidal Zn (Te, Se)/ZnS core/shell quantum dots exhibiting narrow-band and green photoluminescence. *ACS Omega* **2018**, *3*, 6703–6709. [CrossRef] [PubMed]
8. Asano, H.; Omata, T. Design of cadmium-free colloidal II–VI semiconductor quantum dots exhibiting RGB emission. *AIP Adv.* **2017**, *7*, 045309. [CrossRef]

9. Patra, S.K.; Bhushan, B.; Priyam, A. Water-soluble, luminescent ZnTe quantum dots: Supersaturation-controlled synthesis and self-assembly into nanoballs, nanonecklaces and nanowires. *Dalton Trans.* **2016**, *45*, 3918–3926. [CrossRef] [PubMed]
10. Adachi, S. Optical Constants of Crystalline and Amorphous Semiconductors. In *Numerical Data and Graphical Information*; Kluwer Academic Publishers: Boston, MA, USA, 1999.
11. Zhang, H.; Su, Q.; Chen, S. Quantum-dot and organic hybrid tandem light-emitting diodes with multi-functionality of full-color-tunability and white-light-emission. *Nat. Commun.* **2020**, *11*, 1–8. [CrossRef] [PubMed]
12. Wageh, S. Light emitting devices based on CdSe nanoparticles capped with mercaptoacetic acid. *IEEE J. Quantum Electron.* **2014**, *50*, 1–8. [CrossRef]
13. Prontera, C.T.; Pugliese, M.; Giannuzzi, R.; Carallo, S.; Esposito, M.; Gigli, G.; Maiorano, V. Flexible distributed Bragg reflectors as optical outcouplers for OLEDs based on a polymeric anode. *J. Inf. Disp.* **2021**, *22*, 39–47. [CrossRef]
14. Assafli, H.T.; Abdulhadi, A.H.; Nassir, W.Y. Design High Efficient Reflectivity of Distributed Bragg Reflectors. *Iraqi J. Laser* **2016**, *15*, 15–20.
15. Dong, W.J.; Lo, N.-T.; Jung, G.H.; Ham, J.; Lee, J.-L. Efficiency enhancement and angle-dependent color change in see-through organic photovoltaics using distributed Bragg reflectors. *Appl. Phys. Lett.* **2016**, *108*, 103902. [CrossRef]
16. Delbeke, D.; Bockstaele, R.; Bienstman, P.; Baets, R.; Benisty, H. High-efficiency semiconductor resonant-cavity light-emitting diodes: A review. *IEEE J. Sel. Top. Quantum Electron.* **2002**, *8*, 189–206. [CrossRef]
17. Wu, Y.; Yang, J.; Wang, S.; Ling, Z.; Zhang, H.; Wei, B. High-performance white organic light-emitting diodes using distributed Bragg reflector by atomic layer deposition. *Appl. Sci.* **2019**, *9*, 1415. [CrossRef]
18. Tan, X.; Tu, Y.; Deng, C.; von Czarnowski, A.; Yan, W.; Ye, M.; Yi, Y. Enhancement of light trapping for ultrathin crystalline silicon solar cells. *Opt. Commun.* **2018**, *426*, 584–588. [CrossRef]
19. Lohithakshan, L.C.; Geetha, V.; Kannan, P. Single polymer-variable index for the design and fabrication of variable stop band distributed Bragg reflectors. *Opt. Mater.* **2020**, *110*, 110509. [CrossRef]
20. Kitabayashi, T.; Asashita, T.; Satoh, N.; Kiba, T.; Kawamura, M.; Abe, Y.; Kim, K.H. Fabrication and characterization of microcavity organic light-emitting diode with CaF2/ZnS distributed Bragg reflector. *Thin Solid Films* **2020**, *699*, 137912. [CrossRef]
21. Zhang, J.; Zhang, H.; Zheng, Y.; Wei, M.; Ding, H.; Wei, B.; Zhang, Z. Super color purity green organic light-emitting diodes with ZrO2/zircone nanolaminates as a distributed Bragg reflector deposited by atomic layer deposition. *Nanotechnology* **2016**, *28*, 044002. [CrossRef]
22. Martin, L.F.; Patwardhan, A.M.; Jain, S.V.; Salloum, M.M.; Freeman, J.; Khanna, R.; Gannala, P.; Goel, V.; Jones-MacFarland, F.N.; Killgore, W.D. Evaluation of green light exposure on headache frequency and quality of life in migraine patients: A preliminary one-way cross-over clinical trial. *Cephalalgia* **2021**, *41*, 135–147. [CrossRef] [PubMed]
23. Manaf, N.A.A.; Alias, M.S.; Mithani, S.M.; Maulud, M.F.; Yahya, M.R.; Mat, A.F.A. Design and optimization of distributed Bragg reflector for 1310nm vertical cavity surface emitting lasers. In Proceedings of the 2008 IEEE International Conference on Semiconductor Electronics ICSE, Johor Bahru, Malaysia, 25–27 November 2008; pp. 254–258.
24. Zhang, C.; ElAfandy, R.; Han, J. Distributed Bragg reflectors for GaN-based vertical-cavity surface-emitting lasers. *Appl. Sci.* **2019**, *9*, 1593. [CrossRef]
25. Sharhan, A.A. Transfer Matrix Mathematical Method for Evaluation the DBR Mirror for Light Emitting Diode and Laser. *J. Phys. Conf. Ser.* **2020**, *1535*, 1–7. [CrossRef]
26. Poitras, D.; Kuo, C.-C.; Py, C. Design of high-contrast OLEDs with microcavity effect. *Opt. Express* **2008**, *16*, 8003–8015. [CrossRef] [PubMed]
27. Chen, S.; Deng, L.; Xie, J.; Peng, L.; Xie, L.; Fan, Q.; Huang, W. Recent developments in top-emitting organic light-emitting diodes. *Adv. Mater.* **2010**, *22*, 5227–5239. [CrossRef]
28. Varu, H. The Optical Modelling and Design of Fabry Perot Interferometer Sensors for Ultrasound Detection. Ph.D. Thesis, University College London, London, UK, 2014.
29. Tadayon, M.A.; Baylor, M.-E.; Ashkenazi, S. High quality factor polymeric Fabry-Perot resonators utilizing a polymer waveguide. *Opt. Express* **2014**, *22*, 5904–5912. [CrossRef] [PubMed]
30. Thomschke, M.; Nitsche, R.; Furno, M.; Leo, K. Optimized efficiency and angular emission characteristics of white top-emitting organic electroluminescent diodes. *Appl. Phys. Lett.* **2009**, *94*, 59. [CrossRef]
31. Hadjaj, F.; Belghachi, A.; Halmaoui, A.; Belhadj, M.; Mazouz, H. Study of a Fabry-Perot resonator. *Int. J. Phys. Math. Sci.* **2013**, *7*, 1713–1717.
32. Darrow, M.C. Finesse Measurement in Fabry-Perot Interferometers. *Macalester J. Phys. Astron.* **2014**, *2*, 3.
33. Gryshchenko, S.; Demin, A.; Lysak, V. Quantum efficiency and reflection in resonant cavity photodetector with anomalous dispersion mirror. In Proceedings of the 2008 4th International Conference on Advanced Optoelectronics and Lasers IEEE, Alushta, Ukraine, 29 September–4 October 2008; pp. 229–232.
34. Coldren, L.A.; Corzine, S.W. *Diode Lasers and Photonic Integrated Circuits*; John Wiley & Sons Inc: New York, NY, USA, 1995.
35. Yahia, K.Z. Simulation of multilayer layer antireflection coating for visible and near IR region on silicon substrate using MATLAB program. *Al-Nahrain J. Sci.* **2009**, *12*, 97–103.
36. Born, M.; Wolf, E. *Principles of Optics: Electromagnetic Theory of Propagation, Interference and Diffraction of Light*; Elsevier: Amsterdam, The Netherlands, 2013.

37. Shaaban, I.E.; Samra, A.S.; Yousif, B.; Alghamdi, N.; El-Sherbiny, S.; Wageh, S. Cavity Design and Optimization of Hybrid Quantum Dot Organic Light Emitting Devices for Blue Light Emission. *J. Nanoelectron. Optoelectron.* **2020**, *15*, 1364–1373. [CrossRef]
38. Wang, L.-W.; Zunger, A. Pseudopotential calculations of nanoscale CdSe quantum dots. *Phys. Rev. B* **1996**, *53*, 9579. [CrossRef]
39. Li, H. Refractive index of ZnS, ZnSe, and ZnTe and its wavelength and temperature derivatives. *J. Phys. Chem. Ref. Data* **1984**, *13*, 103–150. [CrossRef]
40. Wood, D.L.; Nassau, K. Refractive index of cubic zirconia stabilized with yttria. *Appl. Opt.* **1982**, *21*, 2978–2981. [CrossRef] [PubMed]
41. Rodríguez-de Marcos, L.V.; Larruquert, J.I.; Méndez, J.A.; Aznárez, J.A. Self-consistent optical constants of SiO_2 and Ta_2O_5 films. *Opt. Mater. Express* **2016**, *6*, 3622–3637. [CrossRef]
42. Duttagupta, S.; Ma, F.; Hoex, B.; Mueller, T.; Aberle, A.G. Optimised antireflection coatings using silicon nitride on textured silicon surfaces based on measurements and multidimensional modelling. *Energy Procedia* **2012**, *15*, 78–83. [CrossRef]
43. Cheng, F.; Su, P.-H.; Choi, J.; Gwo, S.; Li, X.; Shih, C.-K. Epitaxial growth of atomically smooth aluminum on silicon and its intrinsic optical properties. *ACS Nano* **2016**, *10*, 9852–9860. [CrossRef] [PubMed]
44. Moerland, R.J.; Hoogenboom, J.P. Subnanometer-accuracy optical distance ruler based on fluorescence quenching by transparent conductors. *Optica* **2016**, *3*, 112–117. [CrossRef]

Article

Clean Electrochemical Synthesis of Pd–Pt Bimetallic Dendrites with High Electrocatalytic Performance for the Oxidation of Formic Acid

Jie Liu [1], Fangchao Li [1], Cheng Zhong [1,2,3,*] and Wenbin Hu [1,2,3]

[1] Key Laboratory of Advanced Ceramics and Machining Technology (Ministry of Education), School of Materials Science and Engineering, Tianjin University, Tianjin 300072, China; jieliu0109@tju.edu.cn (J.L.); lfc2018208170@tju.edu.cn (F.L.); wbhu@tju.edu.cn (W.H.)
[2] Tianjin Key Laboratory of Composite and Functional Materials, School of Materials Science and Engineering, Tianjin University, Tianjin 300072, China
[3] Joint School of National University of Singapore and Tianjin University, International Campus of Tianjin University, Binhai New City, Fuzhou 350207, China
* Correspondence: cheng.zhong@tju.edu.cn

Abstract: Pd–Pt bimetallic catalysts with a dendritic morphology were in situ synthesized on the surface of a carbon paper via the facile and surfactant-free two step electrochemical method. The effects of the frequency and modification time of the periodic square-wave potential (PSWP) on the morphology of the Pd–Pt bimetallic catalysts were investigated. The obtained Pd–Pt bimetallic catalysts with a dendritic morphology displayed an enhanced catalytic activity of 0.77 A mg^{-1}, almost 2.5 times that of the commercial Pd/C catalyst reported in the literature (0.31 A mg^{-1}) in acidic media. The enhanced catalytic activity of the Pd–Pt bimetallic catalysts with a dendritic morphology towards formic acid oxidation reaction (FAOR) was not only attributed to the large number of atomic defects at the edges of dendrites, but also ascribed to the high utilization of active sites resulting from the "clean" electrochemical preparation method. Besides, during chronoamperometric testing, the current density of the dendritic Pd–Pt bimetallic catalysts for a period of 3000 s was 0.08 A mg^{-1}, even four times that of the commercial Pd/C catalyst reported in the literature (about 0.02 A mg^{-1}).

Keywords: electrochemical synthesis; Pd–Pt dendrites; formic acid oxidation; high electrocatalytic performance

1. Introduction

Noble metals are widely used in various important catalytic or electrocatalytic reactions in the fields of energy conversion/storage as well as water pollution such as hydrogen production [1], reduction of oxygen [2], oxidation of organic or inorganic small molecule [3], fuel cells [4], and degradation of organic dyes in water [5]. Among these important applications, fuel cells have been receiving extensive attention due to their ability to directly convert the chemical energy of small-molecule fuel oxidation into electricity [6]. Direct formic acid fuel cells (DFAFCs) are believed to be a promising power generation system owing to their reasonable power density, high electromotive force, and limited fuel crossover [7,8]. As a key component of DFAFCs, electrocatalysts towards formic acid oxidation reaction (FAOR) play a vital role in the development of DFAFCs. Recently, noble metal palladium (Pd) has drawn intensive attention due to its high catalytic activity for FAOR and better resistance to CO poisoning than Pt [9–12]. However, the Pd catalyst is prone to dissolution in acidic media during catalytic reactions, hampering its commercial application [13]. In addition, there is still much room for improvement in the catalytic activity of the Pd catalyst. Therefore, it is particularly eager to develop a highly active and durable catalyst towards FAOR.

It is widely accepted the oxidation of formic acid involves two mechanisms, i.e., the direct oxidation pathway (HCOOH → CO_2 + $2H^+$ + $2e^-$) and the CO oxidation pathway (HCOOH → CO_{ads} + H_2O → CO_2 + $2H^+$ + $2e^-$) [14]. In the direct oxidation pathway, HCOOH molecules directly dehydrogenate to form CO_2 via one or more active intermediates. In the CO oxidation pathway, HCOOH molecules dehydrate to produce CO which depends on the applied potential. The generated CO may be further oxidized to CO_2 or poison the catalyst. Up to date, several strategies have been developed to improve the electrocatalytic properties of Pd. The previous literature has reported that alloys of Pd with other metals (such as Cu [9], Pt [15], Co [16], Ag [17], Au [18], Sn [19], Ni [20], Rh [21], Zn [22], Pb [23], Cr [24], and Ir [25]) can not only enhance catalytic activity, but also improve the corrosion resistance of Pd. In particular in Pd–Pt alloys, the electronegativities and bulk Wigner–Seitz radii of Pt and Pd are similar [26]. The alloying of Pt and Pd will produce a synergistic electronic effect [27]. This effect favors formic acid oxidation via the direct oxidation pathway. Additionally, Pt also exhibits high stability in acidic media due to its chemically inert property [28].

It is generally accepted the morphology of catalysts plays a crucial role in enhancing the catalytic activity. A great deal of work has focused on the synthesis of bimetallic Pt–Pd catalysts with various morphologies. For instance, Guo et al. [15] synthesized three-dimensional (3D) dendritic Pt-on-Pd bimetals on graphene sheets via a facile wet-chemical approach and found that Pt–Pd bimetallic nanodendrites/graphene hybrids exhibited high catalytic activity for the oxidation of methanol. Yuan et al. [29] prepared Pd–Pt random alloy nanocubes in an aqueous solution containing KBr, polyvinyl pyrrolidone, and sodium lauryl sulfate with $PdCl_2$ and K_2PtCl_6 as precursors. Zhang et al. [30] reported different shapes of Pd–Pt alloys by a solvothermal process, such as flowers, dendrites, bars, cubes, and concave cubes, and concluded that the obtained Pd–Pt alloys had an enhanced catalytic activity and CO tolerance towards FAOR. Lu et al. [31] fabricated a reduced graphene oxide/Pt–Pd alloy nanocubes by a facile hydrothermal method. However, the synthesis of most Pd–Pt catalysts with various morphologies commonly uses organic additives, thus requiring additional removal processes of additives. Otherwise, incomplete removal of organic additives limits the active sites of the catalyst and thus negatively affects its performance. Additionally, during conventional electrode preparation, the powder catalysts have to be transferred to the surface of the electrode, which inevitably requires the use of conductive additives and binders. The use of conductive additives and binders will lead to a reduction in the active sites of the catalyst. Therefore, it is highly desired to develop a facile and surfactant-free route to synthesize Pd–Pt nanocatalysts. The electrochemical synthesis method is considered to be an effective catalyst preparation method because of its simplicity, low cost, easy operation, and high purity of the product [32]. However, conventional electrochemical synthesis techniques have limited control parameters, making it difficult to effectively control the morphology of the catalyst. The combination of different modes of electrochemical technologies can effectively expand the range of its control parameters and is an effective method to realize the control of the catalyst morphology. For example, Tian et al. [33,34] synthesized monometallic Pt and Pd tetrahexahedral nanocrystals by the electrochemical technology, which showed a high electrocatalytic activity for the oxidation of formic acid and ethanol. The dendritic Pt and highly dispersed Pd particles also synthesized by a similar electrochemical method and displayed a high electrocatalytic activity [35,36]. However, there are few reports on the synthesis of morphology-controlled Pd–Pt bimetallic catalysts via a clean electrochemical approach. The research on the effect of electrochemical parameters on the structure and morphology of Pd–Pt bimetallic catalysts is very limited. The intrinsic relationship between the microstructure of Pd–Pt bimetallic catalysts and its macroscopic catalytic performance also needs to be further illustrated.

Herein, Pd–Pt bimetallic catalysts with a dendritic morphology were in situ synthesized on the surface of a carbon paper by a facile and clean electrochemical method. Pure Pd particles were firstly electrodeposited on the surface of the carbon paper, and then Pd–Pt bimetallic catalysts with a dendritic morphology were obtained by the periodic square-wave potential (PSWP) treatment of pure Pd particles in an aqueous solution of 0.5 M H_2SO_4 and 5 mM $PdCl_2$. The effects of the frequency and treatment time of the PSWP on the morphology of the bimetallic Pd–Pt catalysts were systematically investigated. The obtained Pd–Pt bimetallic catalyst with a dendritic morphology displayed an outstanding catalytic activity (0.77 A mg^{-1}) and a high stability towards FAOR.

2. Experimental

2.1. Reagents and Materials

Palladium (II) chloride ($PdCl_2$; Adamas Reagent Co., Ltd., Shanghai, China), chloroplatinic acid hexahydrate ($H_2PtCl_6 \cdot 6H_2O$; Shanghai Aladdin Biochemical Technology Co., Ltd., Shanghai, China), sulfuric acid (H_2SO_4; Jiangtian Chemical Technology Co., Ltd., Tianjin, China), formic acid, and anhydrous ethanol (Yuanli Chemical Technology Co., Ltd., Tianjin, China) were of analytical reagent grade. A carbon paper (TGP-H-060) and commercial 10 wt % Pd/C catalysts were purchased from Toray (Toray Industries, Tokyo, Japan) and Shanghai Aladdin Biochemical Technology Co., Ltd. (Shanghai, China), respectively. The deionized water was produced with a Millipore Milli-Q system (18.2 MΩ cm).

2.2. Electrochemical Fabrication of Pd Particles and Pd–Pt Bimetallic Catalysts

Firstly, prior to the electrodeposition of Pd particles, the carbon paper was ultrasonically cleaned with anhydrous ethanol, acetone, and deionized water for 30 min separately. Secondly, Pd particles was electrochemically deposited on the surface of the carbon paper at -0.15 mA cm^{-2} for 45 min in an aqueous solution of 5 mM $PdCl_2$ and 0.5 M H_2SO_4, using a conventional electrochemical cell with a three-electrode system (Ivium Stat, Ivium Technologies, Eindhoven, Netherlands). The cleaned carbon paper served as the working electrode. A platinum plate electrode and a mercurous sulfate electrode (MSE) were used as the counter electrode and the reference electrode, respectively. At last, based on the obtained pure Pd particles, a Pd–Pt bimetallic catalysts/carbon paper electrode was finally synthesized in a two-electrode system by PSWP (NF BP4610, NF Corporation, Yokohama Japan) with different frequencies (10 Hz, 50 Hz, and 90 Hz) and modification times (1 h, 2 h, and 4 h) in an aqueous solution of 0.1 mM H_2PtCl_6 and 1 M H_2SO_4. Correspondingly, the obtained Pd–Pt samples were denoted as PdPts–10 Hz, PdPts–50 Hz (4 H), PdPts–90 Hz, PdPts–1 H, and PdPts–2 H, respectively. The upper and lower limit potentials of the PSWP were 0.6 V and -3.2 V, respectively.

2.3. Characterization of Pd Particles and Pd–Pt Bimetallic Catalysts

The surface morphologies of the Pd particles and the Pt−Pd bimetallic catalysts were investigated by a field-emission scanning electron microscope (Hitachi S-4800, Hitachi, Tokyo, Japan). The structure and composition of the Pt−Pd bimetallic catalysts were investigated by a transmission electron microscope (JEOL 2100F, JELO Ltd., Tokyo Japan) equipped with energy-dispersive X-ray (EDX) analysis. The phase and the crystallinity of the Pd–Pt bimetallic catalysts were analyzed by X-ray diffraction (XRD; Bruker D8 Advanced, Billerica, MA, USA) with CuK_α radiation ($\lambda = 1.5418$ Å). The loading amounts of Pd and Pt particles were obtained by inductively coupled plasma-mass spectrometry (ICP-MS; Agilent 7700, Agilent Technologies, Santa Clara, CA, US).

2.4. Electrochemical Test

All electrochemical tests were performed in a conventional three-electrode system (Ivium Stat, Ivium Technologies, Eindhoven, Netherlands). The obtained Pd–Pt bimetallic catalyst/carbon paper electrode was used as the working electrode. A platinum plate and a saturated calomel electrode (SCE) served as the counter and the reference electrodes, respectively. The cyclic voltammetry (CV) curves of the obtained Pd–Pt bimetallic catalysts were tested in a N_2-saturated aqueous solution of 0.5 M H_2SO_4 at a scan rate of 50 mV s^{-1}. The voltammetric curves of the obtained Pd–Pt bimetallic catalysts towards FAOR were recorded in an aqueous solution of 0.5 M H_2SO_4 and 0.5 M HCOOH at a scan rate of

50 mV s^{-1}. The chronoamperometry curves of the obtained Pd–Pt bimetallic catalysts were recorded in an aqueous solution of 0.5 M H_2SO_4 and 0.5 M HCOOH at 0.15 V for 3000 s.

3. Results and Discussion

Figure 1 displays the SEM images of the pure Pd particles and the Pd–Pt bimetallic catalysts with various morphologies fabricated by PSWP modification with different frequencies in an aqueous solution containing a Pt precursor. The pure Pd particles were electrodeposited on the surface of the carbon paper (Figure 1a), and the average particle size was about 270 nm (Figure 1b). A large number of Pd–Pt particles with a relatively smooth surface were electrodeposited on the surface of the carbon paper, after the pure Pd particles were treated by the PSWP with a frequency of 90 Hz (Figure 1c,d). The size of the Pd–Pt particles was similar to that of the pure Pd particles (Figure 1d). When the modification frequency of the PSWP decreased to 50 Hz, the obtained Pd–Pt bimetallic catalysts showed a well-defined dendritic morphology (Figure 1e). A lot of long secondary dendrites grew on the trunk of the dendrite of the Pd–Pt bimetallic catalysts, and the short tertiary dendrites were also formed on the secondary dendrite arms (Figure 1e,f). As the frequency of the PSWP further reduced to 10 Hz, the Pd–Pt bimetallic catalysts showed an agglomerated morphology with a rough surface (Figure 1g,h). Obviously, the frequency of the PSWP had a significant effect on the morphology of the Pd–Pt bimetallic catalysts. This may be related to the process of adsorption/desorption of oxygen on the surface of Pd and Pt metals during PSWP modification [37]. The oxidation processes of the Pd and Pt surfaces were affected by the upper limit potential of the PSWP, and the dynamic adsorption/desorption process of oxygen on the Pd and Pt surfaces affected the dissolution of Pd and Pt. The lower limit potential of the PSWP was responsible for the deposition of Pd and Pt. When the frequency of the PSWP was high (90 Hz), the square-wave period of the Pd particle surface was relatively short, and thus the modification effect on the surface morphology of Pd particles was limited. As a result, many Pd–Pt particles with a relatively smooth surface were formed (Figure 1c,d), and their morphology was similar to that of pure Pd particles (Figure 1a,b). As the frequency of the PSWP decreased to 50 Hz, the square-wave period increased, resulting in a long period of the dissolution and deposition of metal atoms on the surface of Pd particles. The long-period deposition process caused the diffusion-limited growth of metal ions near the electrode surface. These metal ions tended to diffuse towards the tip of the electrode surface, which finally led to the formation of Pd–Pt bimetallic catalysts with a dendritic morphology (Figure 1e,f). When the frequency of the PSWP reduced to the smallest frequency, i.e., 10 Hz, the PSWP period was longer. The formed dendrites were dissolved, since the dissolution caused by the upper limit potential played a dominant role during the long square-wave period. Consequently, the Pd–Pt bimetallic catalysts displayed an agglomerated morphology with a rough surface (Figure 1g,h).

To further gain insight into the formation mechanism of the Pd–Pt bimetallic catalysts with a dendritic morphology, the surface morphology of the Pd–Pt bimetallic catalysts as a function of the modification time of the PSWP was investigated. Figure 2 shows the SEM images of the Pd–Pt bimetallic catalysts obtained by the PSWP modification with different times. As the modification time increased from 1 h to 4 h, the morphology of the modified Pd–Pt bimetallic catalysts evolved from the agglomerated particles (Figure 2a,b), leaf-like catalysts (Figure 2c,d) to dendritic catalysts (Figure 1e,f). During modification, in the early stage (modification time of 1 h) of the formation of the Pd–Pt bimetallic catalysts, the short modification time had limited effect on the morphology of the Pd–Pt bimetallic catalysts. Only the Pd–Pt bimetallic catalysts with an agglomerated morphology were observed (Figure 2a,b). When the modification time was extended to 2 h, the diffusion of metal ions near the electrode surface was limited, and the metal ions diffused toward the tip of the electrode surface during the deposition process. As a result, the leaf-like catalysts with a protruding texture were formed (Figure 2c,d). As the modification time was further extended to 4 h, the leaf-like catalysts were selectively dissolved, only leaving the protruding texture part. Finally, the Pd–Pt bimetallic catalysts displayed a well-defined dendritic morphology (Figure 1e,f).

Figure 1. SEM images of pure Pd particles (**a**) and the obtained Pd–Pt bimetallic catalysts modified by the periodic square-wave potential (PSWP) with the frequencies of 90 Hz (**c**), 50 Hz (**e**), 10 Hz (**g**) for 4 h in a solution of 0.1 mM H_2PtCl_6 + 1 M H_2SO_4. (**b,d,f,h**) are the high-magnification SEM images of the corresponding catalysts.

Figure 3 displays the XRD spectra of the Pd–Pt catalysts. All the catalysts showed the distinct characteristic diffraction peaks at about 40.1°, 46.6°, 67.9°, 81.7°, and 86.7°, which corresponded to the (111), (200), (220), (311), and (222) lattice planes of the Pd–Pt bimetallic catalysts, respectively (Figure 3a) [38]. This indicated that the obtained Pd–Pt bimetallic catalysts possessed a polycrystalline structure. Figure 3b shows the enlarged (111) peaks of the Pd–Pt bimetallic catalysts with a dendritic morphology. It was found that the (111) peak of the Pd–Pt bimetallic catalysts with a dendritic morphology shifted to the position between the (111) peaks of monometallic Pt (JCPDS 87-0640) and Pd (JCPDS 87-0638). This phenomenon was attributed to the substitution of Pd atoms with Pt in the lattice, which resulted in the expansion of the face-centered cubic lattice [38], indicating the successful formation of Pd–Pt alloy.

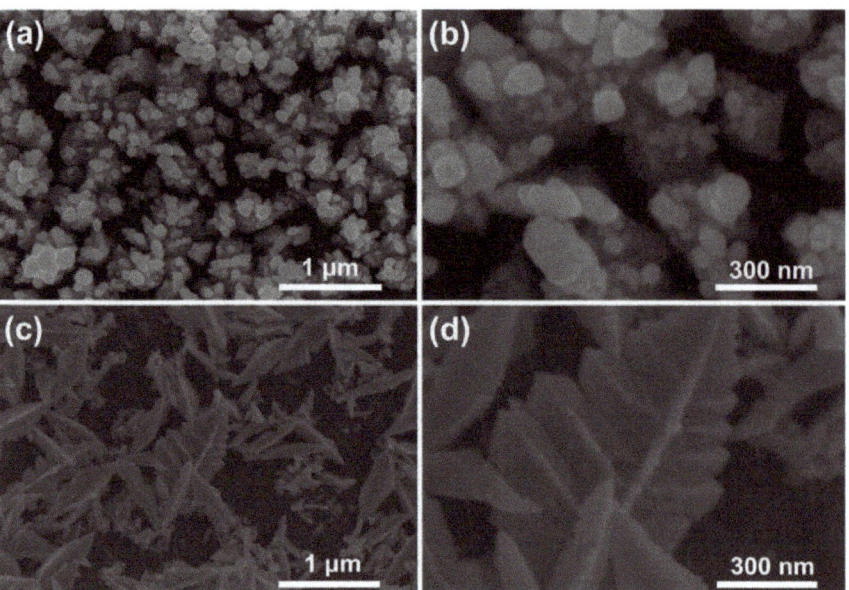

Figure 2. SEM images of the obtained Pd–Pt bimetallic catalysts modified by the PSWP with a frequency of 50 Hz for 1 h (**a**) and 2 h (**c**) in a solution of 0.1 mM H_2PtCl_6 + 1 M H_2SO_4. (**b**,**d**) are the high-magnification SEM images of the corresponding Pd–Pt bimetallic catalysts.

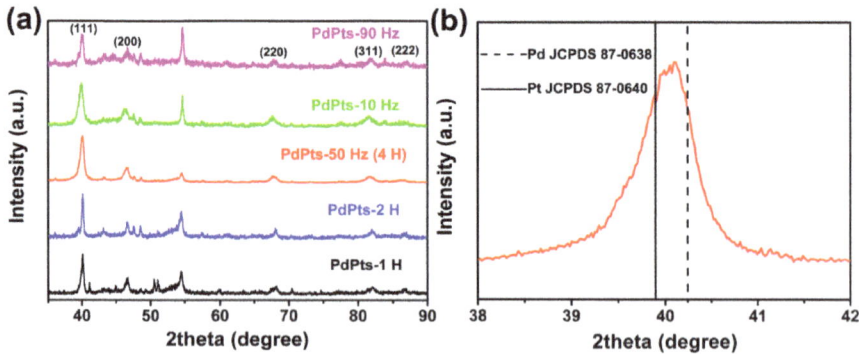

Figure 3. (**a**) X-ray diffraction (XRD) patterns of the obtained Pd–Pt bimetallic catalysts; (**b**) the enlarged XRD spectrum of the Pd–Pt bimetallic catalysts with a dendritic morphology in the 2θ range of 38–42°.

To obtain the detailed structural information of the Pd–Pt bimetallic catalysts with a dendritic morphology, the TEM analysis was conducted. Figure 4a displays the TEM images of the obtained Pd–Pt bimetallic catalysts with a dendritic morphology. It was observed that the long nanothorns grew on the tip of the Pd–Pt dendrites. The length of the nanothorns was about 100 nm. This is coincident with the observed SEM results (Figure 1e,f). Figure 4b,c shows the elemental mapping images of the corresponding Pd–Pt dendrites. The Pd–Pt dendries were composed of Pd (Figure 4b) and Pt (Figure 4c) elements, and these elements were uniformly distributed on the surface of Pd–Pt dendrites. Figure 4d shows the high-resolution TEM (HRTEM) image of the Pd–Pt dendrites. The well-defined lattice fringes of the Pd–Pt dendrites were observed, indicating the high crystallinity of the Pd–Pt dendrites. The interplanar spacing of the Pd–Pt dendrites was 0.223 nm, which matched with the (111) facet of the Pd–Pt phase (confirmed by XRD; Figure 3) [6].

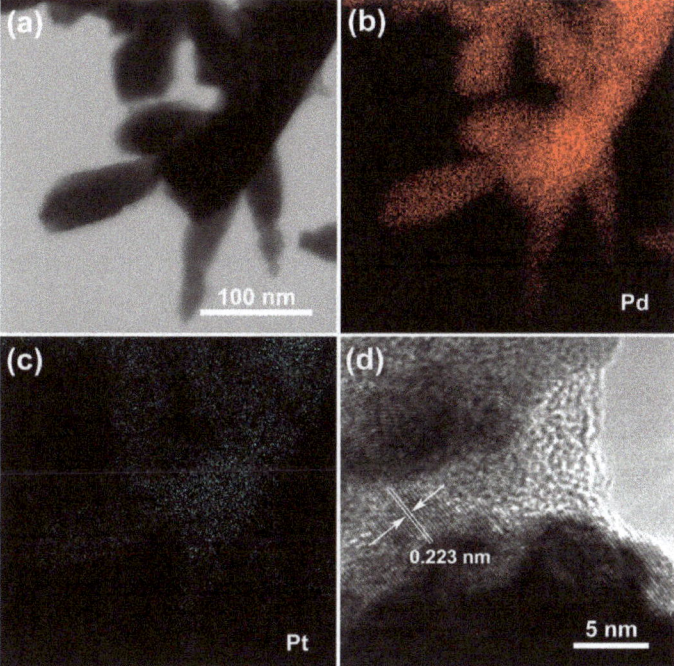

Figure 4. (**a**) TEM image of Pd–Pt dendrites. The elemental mapping images of Pd (**b**) and Pt (**c**) of Pd–Pt dendrites. (**d**) High-resolution TEM (HRTEM) image of Pd–Pt dendrites.

Figure 5 shows the CV profiles of the Pd–Pt bimetallic catalysts tested in a N_2-saturated 0.5 M H_2SO_4 solution. All the CV curves displayed the similar voltammetric features to Pd–Pt polycrystalline. These CV curves exhibited three typical potential regions including the hydrogen adsorption/desorption (−0.20 V to 0.04 V (vs. SCE)), the electric double layer (0.04 V to 0.50 V (vs. SCE)), and the formation/reduction of Pt/Pd oxides (0.50 V to 1.20 V (vs. SCE)) [39]. The multiple peaks of the hydrogen adsorption/desorption of the Pd–Pt bimetallic catalysts indicated that the Pd–Pt bimetallic catalysts possessed a well-developed polycrystalline structure [40,41], which coincided with the XRD results. The electrochemically active surface area (ECSA) can be estimated by integrating a columbic charge associated with reduction peaks of Pd/Pt oxides at about 0.47 V after electric double layer correction, assuming that the charge required for the reduction of the Pd/Pt oxides monolayer was 424 $\mu C\ cm^{-2}$ [42]. The calculated ECSAs of PdPts–10 Hz, PdPts–50 Hz (4 H), PdPts–90 Hz, PdPts–1 H, and PdPts–2 H were 34.43 $m^2\ g^{-1}$, 28.30 $m^2\ g^{-1}$, 33.96 $m^2\ g^{-1}$, 14.15 $m^2\ g^{-1}$, and 31.60 $m^2\ g^{-1}$, respectively. The relatively small specific ECSA of the Pt−Pd bimetallic catalysts with a dendritic morphology can be ascribed to the large dendrite size. This is consistent with the observed SEM results (Figure 1).

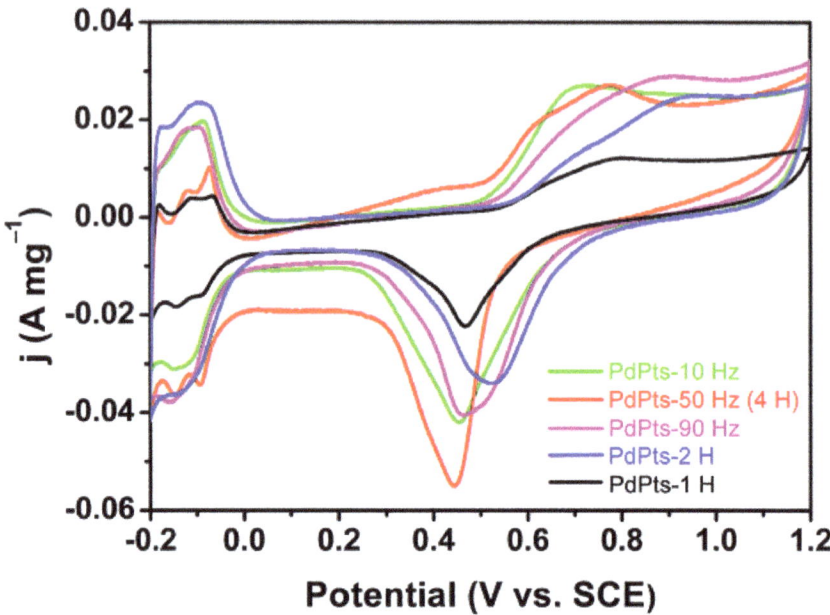

Figure 5. Cyclic voltammetry curves of the obtained Pd–Pt bimetallic catalysts tested in a 0.5 M H_2SO_4 solution at a scan rate of 50 mV s^{-1}, normalized by the Pd–Pt mass.

To evaluate the catalytic activity of the Pd–Pt catalysts, the voltammetry tests of Pd–Pt catalysts towards FAOR were conducted. Figure 6 shows the voltammetric curves of the Pd–Pt bimetallic catalysts towards FAOR. Two well-defined current peaks P_1 and P_2 appeared at about 0.44 V (vs. SCE) and 0.78 V (vs. SCE) (Figure 6a), corresponding to the direct oxidation of formic acid to CO_2 and oxidation of adsorbed CO generated by dehydration of formic acid, respectively [7,14]. The Pd–Pt bimetallic catalysts with a dendritic morphology (PdPt–50 Hz (4 H)) exhibited the highest mass activity (P_1, 0.77 A mg^{-1}) among the obtained Pd–Pt bimetallic catalysts towards FAOR and was almost 2.5 times that of the commercial Pd/C catalyst (0.31 A mg^{-1}) [36]. Furthermore, the Pd–Pt bimetallic catalysts with a dendritic morphology also displayed a higher mass activity compared with the reported Pd-based electrocatalysts towards FAOR (Table 1). Moreover, there was still much room for improvement in the mass activity of this catalyst by further reducing its particle size. Figure 6b shows the voltammetric curves of the Pd–Pt bimetallic catalysts normalized by the ECSA of Pd–Pt catalysts towards FAOR. The specific activity of the Pd–Pt bimetallic catalysts with a dendritic morphology was also much larger than those of the other obtained Pd–Pt bimetallic catalysts. The enhanced specific activity of the Pd–Pt bimetallic catalysts with a dendritic morphology towards FAOR can be ascribed to the large number of unsaturated atoms at the edges of dendrites (Figure 4d). During the preparation of conventional electrodes, the powder catalyst had to be mixed with polymer binders and conductive agents into a catalyst ink, and then the catalyst ink was transferred to the surface of a current collector. The introduction of polymer binders increased the interfacial resistance between the catalyst and the current collector. Besides, the physical transfer of the catalyst caused the agglomeration of the catalyst particles and thus reduced its effective catalytic active sites. These resulted in undesirable side effects on the catalytic activity of the catalyst. On contrary, the Pd–Pt bimetallic catalysts with a dendritic morphology were directly grown on the surface of the carbon paper, and the entire electrochemical synthesis process of the electrode did not involve the use of the binders and the transfer process of the catalyst. Consequently, the direct growth of dendritic Pd–Pt catalysts on the surface of the carbon paper could effectively reduce the interfacial resistance of electrode and maximize the utilization of the effective catalytic active sites of the catalyst, thereby improving the mass activity of the Pd–Pt bimetallic catalysts with a dendritic morphology. Therefore, the enhanced mass activity of the Pd–Pt catalysts with a dendritic morphology was not only attributed to the large number of atomic defects at the edges of dendrites, but also ascribed to the high utilization of active sites caused by the "clean" electrochemical preparation method. In addition, compared with commercial Pd/C catalysts, the electronic effects, caused by the proper downshift of the d-band center of Pd resulting from Pd alloying with Pt, contributed to accelerating the kinetic of FAOR, thus enhancing its catalytic activity [43,44].

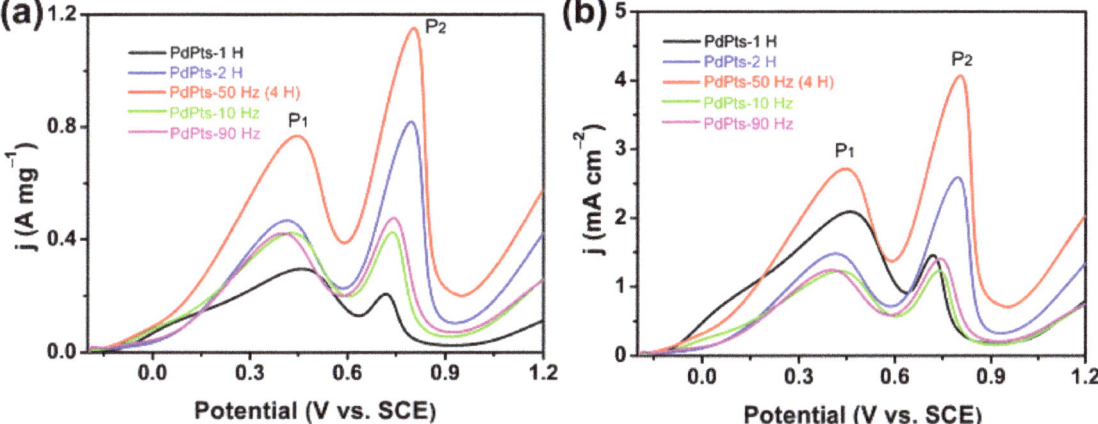

Figure 6. Voltammetric curves of the obtained Pd–Pt bimetallic catalysts in an aqueous solution of 0.5 M H_2SO_4 and 0.5 M HCOOH at a scan rate of 50 mV s^{-1}, normalized by the Pd–Pt mass (**a**) and the Pd–Pt electrochemically active surface area (ECSA) (**b**).

Table 1. Comparison of the mass activity of the Pd–Pt bimetallic catalysts with a dendritic morphology prepared in this work with those of Pd-based electrocatalysts towards formic acid oxidation reaction (FAOR).

Catalyst	Test Protocol	Mass Activity (A mg^{-1})	Reference
Pd–Pt bimetallic catalysts with a dendritic morphology	0.5 M H_2SO_4 + 0.5 M HCOOH, 50 mV s^{-1}	0.77	This work
Pd_1Cu_3/CNTs	0.5 M H_2SO_4 + 0.5 M HCOOH, 50 mV s^{-1}	0.56	[45]
Pd/NS-G	0.5 M H_2SO_4 + 0.5 M HCOOH, 50 mV s^{-1}	0.50	[46]
Pd_3Pt half-shells	0.5 M H_2SO_4 + 0.5 M HCOOH, 50 mV s^{-1}	0.32	[47]
Pd@graphene	0.5 M H_2SO_4 + 0.5 M HCOOH, 50 mV s^{-1}	0.09	[48]
Pd/CN	0.5 M H_2SO_4 + 0.5 M HCOOH, 50 mV s^{-1}	0.20	[49]
PdCuSn/CNFs	0.5 M H_2SO_4 + 0.5 M HCOOH, 50 mV s^{-1}	0.53	[50]
Pt/Pd bimetallic nanotubes with a petal-like surface	0.5 M H_2SO_4 + 0.5 M HCOOH, 50 mV s^{-1}	0.54	[51]
Pd_1Ni_1-NNs/RGO	0.5 M H_2SO_4 + 0.5 M HCOOH, 50 mV s^{-1}	0.60	[52]
PdSnAg/C	0.5 M H_2SO_4 + 0.5 M HCOOH, 50 mV s^{-1}	0.63	[53]
PdSn/C	0.5 M H_2SO_4 + 0.5 M HCOOH, 50 mV s^{-1}	0.17	[53]
PdSnNi/C	0.5 M H_2SO_4 + 0.5 M HCOOH, 50 mV s^{-1}	0.36	[53]
PdSnCo/C	0.5 M H_2SO_4 + 0.5 M HCOOH, 50 mV s^{-1}	0.29	[53]

To assess the stability of the Pd–Pt catalysts, the chronoamperometric testing of the Pd–Pt bimetallic catalysts towards FAOR was conducted. Figure 7 shows the chronoamperometric curves at 0.15 V (vs. SCE) for 3000 s. In the initial stage, a rapid drop of the current density was associated with electric double-layer charging [7]. Subsequently, a slow decay of the current density was observed, which was associated with surface poisoning by intermediates [7]. The current density of the Pd–Pt bimetallic catalysts with a dendritic morphology for a period of 3000 s was 0.08 A mg^{-1}, which was higher than those of PdPts–10 Hz (0.04 A mg^{-1}), PdPts–90 Hz (0.04 A mg^{-1}), PdPts–1 H (0.02 A mg^{-1}), and PdPts–2 H (0.03 A mg^{-1}), even four times that of the commercial Pd/C catalyst reported in the literature (about 0.02 A mg^{-1}) [36]. This indicated that the Pd–Pt bimetallic catalysts with a dendritic morphology possessed an outstanding catalytic activity and a high stability towards FAOR in an acid medium. Therefore, it is possible that the dendritic Pd–Pt catalyst directly electrodeposited on a carbon paper can be applied in direct formic acid fuel cells as a fuel cell anode, methanol fuel cells, and the degradation of organic dyes in water [5,54,55]. Additionally, since the catalyst can be continuously electrodeposited on the surface of a conductive substrate when the conductive substrate (such as carbon paper) is used as a "conveyor belt", the electrochemical synthesis method in this work can realize the continuous industrial production of catalytic electrodes.

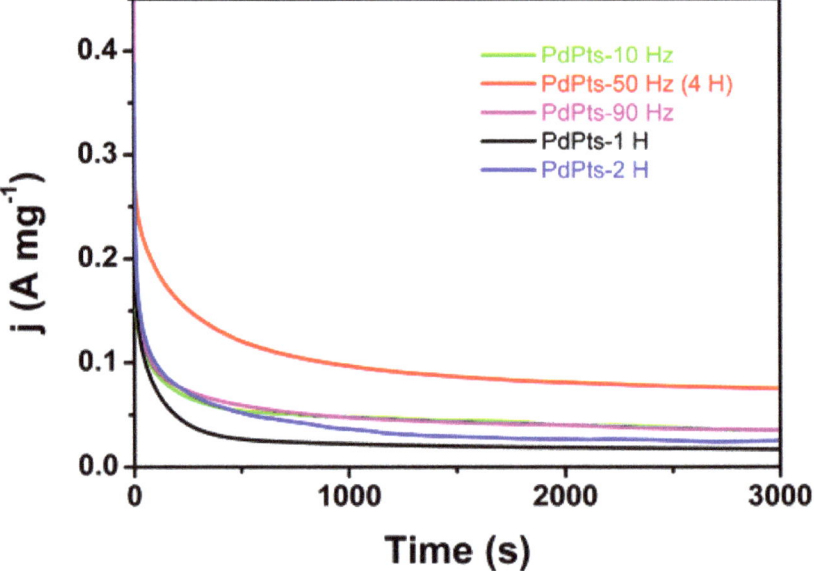

Figure 7. Chronoamperometry curves of the obtained Pd–Pt bimetallic catalysts recorded in an aqueous solution of 0.5 M H$_2$SO$_4$ and 0.5 M HCOOH at 0.15 V (vs. saturated calomel electrode (SCE)) for 3000 s.

4. Conclusions

Pd–Pt catalysts with a dendritic morphology were in situ grown on the surface of a carbon paper via a facile and "green" two-step electrochemical method. The frequency of the PSWP had a significant effect on the morphology of the Pd–Pt bimetallic catalysts. Additionally, as the modification time increased, the morphology of the Pd–Pt bimetallic catalysts evolved from the agglomerated particles, leaf-like catalysts to dendritic catalysts. The obtained dendritic Pd–Pt catalysts displayed an outstanding catalytic activity (0.77 A mg^{-1}) and a high stability towards FAOR. The improved catalytic activity of the Pd–Pt catalysts with a dendritic morphology can be ascribed to the high utilization of its active site and the improved specific activity related to its rough dendritic morphology. The dendritic Pd–Pt catalyst directly electrodeposited on a carbon paper possesses great potential to be applied in direct formic acid fuel cells as a fuel cell anode methanol fuel cells, and the degradation of organic dyes in water.

Author Contributions: Conceptualization, J.L. and F.L.; methodology, F.L.; investigation, J.L. and F.L.; data curation, J.L. and F.L.; writing—original draft preparation, J.L. and F.L.; writing—review and editing, J.L. and C.Z.; supervision, C.Z. and W.H.; project administration, C.Z. and W.H.; funding acquisition, J.L., C.Z. and W.H. All authors have read and agreed to the published version of the manuscript.

Funding: This work was supported by the National Natural Science Foundation of China (Nos. 51771134 and 51801134), National Natural Science Foundation for Distinguished Young Scholar (52125404), Tianjin Natural Science Foundation for Distinguished Young Scholar (18JCJQJC46500), "131" First Level Innovative Talents Training Project in Tianjin, Tianjin Natural Science Foundation (20JCQNJC01130), National Natural Science Foundation of China and Guangdong Province (U1601216), and the National Youth Talent Support Program.

Institutional Review Board Statement: Not applicable.

Informed Consent Statement: Not applicable.

Data Availability Statement: Data are contained within the article.

Conflicts of Interest: The authors declare no conflict of interest.

References

1. Yu, X.-P.; Yang, C.; Song, P.; Peng, J. Self-assembly of Au/MoS$_2$ quantum dots core-satellite hybrid as efficient electrocatalyst for hydrogen production. *Tungsten* **2020**, *2*, 194–202. [CrossRef]
2. Zhao, T.; Luo, E.; Li, Y.; Wang, X.; Liu, C.; Xing, W.; Ge, J. Highly dispersed L10-PtZn intermetallic catalyst for efficient oxygen reduction. *Sci. China Mater.* **2021**, *64*, 1671–1678. [CrossRef]
3. Liu, J.; Liu, B.; Ni, Z.; Deng, Y.; Zhong, C.; Hu, W. Improved catalytic performance of Pt/TiO$_2$ nanotubes electrode for ammonia oxidation under UV-light illumination. *Electrochim. Acta* **2014**, *150*, 146–150. [CrossRef]
4. Huang, L.; Zaman, S.; Wang, Z.; Niu, H.; You, B.; Xia, B.Y. Synthesis and Application of Platinum-based Hollow Nanoframes for Direct Alcohol Fuel Cells. *Acta Phys.-Chim. Sin.* **2021**, *37*, 2009035. [CrossRef]
5. Omidvar, A.; Jaleh, B.; Nasrollahzadeh, M. Preparation of the GO/Pd nanocomposite and its application for the degradation of organic dyes in water. *J. Colloid Interface Sci.* **2017**, *496*, 44–50. [CrossRef] [PubMed]
6. Ren, F.; Wang, H.; Zhai, C.; Zhu, M.; Yue, R.; Du, Y.; Yang, P.; Xu, J.; Lu, W. Clean Method for the Synthesis of Reduced Graphene Oxide-Supported PtPd Alloys with High Electrocatalytic Activity for Ethanol Oxidation in Alkaline Medium. *ACS Appl. Mater. Interfaces* **2014**, *6*, 3607–3614. [CrossRef]
7. Muthukumar, V.; Chetty, R. Electrodeposited Pt–Pd dendrite on carbon support as anode for direct formic acid fuel cells. *Ionics* **2018**, *24*, 3937–3947. [CrossRef]
8. Zhao, Q.; Ge, C.; Cai, Y.; Qiao, Q.; Jia, X. Silsesquioxane stabilized platinum-palladium alloy nanoparticles with morphology evolution and enhanced electrocatalytic oxidation of formic acid. *J. Colloid Interface Sci.* **2018**, *514*, 425–432. [CrossRef]
9. Liu, H.; Adzic, R.R.; Wong, S.S. Multifunctional Ultrathin Pd$_x$Cu$_{1-x}$ and Pt~Pd$_x$Cu$_{1-x}$ One-Dimensional Nanowire Motifs for Various Small Molecule Oxidation Reactions. *ACS Appl. Mater. Interfaces* **2015**, *7*, 26145–26157. [CrossRef]
10. Ding, J.; Liu, Z.; Liu, X.; Liu, J.; Deng, Y.; Han, X.; Zhong, C.; Hu, W. Mesoporous Decoration of Freestanding Palladium Nanotube Arrays Boosts the Electrocatalysis Capabilities toward Formic Acid and Formate Oxidation. *Adv. Energy Mater.* **2019**, *9*, 1900955. [CrossRef]
11. Mazumder, V.; Sun, S. Oleylamine-Mediated Synthesis of Pd Nanoparticles for Catalytic Formic Acid Oxidation. *J. Am. Chem. Soc.* **2009**, *131*, 4588–4589. [CrossRef] [PubMed]
12. Ding, J.; Liu, Z.; Liu, X.; Liu, B.; Liu, J.; Deng, Y.; Han, X.; Hu, W.; Zhong, C. Tunable Periodically Ordered Mesoporosity in Palladium Membranes Enables Exceptional Enhancement of Intrinsic Electrocatalytic Activity for Formic Acid Oxidation. *Angew. Chem.* **2020**, *132*, 5130–5139. [CrossRef]
13. Zhang, S.; Shao, Y.; Yin, G.; Lin, Y. Electrostatic Self-Assembly of a Pt-around-Au Nanocomposite with High Activity towards Formic Acid Oxidation. *Angew. Chem. Int. Ed.* **2010**, *49*, 2211–2214. [CrossRef] [PubMed]
14. Capon, A.; Parsons, R. The oxidation of formic acid at noble metal electrodes Part III. Intermediates and mechanism on platinum electrodes. *J. Electroanal. Chem. Interfacial Electrochem.* **1973**, *45*, 205–231. [CrossRef]
15. Guo, S.; Dong, S.; Wang, E. Three-Dimensional Pt-on-Pd Bimetallic Nanodendrites Supported on Graphene Nanosheet: Facile Synthesis and Used as an Advanced Nanoelectrocatalyst for Methanol Oxidation. *ACS Nano* **2010**, *4*, 547–555. [CrossRef] [PubMed]
16. Morales-Acosta, D.; Ledesma-Garcia, J.; Godinez, L.A.; Rodríguez, H.; Álvarez-Contreras, L.; Arriaga, L. Development of Pd and Pd–Co catalysts supported on multi-walled carbon nanotubes for formic acid oxidation. *J. Power Sources* **2010**, *195*, 461–465. [CrossRef]
17. Lu, Y.; Chen, W. Nanoneedle-Covered Pd−Ag Nanotubes: High Electrocatalytic Activity for Formic Acid Oxidation. *J. Phys. Chem. C* **2010**, *114*, 21190–21200. [CrossRef]

18. Zhang, G.; Wang, Y.; Wang, X.; Chen, Y.; Zhou, Y.; Tang, Y.; Lu, L.; Bao, J.; Lu, T. Preparation of Pd–Au/C catalysts with different alloying degree and their electrocatalytic performance for formic acid oxidation. *Appl. Catal. B Environ.* **2011**, *102*, 614–619. [CrossRef]
19. Shen, T.; Lu, Y.; Gong, M.; Zhao, T.; Hu, Y.; Wang, D. Optimizing Formic Acid Electro-oxidation Performance by Restricting the Continuous Pd Sites in Pd–Sn Nanocatalysts. *ACS Sustain. Chem. Eng.* **2020**, *8*, 12239–12247. [CrossRef]
20. Bao, Y.; Feng, L. Formic Acid Electro-oxidation Catalyzed by PdNi/Graphene Aerogel. *Acta Phys.-Chim. Sin.* **2021**, *37*, 2008031. [CrossRef]
21. Bai, Z.; Yang, L.; Zhang, J.; Li, L.; Lv, J.; Hu, C.; Zhou, J. Solvothermal synthesis and characterization of Pd–Rh alloy hollow nanosphere catalysts for formic acid oxidation. *Catal. Commun.* **2010**, *11*, 919–922. [CrossRef]
22. Zhang, X.; Fan, H.; Zheng, J.; Duan, S.; Huang, Y.; Cui, Y.; Wang, R. Pd–Zn nanocrystals for highly efficient formic acid oxidation. *Catal. Sci. Technol.* **2018**, *8*, 4757–4765. [CrossRef]
23. Li, R.; Hao, H.; Cai, W.-B.; Huang, T.; Yu, A. Preparation of carbon supported Pd–Pb hollow nanospheres and their electrocatalytic activities for formic acid oxidation. *Electrochem. Commun.* **2010**, *12*, 901–904. [CrossRef]
24. Mebed, A.M.; Zeid, E.F.A.; Abd-Elnaiem, A.M. Synthesis and Thermal Treatment of Pd-Cr@Carbon for Efficient Oxygen Reduction Reaction in Proton-Exchange Membrane Fuel Cells. *J. Inorg. Organomet. Polym. Mater.* **2021**, *31*, 3772–3779. [CrossRef]
25. Wang, X.; Tang, Y.; Gao, Y.; Lu, T. Carbon-supported Pd–Ir catalyst as anodic catalyst in direct formic acid fuel cell. *J. Power Sources* **2008**, *175*, 784–788. [CrossRef]
26. Demirci, U.B. Theoretical means for searching bimetallic alloys as anode electrocatalysts for direct liquid-feed fuel cells. *J. Power Sources* **2007**, *173*, 11–18. [CrossRef]
27. Li, X.; Hsing, I.-M. Electrooxidation of formic acid on carbon supported Pt_xPd_{1-x} (x = 0–1) nanocatalysts. *Electrochim. Acta* **2006**, *51*, 3477–3483. [CrossRef]
28. Kang, Y.; Murray, C.B. Synthesis and Electrocatalytic Properties of Cubic Mn−Pt Nanocrystals (Nanocubes). *J. Am. Chem. Soc.* **2010**, *132*, 7568–7569. [CrossRef]
29. Yuan, Q.; Zhou, Z.; Zhuang, J.; Wang, X. Pd–Pt random alloy nanocubes with tunable compositions and their enhanced electrocatalytic activities. *Chem. Commun.* **2010**, *46*, 1491–1493. [CrossRef]
30. Zhang, Z.-C.; Hui, J.-F.; Guo, Z.-G.; Yu, Q.-Y.; Xu, B.; Zhang, X.; Liu, Z.-C.; Xu, C.-M.; Gao, J.-S.; Wang, X. Solvothermal synthesis of Pt–Pd alloys with selective shapes and their enhanced electrocatalytic activities. *Nanoscale* **2012**, *4*, 2633–2639. [CrossRef]
31. Lu, Y.; Jiang, Y.; Wu, H.; Chen, W. Nano-PtPd Cubes on Graphene Exhibit Enhanced Activity and Durability in Methanol Electrooxidation after CO Stripping–Cleaning. *J. Phys. Chem. C* **2013**, *117*, 2926–2938. [CrossRef]
32. Paoletti, C.; Cemmi, A.; Giorgi, L.; Giorgi, R.; Pilloni, L.; Serra, E.; Pasquali, M. Electro-deposition on carbon black and carbon nanotubes of Pt nanostructured catalysts for methanol oxidation. *J. Power Sources* **2008**, *183*, 84–91. [CrossRef]
33. Tian, N.; Zhou, Z.-Y.; Sun, S.-G.; Ding, Y.; Wang, Z.L. Synthesis of Tetrahexahedral Platinum Nanocrystals with High-Index Facets and High Electro-Oxidation Activity. *Science* **2007**, *316*, 732–735. [CrossRef] [PubMed]
34. Tian, N.; Zhou, Z.-Y.; Yu, N.-F.; Wang, L.-Y.; Sun, S.-G. Direct Electrodeposition of Tetrahexahedral Pd Nanocrystals with High-Index Facets and High Catalytic Activity for Ethanol Electrooxidation. *J. Am. Chem. Soc.* **2010**, *132*, 7580–7581. [CrossRef]
35. Fu, W.; Liu, B.; Liu, J.; Han, X.; Deng, Y.; Zhong, C.; Hu, W. Square-Wave Potential-Modified Pt Particles for Methanol and Ammonia Oxidation. *Int. J. Electrochem. Sci.* **2021**, *16*, 210834. [CrossRef]
36. Li, F.; Liu, B.; Shen, Y.; Liu, J.; Zhong, C.; Hu, W. Palladium Particles Modified by Mixed-Frequency Square-Wave Potential Treatment to Enhance Electrocatalytic Performance for Formic Acid Oxidation. *Catalysts* **2021**, *11*, 522. [CrossRef]
37. Liu, J.; Fan, X.; Liu, X.; Song, Z.; Deng, Y.; Han, X.; Hu, W.; Zhong, C. Synthesis of Cubic-Shaped Pt Particles with (100) Preferential Orientation by a Quick, One-Step and Clean Electrochemical Method. *ACS Appl. Mater. Interfaces* **2017**, *9*, 18856–18864. [CrossRef]
38. Lu, Y.; Jiang, Y.; Chen, W. PtPd porous nanorods with enhanced electrocatalytic activity and durability for oxygen reduction reaction. *Nano Energy* **2013**, *2*, 836–844. [CrossRef]
39. Shahrokhian, S.; Rezaee, S. Vertically standing Cu_2O nanosheets promoted flower-like PtPd nanostructures supported on reduced graphene oxide for methanol electro-oxidation. *Electrochim. Acta* **2018**, *259*, 36–47. [CrossRef]
40. Fan, Y.; Liu, P.-F.; Yang, Z.-J.; Jiang, T.-W.; Yao, K.-L.; Han, R.; Huo, X.-X.; Xiong, Y.-Y. Bi-functional porous carbon spheres derived from pectin as electrode material for supercapacitors and support material for Pt nanowires towards electrocatalytic methanol and ethanol oxidation. *Electrochim. Acta* **2015**, *163*, 140–148. [CrossRef]
41. Sun, S.; Zhang, G.; Geng, D.; Chen, Y.; Banis, M.N.; Li, R.; Cai, M.; Sun, X. Direct Growth of Single-Crystal Pt Nanowires on Sn@CNT Nanocable: 3D Electrodes for Highly Active Electrocatalysts. *Chem.-A Eur. J.* **2010**, *16*, 829–835. [CrossRef] [PubMed]
42. Yang, Y.; Du, J.-J.; Luo, L.-M.; Zhang, R.-H.; Dai, Z.-X.; Zhou, X.-W. Facile Aqueous-Phase Synthesis and Electrochemical Properties of Novel PtPd Hollow Nanocatalysts. *Electrochim. Acta* **2016**, *212*, 966–972. [CrossRef]
43. Wang, C.; Peng, B.; Xie, H.-N.; Zhang, H.-X.; Shi, F.; Cai, W.-B. Facile Fabrication of Pt, Pd and Pt−Pd Alloy Films on Si with Tunable Infrared Internal Reflection Absorption and Synergetic Electrocatalysis. *J. Phys. Chem. C* **2009**, *113*, 13841–13846. [CrossRef]
44. Ruban, A.; Hammer, B.; Stoltze, P.; Skriver, H.; Nørskov, J. Surface electronic structure and reactivity of transition and noble metals. *J. Mol. Catal. A Chem.* **1997**, *115*, 421–429. [CrossRef]
45. Zhao, Q.; Wang, J.; Huang, X.; Yao, Y.; Zhang, W.; Shao, L. Copper-enriched palladium-copper alloy nanoparticles for effective electrochemical formic acid oxidation. *Electrochem. Commun.* **2016**, *69*, 55–58. [CrossRef]

46. Zhang, X.; Zhu, J.; Tiwary, C.S.; Ma, Z.; Huang, H.; Zhang, J.; Lu, Z.; Huang, W.; Wu, Y. Palladium Nanoparticles Supported on Nitrogen and Sulfur Dual-Doped Graphene as Highly Active Electrocatalysts for Formic Acid and Methanol Oxidation. *ACS Appl. Mater. Interfaces* **2016**, *8*, 10858–10865. [CrossRef]
47. Yan, X.; Hu, X.; Fu, G.; Xu, L.; Lee, J.-M.; Tang, Y. Facile Synthesis of Porous Pd3 Pt Half-Shells with Rich "Active Sites" as Efficient Catalysts for Formic Acid Oxidation. *Small* **2018**, *14*, e1703940. [CrossRef]
48. Zhang, L.Y.; Zhao, Z.L.; Li, C.M. Formic acid-reduced ultrasmall Pd nanocrystals on graphene to provide superior electocatalytic activity and stability toward formic acid oxidation. *Nano Energy* **2015**, *11*, 71–77. [CrossRef]
49. Yang, L.; Wang, X.; Liu, D.; Cui, G.; Dou, B.; Wang, J. Efficient anchoring of nanoscale Pd on three-dimensional carbon hybrid as highly active and stable catalyst for electro-oxidation of formic acid. *Appl. Catal. B Environ.* **2020**, *263*, 118304. [CrossRef]
50. Zhu, F.; Ma, G.; Bai, Z.; Hang, R.; Tang, B.; Zhang, Z.; Wang, X. High activity of carbon nanotubes supported binary and ternary Pd-based catalysts for methanol, ethanol and formic acid electro-oxidation. *J. Power Sources* **2013**, *242*, 610–620. [CrossRef]
51. Guo, S.; Dong, S.; Wang, E. Pt/Pd bimetallic nanotubes with petal-like surfaces for enhanced catalytic activity and stability towards ethanol electrooxidation. *Energy Environ. Sci.* **2010**, *3*, 1307–1310. [CrossRef]
52. Bin, D.; Yang, B.; Ren, F.; Zhang, K.; Yang, P.; Du, Y. Facile synthesis of PdNi nanowire networks supported on reduced graphene oxide with enhanced catalytic performance for formic acid oxidation. *J. Mater. Chem. A* **2015**, *3*, 14001–14006. [CrossRef]
53. Zhu, F.; Wang, M.; He, Y.; Ma, G.; Zhang, Z.; Wang, X. A comparative study of elemental additives (Ni, Co and Ag) on electrocatalytic activity improvement of PdSn-based catalysts for ethanol and formic acid electro-oxidation. *Electrochim. Acta* **2014**, *148*, 291–301. [CrossRef]
54. Lai, L.; Yang, G.; Zhang, Q.; Yu, H.; Peng, F. Essential analysis of cyclic voltammetry of methanol electrooxidation using the differential electrochemical mass spectrometry. *J. Power Sources* **2021**, *509*, 230397. [CrossRef]
55. Petriev, I.; Pushankina, P.; Lutsenko, I.; Shostak, N.; Baryshev, M. Synthesis, Electrocatalytic and Gas Transport Characteristics of Pentagonally Structured Star-Shaped Nanocrystallites of Pd-Ag. *Nanomaterials* **2020**, *10*, 2081. [CrossRef] [PubMed]

Essay

Molecular Dynamics Simulation of the Oil–Water Interface Behavior of Modified Graphene Oxide and Its Effect on Interfacial Phenomena

Jianzhong Wang [1,*], Suo Tian [1], Xiaoze Liu [2], Xiangtao Wang [1], Yue Huang [1], Yingchao Fu [1] and Qingfa Xu [1]

[1] School of Petroleum Engineering, China University of Petroleum (East China), Qingdao 266580, China; tiasnuotian@163.com (S.T.); wangxt515@163.com (X.W.); huangyue0719@outlook.com (Y.H.); z19020042@s.upc.edu.cn (Y.F.); xu2693253699@163.com (Q.X.)

[2] School of Foreign Languages, Hubei University of Technology, Wuhan 430068, China; lxz3193123682@163.com

* Correspondence: wangjzh@upc.edu.cn

Abstract: Graphene oxide, as a new two-dimensional material, has a large specific surface area, high thermal stability, excellent mechanical stability and exhibits hydrophilic properties. By combining the carboxyl groups on the surface of graphene oxide with hydrophilic groups, surfactant-like polymers can be obtained. In this paper, based on the molecular dynamics method combined with the first nature principle, we first determine the magnitude of the binding energy of three different coupling agents—alkylamines, silane coupling agents, and haloalkanes—and analytically obtain the characteristics of the soft reaction. The high stability of alkylamines and graphene oxide modified by cetylamine, oil, and water models was also established. Then, three different chain lengths of simulated oil, modified graphene oxide–water solution, and oil-modified graphene oxide–water systems were established, and finally, the self-aggregation phenomenon and molecular morphology changes in modified graphene oxide at the oil–water interface were observed by an all-atom molecular dynamics model. The density profile, interfacial formation energy, diffusion coefficient and oil–water interfacial tension of modified graphene oxide molecules (NGOs) at three different temperatures of 300 K, 330 K, and 360 K were analyzed, as well as the relationship between the reduced interfacial tension and enhanced oil recovery (EOR).

Keywords: modified graphene oxide; self-aggregation phenomena; molecular morphology changes; interfacial tension; firstness principle; all-atom molecular dynamics simulations

Citation: Wang, J.; Tian, S.; Liu, X.; Wang, X.; Huang, Y.; Fu, Y.; Xu, Q. Molecular Dynamics Simulation of the Oil–Water Interface Behavior of Modified Graphene Oxide and Its Effect on Interfacial Phenomena. *Energies* **2022**, *15*, 4443. https://doi.org/10.3390/en15124443

Academic Editors: Ioannis F. Gonos, Eleftheria C. Pyrgioti and Diaa-Eldin A. Mansour

Received: 20 May 2022
Accepted: 15 June 2022
Published: 18 June 2022

Publisher's Note: MDPI stays neutral with regard to jurisdictional claims in published maps and institutional affiliations.

Copyright: © 2022 by the authors. Licensee MDPI, Basel, Switzerland. This article is an open access article distributed under the terms and conditions of the Creative Commons Attribution (CC BY) license (https://creativecommons.org/licenses/by/4.0/).

1. Introduction

Polymers, surfactants, and alkalis are the main repellents in current chemical flooding, but a single chemical can only improve the oil displacement efficiency or sweep efficiency. Combination flooding [1] has synergistic effects, with surfactants significantly reducing the interfacial tension at the oil–water interface and increasing the number of capillary numbers. Alkalis injected into the reservoir can chemically react with organic acids in the reservoir, thereby reducing adsorption losses. However, the traditional tertiary oil recovery chemical agents, which can improve the level of recovery, are limited, The nanoparticles are uniform in size and can form compact, well-structured monolayers at the water/non-aqueous phase interface. The emulsions are very stable under high temperatures and high-salt reservoir conditions [2]. Nanoparticle-stabilized emulsions have a high viscosity, which can help manage migration rates during oil transport and provides a viable method of pushing highly viscous oil out of the subsurface, relative to surfactants with a high retention on reservoir rocks [3]. Some oil fields in the geological reserves still constitute 50% of the unswept region, and people are in urgent need of a breakthrough in conventional chemical agents to significantly improve recovery factors [4].

Nanoparticles have the advantages of a large specific surface area and small size and have some special properties different from those of conventional chemical agents. A large number of scholars in China and abroad have carried out a series of theoretical, experimental, and simulation work on the influence of the concentration, size, and charge of nanoparticles on the oil recovery factor [5–7]. Wang et al. [8] found that oil droplets could be spontaneously detached from the solid surface when the charge of the nanofluid composed of charged nanoparticles reached a certain critical value. The modified nanoparticles can effectively reduce the interfacial tension and improve the carrying capacity of water relative to the oil phase. Luo et al. [9] used molecular dynamics to study the self-assembly behavior of SiO_2 nanoparticles at the oil–water interface and found that the modified nanoparticles could effectively reduce the interfacial tension and improve the carrying capacity of water relative to the oil phase. Jia et al. [10] used experimental methods and found that amphiphilic graphene flake nanofluids could form a solid interfacial film, which reduced the interfacial tension at the oil–water interface. Through this mechanism, the oil droplets on the rock surface were desorbed. Compared with conventional chemical flooding, in this agent, the oil displacement efficiency of nanofluids was increased by two times.

At present, the research on nanoparticles is mainly focused on experimental aspects [11–13]. The simulation means for modified nanoparticles are still at the stage of exploration and realization [14,15]. This study selected alkyl-modified graphene oxide as the research object and used the all-atom molecular dynamics simulation method [16] to study the diffusion of modified nanoparticles (NGOs) in the aqueous phase and the self-aggregation phenomenon at the oil–water interface. We also established two different models through the visualization software Materials Studio. Model I analyzed the dispersion nature of nanoparticles in the aqueous phase, Model I analyzed the dispersion properties of nanoparticles in the water phase, while Model II observed the molecular configuration of nanoparticles at the oil–water interface and the self-aggregation phenomenon of nanoparticles at the oil–water interface. The interaction of modified nano molecules on the oil–water phase at three temperatures was investigated according to the constructed models, and finally, the effect of modified graphene oxide on the interfacial tension at different temperatures was revealed.

2. Models and Methods

Molecular dynamics simulation is commonly used method for the software, Material Studio, Lammps, Amber, etc. This visualization software, with built-in rich algorithms, a powerful interactive interface, and multi-scale and multi-functional modules, is widely used in the field of molecular property simulation. The properties of the oil–water interface [17], the aggregation pattern of the solution [18], and the wettability of oil droplets on the solid surface [19] were investigated by domestic and foreign scholars under the action of chemical flooding systems. The simulation process is chosen from all-atom molecular dynamics simulation, which has the advantage of a high accuracy compared to dissipative molecular dynamics simulation [20].

2.1. Model Construction

The simulations were completed with Materials Studio (MS) software, and the simulation process was carried out using the COMPASS force field [21]. Firstly, the simulations established three crude oil systems with different molecular compositions of hexane, heptane, and isooctane, as shown in Figure 1.

The binding energies required for the reaction of the three different coupling agents with graphene oxide were obtained according to the first principle [22], and the parameters are shown in Table 1. The binding energy of haloalkane with graphene oxide was +0.47, which means that the reaction of haloalkane with graphene oxide was not easy to carry out; the reaction conditions were harsh, and the resulting structures were unstable. However, the binding energies of −1.96 and −1.68, for alkylamines and silane coupling agents with graphene oxide, respectively, indicate that the reaction process of alkylamines and silane coupling agents with graphene oxide is easier; the reaction conditions are milder and the resulting products have a good structural stability.

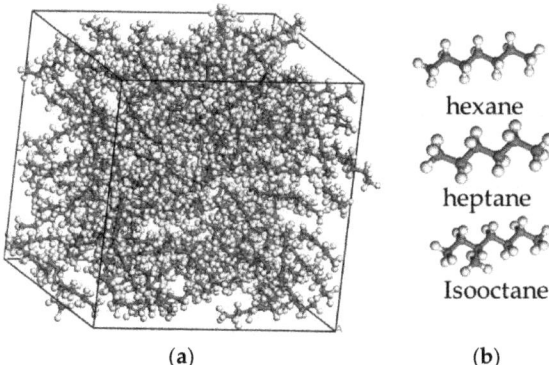

(a) (b)

Figure 1. Molecular configuration of oil droplet composition (grey—carbon, white—hydrogen): (**a**) oil phase box; (**b**) oil phase composition.

Table 1. The binding energy of different modifiers to graphene oxide.

Material	Alkylamines and Graphene Oxide	Silane Coupling Agents and Graphene Oxide	Halothane and Graphene Oxide
Binding of energy/eV	−1.96	−1.68	0.47

To investigate the effect of graft length on the properties of graphene oxide, the binding energies of thirteen, sixteen, and nineteen alkylamines were separately investigated. As shown in Table 2, the binding energies gradually decreased with increasing alkylamine carbon chain length, indicating that, as the alkylamine carbon chain length increased, the reaction proceeded with more ease, and the structure of the resulting modified graphene oxide products became more stable.

Table 2. The binding energy of different graft chain lengths.

Material	Thirteen Alkyl	Sixteen Alkyl	Nineteen Alkyl
Binding energy/eV	−1.46	−1.68	−2.52

To investigate the interaction between water molecules, the reservoir surface, and modified graphene oxide, the adsorption energy of water molecules on the surface of the modified graphene oxide was simulated.

As shown in Table 3, with 13 alkylamines, the adsorption energy of the modified water molecule is −1.06; with 16 alkylamines, the modified graphene oxide adsorption energy is −0.94; and with 19 alkyls, the modified graphene oxide adsorption energy is −0.76. The adsorption of water molecules on both types of modified graphene oxide is an exothermic process, and the surface of the modified product is strongly chemisorbed. Additionally, the octadecylamine-modified graphene oxide had the weakest interaction with water compared to the other chain lengths.

Table 3. The adsorption energy of water molecules with modified graphene oxide.

Material	Thirteen Alkyl	Sixteen Alkyl	Nineteen Alkyl
Adsorption energy/eV	−1.06	−0.94	−0.76

An analysis of the change in binding energy and adsorption energy shows that as the graft chain length increases, the exothermic reaction is enhanced and the binding energy increases. As the graft chain length increases, the adsorption is a spontaneous process and the adsorption energy decreases. On balance, the final choice was to choose s with a relatively smooth reaction process, a high adsorption energy of the modified molecules with water, and a better reduction in interfacial tension.

By oxidizing a thin layer of graphene, structures containing -COOH and -OH on the surface could be obtained. In this study, alkyl long-chain groups were grafted onto the surface of graphene oxide using the azide chemical reaction of cetylamine, as shown in Figure 2.

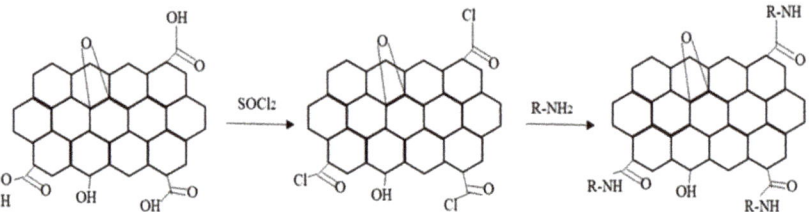

Figure 2. Azide chemical reaction process.

A thin-layer, graphene oxide model with a diameter of 1.8 nm was constructed, and seven cetylamine long chains were grafted on the unilateral side of the graphene oxide model to obtain partially alkyl-modified graphene oxide nanoparticles (NGOs). The NGOs before and after modification are shown in Figure 3.

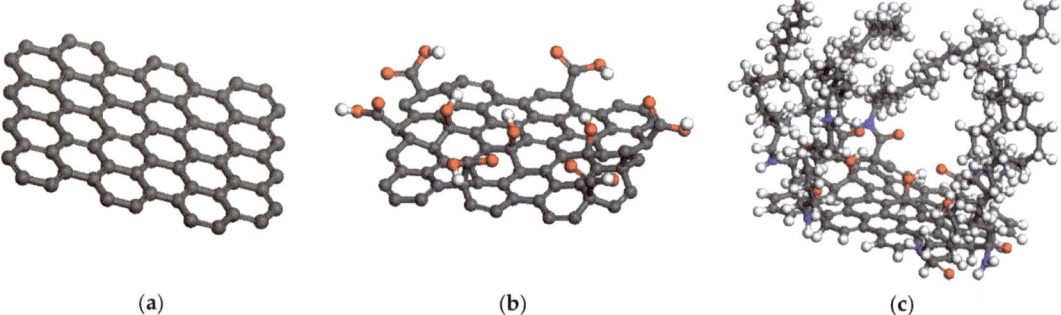

Figure 3. Modified graphene oxide conformation (gray—carbon atoms; red—oxygen atoms; white —hydrogen atoms; blue—nitrogen atoms): (**a**) Thin layer of carbon; (**b**) GO; (**c**) NGOs.

To investigate the dispersion properties of nanoparticles in water, a water–NGO miscible system was built. A square simulation box with a size of 21.92 Å × 21.92 Å × 21.92 Å was created by "amorphous cells tools". The mixed solution system contained 56.4% water and 42.6% NGOs. The initial constructed model is shown in Figure 4a. To investigate the self-aggregation of nanoparticles at the oil–water interface, a columnar simulation box of 26.25 Å × 26.25 Å × 777.61 Å was constructed by the "build layer" tool. To eliminate the influence of periodic boundaries, a thickness of 10 was added above the oil model. The initial model of the vacuum layer is shown in Figure 4b, the molecular dynamics simulation was performed using the "forcite tools", and the simulation parameters are shown in Table 4.

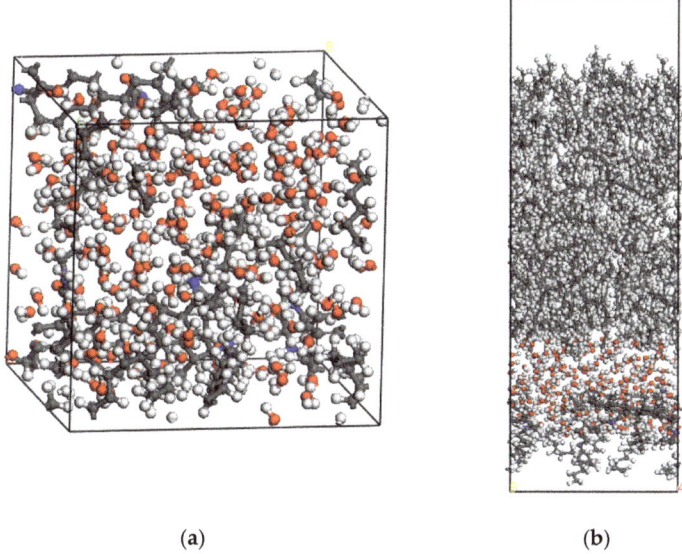

(a)   (b)

Figure 4. Schematic diagram of the initial model: (**a**) Model I (**b**) Model II.

Table 4. Model II force field parameters.

Summation Method	Cutoff Distance	Long-Range Correction	Buffer Width
Atom-based	12.5 Å	YES	0.5 Å

2.2. Mathematical Model Construction

This simulation used an all-atom molecular dynamics simulation approach, and the Newton's Equation of motion (1) was used during the simulation:

$$F_i(t) = m_i a_i(t) \tag{1}$$

where: F_i is combined force on the molecule, m_i is relative molecular mass, a_i is the acceleration of the molecule. The force on the atom can be obtained from the potential energy concerning the position in the coordinate system as Equation of motion (2):

$$-\frac{\partial V}{\partial r_i} = m_i \frac{\partial^2 r_i}{\partial t^2} \tag{2}$$

Because there are no charged molecules in the system, the potential energy function [23] discards electrostatic interactions and constructs a potential energy function that includes inter-bonding atomic interactions and van der Waals forces, with the specific functional relationship shown in Equation (3):

$$V = \sum_{bonds} k_r(r-r_{eq})^2 + \sum_{angles} k_\theta(\theta-\theta_{eq})^2 + \sum_{dihedral} k_\phi(1+\cos[n\phi-\gamma]) + \sum_{i<j}^{atoms} \varepsilon_{ij}\left[\left(\frac{r_m}{r_{ij}}\right)^{12} - 2\left(\frac{r_m}{r_{ij}}\right)^6\right] \tag{3}$$

where: k_r is force constant, r_{eq} is equilibrium bond length, θ_{eq} is equilibrium bond angle, ε_{ij} is van der Waals force constant between atoms, r_m is the minimum distance between atoms, and r_{ij} is the distance between atoms at equilibrium.

The integration process uses the Verlet (1967) integral equation of motion algorithm. The advantages of the Verlet integrator are that it has only one energy evaluation per step, requires only a modest amount of memory, and allows relatively large time steps to be used. The Verlet velocity algorithm overcomes the disadvantages of the Verlet step-over method, which is not synchronous. The Verlet velocity algorithm is provided in Equations (4)–(6):

$$r(t + \Delta t) = r(t) + \Delta t v(t) + \frac{\Delta t^2 a(t)}{2} \quad (4)$$

$$a(t + \Delta t) = \frac{f(t + \Delta t)}{m} \quad (5)$$

$$v(t + \Delta t) = v(t) + \frac{1}{2}\Delta t[a(t) + a(t + \Delta t)] \quad (6)$$

where: $r(t)$ is relative position, $v(t)$ is relative velocity, and $a(t)$ is the acceleration of the atom.

2.3. Simulation Details

For Model I, 500 ps of canonical tethered MD simulations were carried out with a simulated temperature of 300 K. For Model II, 500 ps of isothermal isobaric (NPT) MD simulations were first carried out with a pressure level of 0.1 MPa, followed by 500 ps of canonical tethered MD simulations, with three simulated temperatures of 300 K, 330 K, and 360 K selected. The Andersen method [24] was used for temperature control; the Be-redden method [25] was used for pressure control; the Coulomb interaction was calculated using the Ewald method [26]; the truncation radius was 1 nm; the step size was 1.0 fs; and every 100 ps the system trajectory information was recorded once.

3. Characterization of Molecular Dynamics Simulation Results

3.1. Visualization Characterization

The visualization results of Model I are shown in Figure 5. The left panel shows that, after a 500 ps canonical system synthesis simulation, the nanoparticles gradually leaned towards the center through the attraction of their own branched chains and the repulsion of the branched chains by water. The right panel of Figure 5 shows that the nanoparticles are well-dispersed among themselves and show different characteristics from conventional surface-active multimolecular aggregation.

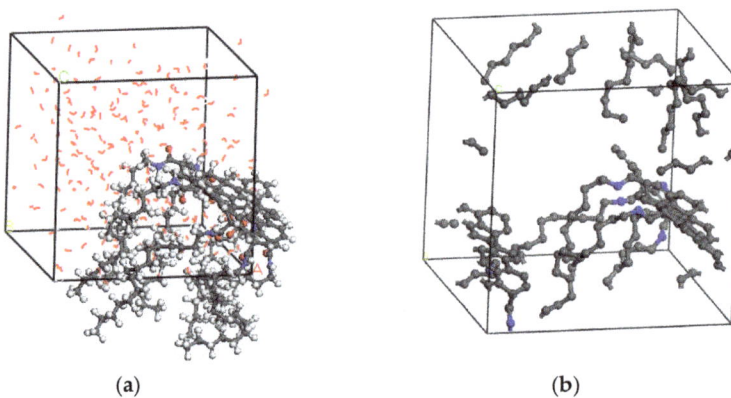

Figure 5. Schematic diagram of the initial model: (**a**) Dispersed morphology, (**b**) NGO distribution.

The visualization model of Model II is shown in Figure 6a: the simulation process of NGOs goes through two processes. The first process is the transport of water solution towards the oil–water interface. The second process is the spontaneous orientation of the lipophilic end towards the oil phase, and the spontaneous orientation of the hydrophilic end towards the water phase. The regular self-aggregation at the oil–water interface is shown in Figure 6b.

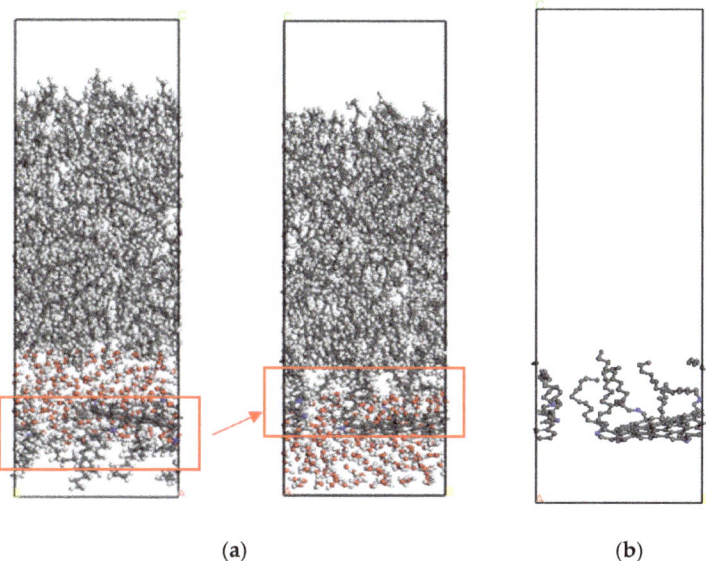

Figure 6. Self-aggregation equilibrium configuration of the system: (**a**) Self-aggregation process, (**b**) self-assembled membrane configuration.

3.2. Density Profile

The alkyl groups in the modified molecules are mixed with the oil phase, and the GO molecules are distributed on the water phase interface because of the polar group -OH. As shown in Figure 6, the oil and water phases are regularly separated by the modified nanoparticle NGOs. To more intuitively observe the self-assembled structure of the NGOs in the oil–water interface layer and the distribution of oil, water, alkyl, and GO molecules at the interface, the density distribution curves of various molecules along the Z-axis direction were calculated, as shown in Figure 7. The nanoparticle NGOs formed a more disordered self-assembled interfacial film structure. An analysis of the density profiles showed that a distinct oil–water interface at 35 Å formed after being distributed, and the nanoparticles were mainly found in the interfacial range of 15 Å–45 Å.

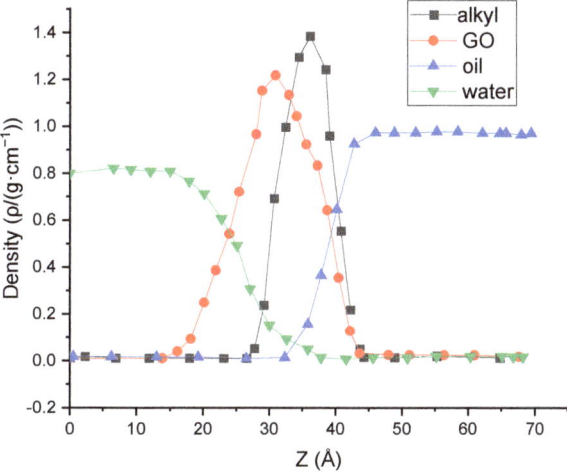

Figure 7. The density of the components in the oil–water and nanoparticle NGO systems.

As shown in Figure 7, the systems appearing along the Z-axis direction are the water phase and NGO–oil phase, with alkyl groups mainly distributed on the oil side. The graphene oxide thin layers are mainly distributed on the water side, forming an interfacial film system at the oil–water interface.

3.3. Interfacial Formation Energy

In order to investigate the stability of the interface in the presence of NGOs at three temperatures, the interaction energy between nanoparticles and water molecules was calculated, the calculation process is shown in Equation (7):

$$\Delta E = \frac{(E_{Nano} + E_{Water}) - E_{Total}}{V_{Nano}} \qquad (7)$$

where: ΔE is interaction energy per unit volume; E_{Total} is the total energy of the nanoparticle and water phase system; E_{Nano} is the energy of the nanoparticle; E_{Water} is the energy of the water phase; and V_{Nano} is occupied volume of the nanoparticle.

From Table 5, it can be observed that the interaction energy between nanoparticles and water becomes smaller as the temperature increases, indicating that an increasing temperature can reduce the resistance to movement between nanoparticles and water, making them easier to transport to the oil–water interface.

Table 5. The interaction energy of nanoparticles with water per unit volume.

Free Energy \ Temperature	300 K	330 K	360 K
E_{Total} (KJ × mol^{-1} × nm^{-3})	374.65	312.84	281.32

3.4. Diffusion Coefficient

The aggregation pattern of each molecule significantly affects the microstructure of the oil–water emulsion system. The nanoparticulate NGOs are spontaneously transported from the aqueous phase to the oil–water interface, and the temperature resistance and diffusion rates of the NGOs are analyzed by comparing the mean square displacement (MSD), as shown in Equation (8), at the three temperatures of 300 K, 330 K, and 360 K set by the model. The mean square displacement can be characterized by a diffusion coefficient (D) related to the simulation time [27]:

$$D = \frac{1}{6N} \lim_{t \to \infty} \frac{d}{dt} \sum_{i=1}^{N} \left\{ [r_i(t) - r_i(0)]^2 \right\} \qquad (8)$$

where D is the diffusion coefficient of the molecule, N is a molecular term of diffusion in the system, $r_i(t)$ is the position of the molecule at the moment, and the differential term is the ratio of mean square displacement to time.

The calculation shows that the mean squared displacement MSD is 7.43 Å at 300 K, 11.98 Å at 330 K, and 18.64 Å at 360 K. As the temperature increases, the NGO diffusion coefficient becomes larger. These results mainly originate from the interaction of the nanoparticle NGOs with the oil phase. The interaction of the surface-oxidized GO molecules with the aqueous phase was much larger than that with the oil phase, and the interaction with the oil phase was greatly enhanced by the surface-grafted cetylamine, so the transportability of the NGOs along the Z-axis was much larger than that of the nanoparticles in the X and Y directions. The results show that the surface-modified alkylamine graphene oxide is highly susceptible to aggregation towards the oil–water interface.

The NGOs move under the combined action of water and oil [28]. As can be seen from Figure 8, the higher the temperature, the greater the slope of the nanoparticle dynamic diffusion curve. It can be observed that the free energy of the mixed-phase is increased and the relative intermolecular displacement rate is expanded under the action of a high temperature, while the NGOs are found to have a good temperature resistance according to the molecular equilibrium conformation.

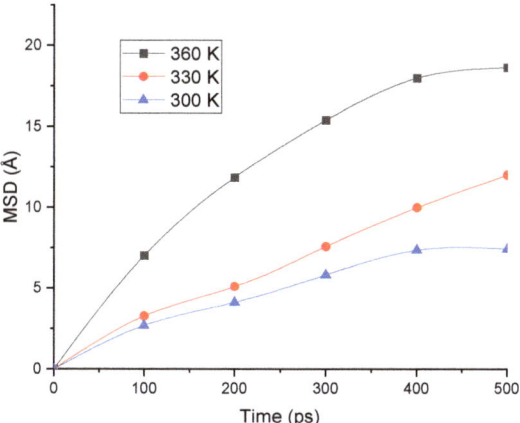

Figure 8. MSD curves for NGOs at different temperatures.

3.5. Interfacial Tension

The oil phase is divided from the water phase by NGOs, forming a clear oil–water interface. Conventional surfactants can be used to improve recovery by reducing the surface tension at the oil–water interface. Numerous experiments demonstrated that modified 2D nanomaterials can significantly improve the recovery rate of cores. In this paper, the interfacial tensions at the oil–water interface, and at the oil–NGO–water interface at three temperatures of 300, 330, and 360 K, were calculated based on Equation (9) [29], and the results are shown in Figure 9:

$$\gamma = \frac{1}{2} L_z \left[p_{zz} - \frac{1}{2}(p_{xx} + p_{yy}) \right] \tag{9}$$

where L_z denotes the length of the system in the z-axis direction. p_{xx}, p_{yy}, p_{zz} are denoted by the pressure tensor in the x, y, and z directions, respectively.

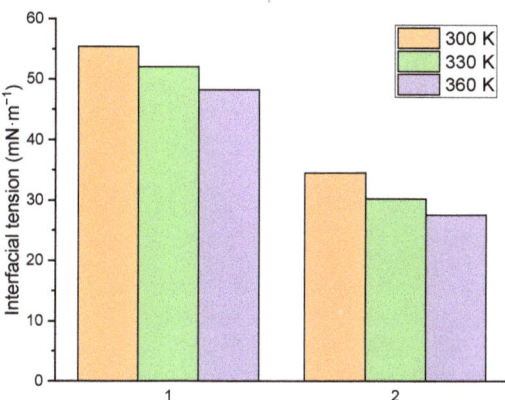

Figure 9. (1) Change in interfacial tension without and (2) with the addition of NGOs.

In order to compare the effect of conventional surfactants with modified graphene oxide by reducing interfacial tension, the oil–water surface tension of three systems (disodium laureth sulfosuccinate DLS, disodium cocoate monoethanolamide sulfosuccinate DMSS and modified graphene oxide) was measured. The results are shown in Table 6, and the modified graphene oxide was the most effective in reducing interfacial tension.

Table 6. Comparison of different surfactants for reducing interfacial tension (mN·m^{-1}).

Time (Min) / Surfactant	DLS	DMSS	NGO
20	69	54	49
40	52	42	36
60	32	32	30

From Table 7, the interfacial tension in both systems decreases with increasing temperature, and the interfacial tension of the system without the addition of NGOs decreases by 7.2 mN·m^{-1} with increasing temperature. The analysis showed that the modified graphene oxide could still significantly reduce the interfacial tension between oil and water at 360 K, showing an excellent temperature resistance. The decrease in interfacial tension was observed at all three temperatures with the addition of NGOs, as shown in Table 7, and it was found that the NGOs could be excellent surfactant substitutes.

Table 7. Variation of interfacial tension (mN·m^{-1}) at different temperatures.

Free Energy / Temperature	300 K	330 K	360 K
$E_{Total}/\left(KJ \times mol^{-1} \times nm^{-3}\right)$	374.65	312.84	281.32

In the process of tertiary oil recovery, the remaining oil is mainly subject to the combined effect of pressure gradient force, surface tension, cohesive force [30], oil drops and oil films, which are the main methods of maintaining oil in the pore space. By using interfacial tension as an important parameter to describe the nature of the oil–water interface, this paper analyzes the force of oil droplets and oil films in the nanopore space to obtain the mechanism of tertiary oil recovery to improve the recovery factor.

The cohesive force: When there is relative motion between the oil droplets and the solid surface, a force that blocks this motion occurs, and a force of this nature is known as the cohesive force. Pressure gradient force is the constant velocity in the flow through a small orifice compared with the value of change in pressure per unit time. Surface tension can be considered as the contraction force acting on the interface of a unit length of liquid.

It can be observed from Figure 10 that, after NGOs were added, the surface tension (orange line in Figure 10) at the oil drops and oil films in the remaining oil becomes less intense, resistance to the three recovery processes and the kinetic energy required at the injection end decrease, and the remaining oil is more easily displaced.

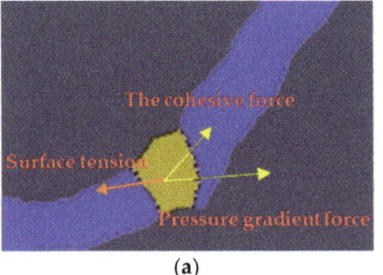 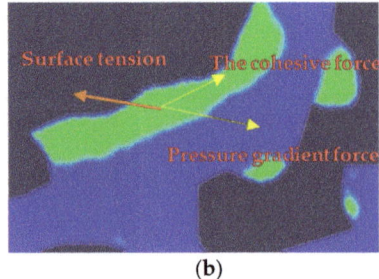

(a) (b)

Figure 10. Vector diagram of forces on oil droplets and oil film: (**a**) Force on oil droplets, (**b**) oil film stresses.

4. Conclusions

The binding energies of alkylamines, silanes, and haloalkanes were compared, and the most stable binding was found for alkylamines. Using this substance as the object of the study, it was found that the grafted nanoparticles were uniformly dispersed in the aqueous solution without agglomeration. The lipophilic end aggregated towards the center of the molecule, showing a dispersion characteristic different from that of conventional surfactants, which form micelles in solution. After surface grafting of graphene oxide, there are a large number of polar hydroxyl groups on the surface layer, which exhibit a strong hydrophilicity, while a large number of alkyl groups exist on one side, showing strong lipophilicity. Unlike conventional surfactants, a single NGO can exhibit the characteristics of multiple surfactant molecules at the oil–water interface.

The simulations show that there is a self-aggregation phenomenon of NGOs at the oil–water interface, specifically the two processes of transport to the oil–water interface and regular at the oil–water interface. After these two stages, a disordered monolayer interfacial film can be formed at the oil–water interface, and the interfacial film can make the oil–water interface irregular, improving the carrying capacity of water to the oil phase. After the addition of NGO nanoparticles, the increase in temperature reduces the free energy of the interfacial layer, which reduces the resistance between the nanoparticles and the water phase, and the NGOs can be dispersed in the oil–water interface faster, thus accelerating the decrease in the interfacial tension between oil and water. Less energy is required at the injection end, thus making it easier to displace oil drops in small pores and oil films adsorbed on rock surfaces in tertiary oil recovery.

Author Contributions: Conceptualization, J.W. and S.T.; methodology, S.T.; software, S.T.; validation, X.W., Y.H. and Y.F.; formal analysis, S.T.; investigation, Q.X.; resources, X.L.; data curation, X.L.; writing—original draft preparation, S.T.; writing—review and editing, S.T.; visualization, J.W.; supervision, J.W.; project administration, J.W.; funding acquisition, J.W. All authors have read and agreed to the published version of the manuscript.

Funding: The APC was funded by [Wang, J.].

Institutional Review Board Statement: Not applicable.

Informed Consent Statement: Not applicable.

Data Availability Statement: Not applicable.

Conflicts of Interest: The authors declare no conflict of interest.

References

1. Li, J.B.; Niu, L.W.; Lu, X.G. Performance of ASP compound systems and effects on flooding efficiency. *J. Pet. Sci. Eng.* **2019**, *178*, 1178–1193. [CrossRef]
2. Mandal, A.; Bera, A.; Ojha, K. Characterization of Surfactant Stabilized Nanoemulsion and Its Use in Enhanced Oil Recovery. In Proceedings of the SPE International Oilfield Nanotechnology Conference and Exhibition, Noordwijk, The Netherlands, 12 June 2012.
3. Qiu, F.D.; Mamora, D. Experimental Study of Solvent-Based Emulsion Injection to Enhance Heavy Oil Recovery in Alaska North Slope Area. In Proceedings of the Canadian Unconventional Resources and International Petroleum Conference, Calgary, AB, Canada, 19 October 2010.
4. Hanam, S.; Hyuntae, K.; Geunju, L.; Jinwoong, K.; Wonmo, S. Enhanced oil recovery using nanoparticle-stabilized oil/water emulsions. *Korean J. Chem. Eng.* **2014**, *31*, 338–342.
5. Irfan, S.A.; Shafie, A.; Yahya, N.; Zainuddin, N. Mathematical Modeling and Simulation of Nanoparticle-Assisted Enhanced Oil Recovery—A Review. *Energies* **2019**, *12*, 1575. [CrossRef]
6. Maurya, N.K.; Mandal, A. Investigation of synergistic effect of nanoparticle and surfactant in macro emulsion based EOR application in oil reservoirs. *Chem. Eng. Res. Des.* **2018**, *132*, 370–384. [CrossRef]
7. He, Y.F.; Liao, K.L.; Bai, J.M.; Fu, L.P.; Ma, Q.L. Study on a Nonionic Surfactant/Nanoparticle Composite Flooding System for Enhanced Oil Recovery. *ACS Omega* **2021**, *6*, 11068–11076. [CrossRef]
8. Wang, J.H.; Wu, H.A. Enhanced oil droplet detachment from solid surfaces in charged nanoparticle suspensions. *Soft Matter* **2013**, *9*, 7974–7979. [CrossRef]
9. Luo, F.C.; Ding, B.; Wang, P.H. Molecular dynamics simulation of adsorption properties of alkane-modified SiO_2 nanoparticles at oil/water interface. *J. China Univ. Pet.* **2015**, *39*, 130–137.

10. Jia, H.; Huang, P.; Han, P.G.; Wang, Q.X.; Wei, X. Synergistic effects of Janus graphene oxide and surfactants on the heavy oil/water interfacial tension and their application to enhance heavy oil recovery. *J. Mol. Liq.* **2020**, *314*, 113791–113800. [CrossRef]
11. Wasan, D.T.; NIikolov, A.D. Spreading of nanofluids on solids. *Nature* **2003**, *423*, 156–159. [CrossRef]
12. Kondiparty, K.; Nikolov, A.D.; Wasan, D.; Liu, K.L. Dynamic Spreading of Nanofluids on Solids. Part I: Experimental. *Langmuir* **2012**, *28*, 14618–14623. [CrossRef]
13. Chen, H.J.; Di, Q.F.; Ye, F.; Gu, C.Y. Numerical simulation of drag reduction effects by hydrophobic nanoparticles adsorption method in water flooding processes. *J. Nat. Gas Sci. Eng.* **2016**, *35*, 1261–1269. [CrossRef]
14. Bellussi, F.M.; Laspalas, M.; Chiminelli, A. Effects of Graphene Oxidation on Interaction Energy and Interfacial Thermal Conductivity of Polymer Nanocomposite: A Molecular Dynamics Approach. *Nanomaterials* **2021**, *11*, 1709. [CrossRef] [PubMed]
15. Feng, Y.; Hou, J.R.; Yang, Y.L.; Wang, D.S.; Wang, L.K. 2-D Molecular dynamics simulation of the microscopic oil repulsion mechanism at the oil-water interface of nano black card. *Oilfield Chem.* **2022**, *51*, 1292–1305.
16. Tsourtou, F.D.; Peroukidis, S.D.; Peristeras, L.D. The phase behaviour of cetyltrimethylammonium chloride surfactant aqueous solutions at high concentrations: An all-atom molecular dynamics simulation study. *Soft Matter* **2022**, *18*, 1371–1384. [CrossRef]
17. Li, N.; Sun, Z.Q.; Pang, Y.H. Microscopic mechanism for electrocoalescence of water droplets in water-in-oil emulsions containing surfactant: A molecular dynamics study. *Sep. Purif. Technol.* **2022**, *289*, 120756–120769. [CrossRef]
18. Liu, Z.N. MD Simulation-Based Study of the Aggregation Behaviour of Alkyl Benzene Sulfonate Surfactants at the Oil/Water Interface. Master's Thesis, Northeastern Petroleum University, Daqing, China, 2022.
19. Zhong, J. Molecular Simulation Study of Reservoir Wettability Formation and Its Regulation. Master's Thesis, China University of Petroleum (East China), Qingdao, China, 2016.
20. Deng, X.J. Simulation Study of the Micelle Aggregation Behavior of Baryonic Surfactants in Solution. Master's Thesis, China University of Petroleum, Beijing, China, 2021.
21. Kondratyuk, N.D.; Pisarev, V.V. Calculation of viscosities of branched alkanes from 0.1 to 1000 MPa by molecular dynamics methods using COMPASS force field. *Fluid Phase Equilibria* **2019**, *498*, 151–159. [CrossRef]
22. Gao, M.K.; Wei, S.Z. First-principles study of the electronic properties of the (Fe, Cr)$_7$C$_3$/MoC interface. *Mater. Guide* **2022**, *36*, 87–92.
23. Shan, D.D.; Zhu, J.G.; Shao, Q.F. Numerical simulation study of nanodrop wetting on graphene oxide surface. *Technol. Innov.* **2021**, *23*, 101–102, 105.
24. Ström, B.; Rota, A.; Linde-Forsberg, C. In vitro characteristics of canine spermatozoa subjected to two methods of cryopreservation. *Theriogenology* **1997**, *48*, 247–256. [CrossRef]
25. Sepehri, M.; Moradi, B.; Emamzadeh, A. High Thickness Uniformity Vaporization Source. U.S. Patent 10805847, 22 September 2005.
26. Petersen, H.G. Accuracy and efficiency of the particle mesh Ewald method. *J. Chem. Phys.* **1995**, *103*, 3668–3679. [CrossRef]
27. Chen, Z.L.; Xu, W.R.; Tang, L.D. *Theory and Practice of Molecular Simulation*, 1st ed.; Chemical Industry Press: Beijing, China, 2007; pp. 62–115.
28. Chanda, J.; Bandyopadhyay, S. Molecular Dynamics Study of Surfactant Monolayers Adsorbed at the Oil/Water and Air/Water Interfaces. *J. Phys. Chem. B* **2006**, *110*, 23482–23488. [CrossRef] [PubMed]
29. Li, K.M.; Liu, Y.L.; Yan, H. Molecular dynamics simulation of self-assembly behavior at SDBS/BMAB oil-water interface. *J. Liaocheng Univ.* **2019**, *32*, 623–627.
30. Liu, X.Z.; Yang, Z.P.; Hui, X.Z. Analysis of the force on the remaining oil in the small hole and improvement of the mathematical model. *Daqing Pet. Geol. Dev.* **2014**, *33*, 77–82.

MDPI
St. Alban-Anlage 66
4052 Basel
Switzerland
www.mdpi.com

MDPI Books Editorial Office
E-mail: books@mdpi.com
www.mdpi.com/books

Disclaimer/Publisher's Note: The statements, opinions and data contained in all publications are solely those of the individual author(s) and contributor(s) and not of MDPI and/or the editor(s). MDPI and/or the editor(s) disclaim responsibility for any injury to people or property resulting from any ideas, methods, instructions or products referred to in the content.

www.ingramcontent.com/pod-product-compliance
Lightning Source LLC
LaVergne TN
LVHW070426100526
838202LV00014B/1535